T5-CPW-901

THOMAS
HUNT
MORGAN

THOMAS HUNT MORGAN

The Man and His Science

GARLAND E. ALLEN

PRINCETON UNIVERSITY PRESS
Princeton, New Jersey

Published by Princeton University Press, Princeton, New Jersey
In the United Kingdom: Princeton University Press, Guildford, Surrey

Library of Congress Cataloging in Publication Data will be found on the last printed page of this book

Publication of this book has been aided by the Louis A. Robb Fund of Princeton University Press

This book has been composed in linotype Times Roman

Clothbound editions of Princeton University Press books are printed on acid-free paper, and binding materials are chosen for strength and durability

Printed in the United States of America by
Princeton University Press, Princeton, New Jersey

To Tove Mohr
Doctor, Scientist, Historian, Friend
whose ideas and warmth have been
a source
of inspiration in this study of the
early years of genetics

CONTENTS

LIST OF ILLUSTRATIONS

PREFACE

In 1866 Gregor Mendel (1822-1884), a teacher and monk of the Augustinian Monastery in Brünn (then part of the Austro-Hungarian empire, now Brno, in the Moravian region of Czechoslovakia), published what seemed to be an ordinary paper on plant hybridization. Mendel had read his paper the previous year at the February and March sessions of the Brünn Natural History Society to a largely uninterested audience. When the paper was published in the society's *Proceedings* for 1866, there was virtually no response, even though the journal was received by all major scientific libraries in the world. It was only after some thirty-five years, in 1900, that Mendel's work was "rediscovered," and the laws of heredity that he formulated were recognized as a bold and revolutionary step in the history of biology. Historians have long pondered the question of why Mendel's work was not appreciated in his own period, yet so much more readily accepted a generation later.

In the same year that Mendel published his paper in the *Proceedings of the Brünn Natural History Society*, Thomas Hunt Morgan (1866-1945) was born to Charlton Hunt (1839-1912) and Ellen Key Howard Morgan (1840-1925) in Lexington, Kentucky. It was to be Thomas Hunt Morgan and his closely knit group of young colleagues who, in the period after 1910, would give a physical basis to Mendelian theory by demonstrating the structural relationship between genes and chromosomes. Although Morgan was not among those who initially rediscovered Mendel's work, or publicized it between 1900 and 1910, it would be his role to exploit the new vistas of heredity that Mendel had opened up.

The central position of genetics in the history of twentieth-century biology can hardly be disputed. Beginning with the rediscovery of Mendel's laws in 1900, through elucidation of the chromosome theory of heredity between 1910 and 1920 by Morgan and his group, to the development of the Watson and Crick model of deoxyribonucleic acid (DNA) in the 1950s, genetics has become a dominant area of modern biology. It contributed significantly to cytology by showing that the rod-shaped chromosomes, found in the nucleus of most cells, could be regarded as the material bearers of heredity. It provided a sound basis to Darwin's theory of natural selection, which had previously suffered from the lack of any coherent understanding of either heredity or its counterpart, variation. It revolutionized the study of biochemistry by suggesting that genes controlled the production of proteins, many of which, as enzymes, guided the cells' myriad metabolic pathways. It contributed a significantly new viewpoint to the study of embryology by showing that genes influence specific changes during development and in some manner still largely unknown can be turned on or off at appropriate times. It opened up new insights in medicine, from the study of genetic diseases to the investigation of cancer as a change in genetically controlled processes. Just as Darwin's theory became a focus for relating many disparate areas of biology in the nineteenth century, so genetics came to occupy a similar position in the twentieth.

The present work focuses on the personal and scientific work of a leader in establishing the modern science of heredity, Thomas Hunt Morgan. The book can be called a "scientific biography" for two reasons. It is the story of the life of a single individual and, hence, is biography in the usual sense of the word. But it is also the story of the growth of a field of science and, hence, goes beyond the career of any one man or woman. In many ways, Morgan serves historically as a vehicle through which to describe the development of the whole field of genetics during the first several decades of the twentieth century. Because his interests were so wide-ranging, Morgan's career provides a valuable case study of the interrelationships between genetics and other areas of the biological, and even the physical sciences. Thus, the major purpose of the present study is not so much to describe Morgan's personal life for its own sake (though that is, of course, interesting), but to analyze historically some important aspects of the growth of the science of heredity in the early twentieth century.

My own view of history is, in many ways, quite opposite to that which characterizes most biographies. By and large, I do not believe that history is made by great individuals. Individuals do, of course, matter. To some degree their own unique personalities do affect the course of events. But I do not see individuals initiating or perpetuating broad-scale historical changes. The science of heredity became important in the early twentieth century not because Gregor Mendel had written a paper in 1866, or because Thomas Hunt Morgan discovered a white-eyed fruit fly in one culture bottle in 1910. It developed because of a confluence of historical forces—economic, social, scientific—which not only conspired to create more interest in heredity itself but also made resources available for promoting genetic research. The rise of the science of genetics cannot be understood out of context of changing economic patterns in the world at the end of the nineteenth and beginning of the twentieth centuries. The development of European empires leading up to World War I and the rapid growth of American science, industry, and agriculture in the post-Civil War period, all contributed significantly to the rise of genetics as a prominent science in the early twentieth century. Although the present book cannot boast that it delineates all of these interactions in depth, it is my hope that they can be sketched clearly enough to show the context in which genetics in general, and Morgan's work in particular, developed. If there is any value to the study of history, other than as what Voltaire called an "idle curiosity," it is in the underlying, and continually operative forces that have shaped the past and continue to shape the present. I believe that such forces, or patterns, exist; it is the historian's job to analyze how they operate. Along with Marx, I believe that the function of studying history is ultimately to be able to change the future.

The conception, and outline, of the present work were laid down long before I came to see history in terms of broad, interacting forces. Hence, the story of Morgan's work in genetics is not set in as wide-ranging a framework as I might attempt if I were writing the book today. However, I have tried to sketch as much as possible of the background, external factors (social, political, economic) that influenced the whole field of heredity and Morgan's work. The obvious influences of large-scale events such as World War I on scientific development can be traced directly and unmistakably. Less obvious influences, such as the shift in Western economic systems from laissez-faire monopoly to more regulated capitalism, and attendant social changes, are more difficult to trace in a work focused on the life of a single individual. Another study of the history of twentieth-century genetics, more broadly conceived, will have to trace such contextual developments in greater depth.

I first became interested in Morgan's life and work in 1963, while still a somewhat fumbling graduate student in the history of science. As I was casting about for a dissertation topic, the idea of working on Morgan came to me through a stroke of luck. In August of 1963 I was visiting one of my undergraduate biology professors at the University of Louisville (Kentucky). In discussing possible dissertation topics he noticed my singular lack of concrete ideas. I vaguely knew I was interested in the area of the history of evolution and genetics, but that's about as far as it went. He suggested that I might be interested in the work of Morgan, whose personal papers he claimed had just been placed in the old Morgan family home in Lexington, Kentucky. As Lexington is only a short drive from Louisville, I went down for a visit during the following week. To my dismay, there were no letters by or about T. H. Morgan at the house, nor, for that matter, at the University of Kentucky Library (since 1963 a number of Morgan family letters have been placed in the university archives). Still the drive to Lexington, through miles of rolling bluegrass country, had been beautiful and had provided some time to think about the possibilities of a research project in the history of twentieth-century genetics. Although I had barely managed to obtain a C in a college genetics course, my interests in biology had always centered more in the area of heredity and evolution than anywhere else. Thus, with a little regional chauvinism to bolster my decision, I soon settled on Morgan as a focus for my dissertation.

As research on Morgan's work in genetics and evolution unfolded, the fortuitous nature of my choice became more and more apparent. Not only was Morgan responsible for elucidating one of the fundamental concepts of twentieth-century biology, what came to be known as the "chromosome theory of heredity" or "the theory of the gene," he was also part of a vanguard movement among younger biologists during the first decades of the century to emancipate biology from its overweening attachment to the methods and aims of nineteenth-century natural history. Thus, through his work in embryology, as well as in genetics and evolution, Morgan helped to fashion an experimental, quantitative methodology which aimed to place biology on an equal footing with physics and chemistry as a quantitative, rigorous science. More than I had initially realized, Morgan proved to be an advantageous focal point for viewing many of the changes occurring in the development of biology as a whole during the early to middle twentieth century. A significant indicator of this new role for biology in general and genetics in particular is the fact that Morgan was the first investigator outside of a strictly physiological discipline to be awarded the Nobel Prize for Physiology and Medicine (1933).

Both as biography and history of science, this study of T. H.

Morgan is only a beginning. It raises more questions than it answers, especially about the history of biology in the twentieth century. My hope is that it will help to focus some of the important questions for other historians of biology to pursue in depth. Topics such as the relationship between embryology and genetics, the growth of eugenics and theories of the genetic basis of human and animal behavior, the development of population genetics, and genetics and biochemistry are but a few areas which have only been touched upon, but not developed, in the present volume. If this book serves any long-lasting function, it will be more in the questions it asks, than the answers it provides.

ACKNOWLEDGMENTS

This book has been so long in the making that it is impossible to acknowledge all the people, ideas, or circumstances that have in one way or another contributed to its development. I have been fortunate in being able to draw upon many resources—intellectual, emotional, and financial—from many people; for that I wish to express at least a small measure of gratitude.

I am especially grateful to the many librarians and archivists who gave me access to their collections of letters and other unpublished as well as published materials. Chief among these is Jane Fessenden and the staff of the Marine Biological Laboratory, Woods Hole, Massachusetts. They not only made available for my use in St. Louis a collection of reprints of Morgan's published writings, but went out of their way to provide a place conducive to work during the many summers in which I was engaged in the research and writing of this work. Few words can express my appreciation for their help and cooperative spirit. Furthermore, the library of the MBL itself houses a remarkable collection of journals and individual reprints, organized in an easily accessible and logical way. The library of the American Philosophical Society, and its director, Dr. Whitfield J. Bell, Jr., have provided much information and encouragement over the years. Dr. Bell's interest in the history of genetics has helped to make the APS library a central archival source for the history of genetics in the United States. Dr. Judith Goodstein, archivist at the California Institute of Technology, was most gracious in providing access to, and guidance through, both the Morgan and the Sturtevant papers. In addition, I have received considerable help from the staffs at various libraries where it was necessary to seek out Morgan papers in the collections of other figures: the Firestone Library at Princeton University (the E. G. Conklin papers), the Sterling Library, Yale University (the Ross G. Harrison papers), the Library of Congress, Washington, D.C. (the Jacques Loeb papers), the University of Illinois Library, Champaign-Urbana (the Frank Zeleny papers), the Margaret I. King Library, the University of Kentucky, Lexington (the Hunt-Morgan family papers), the Bancroft Library, University of California at Berkeley (the Goldschmidt papers), and the Stanford

University Libraries, Stanford, California (the David Starr Jordan papers).

Collections in private hands were graciously opened up to me by their owners. Mrs. Nancy Wilson Lobb, of South Hadley, Massachusetts, was extremely hospitable in allowing me to examine all the letters in her possession between Morgan and her father, E. B. Wilson. Miss Adaline Wheeler of Boston was most helpful in providing access to the extensive correspondence files of her father, William Morton Wheeler. I am also grateful to Howard Ensign and Mary Alice Adams, at the time at Harvard's Museum of Comparative Zoology, for introducing me to Miss Wheeler, and providing help in locating materials in the W. M. Wheeler files.

I was fortunate to have the help of a number of Morgan's former students and associates, whose "oral history" provided me with exciting, and otherwise rarely available reminiscences for historical analysis. Among these I am particularly grateful to Dr. Tove Mohr, of Fredrikstad, Norway, to whom this book is dedicated. She first met Morgan in 1918 when her husband spent a postdoctoral year in Morgan's laboratory in New York. Dr. Mohr has freely given of her time, documents (letters, notebooks, diaries), and most of all, friendship. Without her continuing inspiration, this book might have progressed even more slowly than it did. Also of particular value were the conversations I had with the late A. H. Sturtevant, H. J. Muller, and L. C. Dunn. Professor Dunn in particular shared many of his views on the history of genetics and encouraged me from the beginning to write a detailed study of Morgan and his work. Others who have freely shared their reminiscences, or have helped to correct my incomplete or mistaken views about Morgan and his group, were George MacGinitie, Phoebe Sturtevant, Curt Stern, and Alexander Weinstein. In particular, Dr. Weinstein has been most helpful with his precise memory and his strong concern for historical accuracy.

I have greatly appreciated the help provided by various members of Morgan's family. I especially wish to thank Dr. Isabel Morgan Mountain, Morgan's youngest daughter, for her longstanding interest in this project, as well as for the concrete help she has provided at so many stages along the way. She made available to me her father's letters to her, copies of his own books with marginal comments, numerous photographs, reminiscenes, and most of all her frankness in reviewing and criticizing earlier stages of this manuscript. She has been a good friend as well as a valuable colleague. Howard K. Morgan has also been most helpful with personal reminiscences, given freely into a tape recorder on a lovely mid-winter afternoon at his home in Florida. He and his wife also provided a copy of intriguing movie shots of T. H. Morgan in the late 1930s. Lilian Morgan Scherp and Edith Morgan Whitaker kindly reviewed the later stages of this manuscript and helped to correct errors of fact and citation. All, in their various ways, provided me with a very important glimpse into Morgan's family life.

A number of colleagues have provided necessary professional advice along the way. Everett Mendelsohn and Ernst Mayr were especially helpful with encouragement and hard-nosed criticism, respectively, during preparation of my doctoral dissertation on Morgan's genetic and evolutionary theories. Frederick Churchill has made many valuable points regarding issues in the history of evolutionary and embryological theory in the late nineteenth and early twentieth centuries. In particular, he brought to my attention the existence of the Morgan-Driesch correspondence at Leipzig and introduced me to Dr. Reinhard Mocek, who graciously supplied a microfilm of those letters. I have had many useful conversations over the years with William Coleman about problems in the history of science in general, as well as in the history of biology in particular. William Provine read the final draft of the manuscript carefully and offered a number of important and valuable suggestions.

Many friends and associates have helped formally and informally with the progress of this work. I owe an enormous debt of gratitude to my longstanding friend, colleague, and frequent collaborator, Jeffrey J. W. Baker. He graciously offered his skill as a writer and editor in preparing the final draft of the manuscript. My two close friends Jonathan Hake and Randy Bird have not only discussed with me many aspects of the history of science in general, and this work in particular, but through their sincere though often irreverent senses of humor, have kept me from ever taking myself, or my written word, too seriously. My present collaborator Barry Mehler has helped me see some important interrelations between the history of genetics, and American social history in the late nineteenth and early twentieth centuries. Barbara Westergaard of Princeton University Press, and Josh Tolin, formerly of Washington University, have assisted enormously in the technical details of editing, checking for accuracy footnotes, references, and translations. And last, my family—Susan, Tania, and Carin—each in her own way, have provided support and encouragement whose effects reside indelibly on every page.

Four National Science Foundation grants, between 1967 and 1973, aided greatly in the completion of this work. In particular, a sabbatical year (1972-1973) provided the necessary leisure to complete a first draft of the entire manuscript. Several small grants from the graduate school of Washington University aided in various stages of typing (and retyping!) the manuscript. Without this considerable material support, it would have been impossible to collect much of the primary source material and organize it into an integrated whole.

St. Louis, Missouri
January 15, 1977

Garland Allen

THOMAS
HUNT
MORGAN

CHAPTER I

Background and Early Life: The Formation of a Personality (1866-1886)

In his 1909 book, *Grosse Männer*, the famous German chemist Wilhelm Ostwald classified scientific men into two categories, romantic and classic. To the romantic, ideas come quickly, one on the heels of the other, with each idea having to find quick expression in order to make room for the next. According to Ostwald the romantic is an impatient person, who cannot stay with any one idea too long. The classic, on the other hand, is primarily concerned with the perfection of what already exists. He or she wants to consider ideas carefully, in order to set them in proper relation to one another and to the main body of scientific ideas. The classic tends to work over a subject exhaustively. It has been said, Ostwald comments, that the romantic revolutionizes, while the classic builds from the ground up.[1]

In 1936, the American embryologist Ross G. Harrison used Ostwald's classification to contrast the life styles of two of his close friends: Edmund Beecher Wilson (1856-1939) and Thomas Hunt Morgan. "Wilson is a striking example of a classic," he wrote, "and it is interesting to note that for many years his nearest colleague and closest friend [Morgan] was an equally distinguished romantic."[2]

Harrison's comparison is a useful one and provides a vantage point for viewing T. H. Morgan's life and work. E. B. Wilson represents a perfect example of the restrained and conservative, though dynamic, life style in science. All his life his work was devoted to cytology, the study of cell structure and function. Although trained as an embryologist, he was more concerned with how individual cells were put together, how they developed during the life of the individual, how they evolved during the history of the species, and how their parts functioned as an integrated whole.

1 Wilhelm Ostwald, *Grosse Männer* (Leipzig: Akademische Verlagsgesellschaft, 1909).

2 Ross G. Harrison, "Response on Behalf of the Medallist," *Science* 84 (1936): 565-567; quotation, p. 566.

His great published work, *The Cell in Development and Heredity*, took nearly a decade to write, went through three editions (beginning in 1896), and contained a massive but carefully organized body of material relating to all aspects of cell biology.[3]

Thomas Hunt Morgan, in contrast, moved from problem to problem, from idea to idea. He often left the working out of details suggested by his ideas to other people. Like Wilson, Morgan was trained as an embryologist. Yet, during the course of his career he became interested successively in the problems of regeneration, evolution, sex determination, heredity, and, finally, embryology once again. In all of these areas, Morgan had provocative or even unorthodox ideas. So powerful and active was his imagination, that his patience and persistence were unable to sustain his interest for extended and uninterrupted periods. Dropping one problem and picking up another, he would frequently come back to the first problem at a later date with fresh ideas or a new zest for the old approach. New ideas and interests constantly drove him on from one experiment to another, from one problem to another. A complex man, Morgan's approach to science implied a restless and adventuresome spirit.

Like all classifications, however, Ostwald's dichotomy between classic and romantic does not tell the whole story. Although Morgan and Wilson differed in personality, they also had much in common. They came from similar class backgrounds and matured intellectually in the same era of biological history. They held similar viewpoints about the nature of science and the critical methods of studying biology. Both contributed, in their different ways, to the upheaval in theories of heredity, evolution, and embryology that characterized the changing patterns of biological explanation in the twentieth century.

BACKGROUND[4]

At 201 North Mill Street in Lexington, Kentucky, stands Hopemont, the now restored Hunt-Morgan house, and the property of

[3] E. B. Wilson, *The Cell in Development and Heredity* (New York: Macmillan Co., 1896; 2d ed., 1897; 2d ed. rev., 1900; 3rd ed., 1925; 3rd ed. rev., 1928). A reprinted version of the first edition, with a lengthy introduction by H. J. Muller, has been made available by the Johnson Reprint Corporation, New York (1966).

[4] The most complete published accounts of Morgan's life to date include: Ian Shine and Sylvia Wrobel, *Thomas Hunt Morgan* (Lexington, Ky.: University of Kentucky Press, 1976); A. H. Sturtevant, "Thomas Hunt Morgan,"

the Blue Grass Trust for Historic Preservation (see figure 1). It was in this house on September 25, 1866, that Thomas Hunt Morgan was born. Hopemont had been built in 1814 by the wealthy Lexington merchant, John Wesley Hunt (1773-1849), reputed to have been Kentucky's first millionaire. The house was designed in the then fashionable late Georgian style, examples of which can be seen throughout many cities and towns in the South. A magnificent circular stairway, large rooms with high ceilings, and an intricate fanlight doorway are some of the most obvious characteristics of the style of architecture popular among the rising bourgeoisie of the early nineteenth century. The house was located in

Figure 1. T. H. Morgan's birthplace, Hopemont, built in Lexington, Kentucky in 1814 by his paternal great-grandfather John Wesley Hunt. Having been restored and converted into a museum, the house is now open to the public. (Courtesy of the Blue Grass Trust for Historic Preservation, Lexington, Kentucky.)

Biographical Memoirs, National Academy of Sciences 33 (1959): 283-325 (including a complete bibliography of Morgan's scientific writings); and G. E. Allen, "Thomas Hunt Morgan," *Dictionary of Scientific Biography* (New York: Charles Scribner's Sons, 1974), vol. 9, pp. 515-526.

what was at the time one of Lexington's most pleasant areas: in front of it was College Square, the campus of Transylvania University, and just across the street is the house in which Henry Clay had been married in 1799.

These surroundings, genteel and gracious if not extravagant, help explain T. H. Morgan's social ease and indifference to formal social life. His father's mother, Henrietta Hunt, was John Wesley Hunt's daughter. The Hunt-Morgan family line was itself a distinguished one and matched the physical surroundings with which it became associated in Lexington (see appendix 1).

John Wesley Hunt was born of English ancestry in Trenton, New Jersey, in August 1773. Left motherless at the age of fifteen, he came west with his brother and settled in the then frontier town of Lexington sometime around 1794. Here the two brothers opened a store which soon began to prosper. Hunt's interests rapidly widened to include the manufacture of hemp products, the importing and exporting of goods of various kinds, farming, the breeding of race horses, land speculation, founding of the Lexington & Ohio Railroad, and a variety of other enterprises.[5] He rapidly amassed a fortune unprecedented in Kentucky (on his death in 1849, his estate was valued at over $800,000), being at one time associated with John Jacob Astor and other capitalists of the eastern financial establishment.[6] Hunt was a liberal, a Whig, a supporter of Henry Clay, and also a staunch and longtime friend and trustee of Transylvania University, to which he often made financial contributions. He was also a founder of the Kentucky Lunatic Asylum, one of the earliest in America, and was for a time the postmaster of Lexington.

The Morgan family had originally come to North America in 1636 from Wales. At that time two brothers, Miles Morgan and James Morgan went first to Boston and then removed to Connecticut. The Miles Morgan branch of the family remained in New England and the East, while the James Morgan branch eventually moved south, first to Tennessee, and then to Huntsville, Alabama.

[5] William H. Perrin, *History of Fayette County, Kentucky with an Outline Sketch of the Blue Grass Region by Robert Peter* (Chicago: Baskin and Company, 1882), p. 80.

[6] See Burton Milward, *The Hunt-Morgan House* (Lexington, Ky.: The Foundation for the Preservation of Historic Lexington in Fayette County, 1955), p. 9; for more details on the financial career of John Wesley Hunt, see E. Polk Johnson, *A History of Kentucky and Kentuckians* (Chicago and New York: Lewis Publishing Co., 1912), p. 1,488.

A descendant of James, Calvin Morgan (1801-1854) married Henrietta (1823-1891), daughter of John Wesley Hunt.

In 1849, after the death of John Wesley Hunt, Calvin and Henrietta Morgan moved to Hopemont in Lexington. Their two most illustrious offspring were John Hunt Morgan (1825-1864), later to become a famous general in the Confederate Army (he was the leader of the guerilla band known as "Morgan's Raiders")[7] and Charlton Hunt Morgan (1839-1912), the father of Thomas Hunt Morgan. Charlton Hunt Morgan was born at Hopemont and educated in Lexington, graduating from Transylvania University in 1858. In 1859, at the age of twenty, he was appointed by President James Buchanan as the United States consul to Messina, Sicily. Here he became sympathetic to the cause of Italian liberation from Austria and became acquainted with the great Italian fighter, Giuseppe Garibaldi (1807-1882). Garibaldi's tactics were surprisingly similar to John Hunt Morgan's—both involved a band of volunteers using guerilla-like raids to strike at and overwhelm superior forces. Charlton Hunt Morgan assisted Garibaldi's cause by serving as a treasurer of the Garibaldi Fund. It was through Charlton Hunt Morgan's recommendation that the United States became the first nation to recognize Garibaldi's de facto government in May of 1860.[8] Resigning his consulship in 1861, Charlton Morgan was appointed secretary of the Southern Commission, with headquarters in London, England. Returning to the United States with special dispatches to the secretary of state, he retired from the diplomatic service and entered the Confederate Army, being made an aide-de-camp to General Arnold Elzey at the first battle of Manassas (Bull Run). Wounded and captured several times by the Union forces, he was finally held in the Ohio Penitentiary at Columbus in solitary confinement, and later at Fort Delaware until March 1865, when he was exchanged through special personal intercession between the archbishop of Pennsylvania and the U.S. secretary of war, Edwin M. Stanton (1814-1869).

Soon after his release in 1865, he married Ellen Key Howard, from an old aristocratic family of Baltimore, Maryland. Charlton then returned to military duty and remained with the last remnants of the Confederate Army that escorted President Jefferson Davis in his final flight, which ended at Washington, Georgia in early

[7] Milward, *The Hunt-Morgan House*, p. 11; see also Cecil F. Holland, *Morgan and his Raiders* (New York: Macmillan Co., 1942), p. 26.

[8] Perrin, *History of Fayette County*, p. 661.

May 1865. At the close of the war he returned to Lexington, where his first child, a son, Thomas Hunt Morgan, was born on September 25, 1866. From 1868 to 1869, Charlton was clerk of Beck's Senate Committee on Transportation Routes to the Sea, and was then appointed to the position of tobacco weigher at Louisville, Kentucky, a post he held until 1871. From that time on he remained in Lexington as steward of the Eastern Lunatic Asylum, a newer version of the institution originally founded by his grandfather, John Wesley Hunt.

When Charlton Hunt Morgan married Ellen Key Howard on December 7, 1865, still another prominent line was brought into the Morgan family heritage. Ellen Key Howard was the granddaughter (on her father's side) of John Eager Howard (1752-1827), a colonel in the Revolutionary Army and later governor of Maryland from 1788 to 1791. She was also a granddaughter (on her mother's side) of Francis Scott Key, and thus (through Catherine Grosch, who married John Wesley Hunt) a third cousin to Charlton. Charlton and Ellen Morgan had three children, Thomas Hunt, Charlton Hunt, Jr. (born in 1869), and Ellen Key Howard (born in 1873). An early family portrait is shown in figure 2.

THE CULTURAL MILIEU OF MORGAN'S EARLY YEARS

For the half century preceding T. H. Morgan's birth, Lexington, Kentucky had been a prosperous agricultural and social center. Dating from 1775, the city is situated in the rolling countryside of the bluegrass region of central Kentucky. Travelers and residents in the early nineteenth century described Lexington as "nearly central of the finest and most luxuriant country perhaps on earth," a "very beautiful body of land," and one of the "most fertile points of Kentucky."[9] Lexington also boasted the first college west of the Allegheny Mountains, Transylvania Seminary, founded in 1785 and opened in 1790, later known as Transylvania University and now Transylvania College. After the pressures of Indian warfare died down, Transylvania attracted people of considerable cultivation and learning from many parts of the East. As one observer wrote, "Men of high attainments in every branch of human knowledge" settled in the state "and brought with them

[9] Walter W. Jennings, *Transylvania, Pioneer University of the West* (New York: Pageant Press, 1955), p. 11.

their information and liberal views."[10] Lexington soon came to be known as the "Athens of the Western states."[11]

These growing educational and cultural advantages were made possible by the increasingly important position of Lexington as an agricultural and commercial center. During the early part of the nineteenth century manufactured and agricultural products, as well as travelers, passed through Lexington in increasing numbers on

Figure 2. Portrait of the Morgan family about 1874 (T. H. Morgan is in the middle, front). Shown from left to right are: Ellen Key Howard Morgan, Charlton Hunt Morgan, T. H. Morgan, Ellen Key Howard Morgan, and Charlton Hunt Morgan, Jr. (From the A. H. Sturtevant Papers, Courtesy of the Archives, California Institute of Technology.)

their way west and east as well as north and south. Transylvania remained the only university in Kentucky for almost a century. After the Civil War, however, in 1865, a state-sponsored university, the State College of Kentucky (now the University of Kentucky) was founded, giving the city two institutions of higher learning during T. H. Morgan's formative years.

During the course of the nineteenth century Lexington meta-

[10] The words are those of William Darby, writing about 1816, quoted in Jennings, *Transylvania*, p. 65.

[11] Ibid.

morphosed from a frontier town to one with a distinctly genteel and southern mentality. Before the Civil War most wealthy families owned slaves, and a good deal of the productive agricultural and commercial work was carried out by slaves or low-paid hired hands.

None of this glamour was to last, however. With the war between the states and the demise of the southern cause, social, political, and economic life in Lexington, as well as elsewhere in the South, underwent a change. Never committed strongly to the cause of slavery,[12] the Morgan family did not feel so acutely the effects of the war. They retained a friendliness and congeniality that seemed to carry them with ease through difficult times.

MORGAN'S EARLY CHILDHOOD

Early in his life, T. H. Morgan developed a strong interest in the out-of-doors and in the natural surroundings of Lexington, with its hills, meadows, and rivers. Young Morgan loved to roam through the surrounding countryside, collecting birds, eggs, fossils, and various other relics of nature. This interest emerged even before he reached the age of ten, and he often led groups of smaller boys on collecting trips in the Lexington territory.[13] He spent many summers with his Baltimore cousins in Oakland, in the mountains of western Maryland. In Oakland, as in Lexington, Morgan fished for crayfish and other aquatic specimens, and thus became familiar with a wide variety of living forms.

The 1870s was a period of great expansion of the American railway system. As a result, both in Maryland and in Kentucky,

[12] As a girl of eighteen, T. H. Morgan's mother even described the excitement surrounding the passage of slaves through Lexington on the underground railroad.

[13] W. H. Stephenson, "Thomas Hunt Morgan: Kentucky's Gift to Biological Science," *Filson Club Historical Quarterly* 20 (1946): 97-106, especially p. 98. See also "Some Chronological Notes on the Life of Thomas Hunt Morgan," typewritten manuscript in the T. H. Morgan Papers, Archives of the California Institute of Technology, Pasadena (hereafter referred to as Morgan Papers, Caltech). This manuscript was possibly prepared by Morgan's wife, Lilian, for there are penciled marginalia in her hand. It also follows very closely, both in overall organization and even in phraseology a mimeographed manuscript, "Notes on Thomas Hunt Morgan's Life," by his son Howard Key Morgan, prepared on the occasion of the dedication of the Thomas Hunt Morgan Intermediate School in Seattle, Washington.

railway cuts exposed large amounts of sedimentary rock. Morgan became particularly intrigued with studying these rocks, especially the fossil contents. His interest in geology and paleontology found expression during his college years, when Morgan spent several summers in the Kentucky mountains working for the U.S. Geological Survey.[14]

This early interest in natural history stayed with Morgan all his life. Despite the fact that in later years the laboratory took increasing amounts of his time, he always loved to get out and find organisms in their natural habitats. He knew a surprising number of species of many different higher taxa, and he was remarkably skillful at keeping them alive and in good condition in the laboratory. A remarkably gentle man, Morgan knew how to coax the scientific answers he was seeking from the living organism. During the course of his professional career, he published results on over fifty kinds of animals and at least one plant.[15]

EARLY EDUCATION

After attending the public schools of Lexington, Morgan was admitted in 1880, at the age of sixteen, to the Preparatory Department of the State College of Kentucky at Lexington. The Preparatory Department provided stiff training in what amounted to the final two years of high school.[16] At the time of Morgan's matriculation, the State College of Kentucky had 300 students and ten teachers. Of the latter, only one, the president, James Kennedy Patterson (1833-1922), held the Ph.D. degree.[17]

Patterson's relationship with his faculty as well as with students was one of benevolent despotism. He enforced strictly the 189 rules laid down by the board of trustees, and faculty members who disagreed strongly with his policies were often summarily

[14] Stephenson, "Thomas Hunt Morgan," p. 98; also, H. K. Morgan, "Notes on Thomas Hunt Morgan's Life," p. 1.

[15] Sturtevant, "Thomas Hunt Morgan," pp. 297-298.

[16] A scarcity of good high schools in the state had made necessary the establishment of the Preparatory Department in 1880. Morgan was thus among its first students. James F. Hopkins, *The University of Kentucky: Origins and Early Years* (Lexington, Ky.: University of Kentucky Press, 1951), p. 129.

[17] Ibid., chaps. 7 and 8. A detailed biographical account of Patterson's life is Nabel Hardy Pollitt, *A Biography of James Kennedy Patterson* (Louisville, Ky.: Westerfield-Bonte, 1925).

forced to resign.[18] The autocratic nature of the institution is further reflected in some of its "Rules for Students." For example, rule 75 states: "All deliberations or discussions among students having the object of conveying praise or censure or any mark of approbation or disapprobation toward the the college authorities, are strictly forbidden." Rule 100 states that: "All permits to be absent from any duty, or from quarters during study hours must have the approval of the President." And rule 129 states: "Students are forbidden to take or have in their quarters any newspapers or other periodical publications without special permission from the President. They are also forbidden to keep in their rooms any books, except textbooks, without special permission from the President."[19] It is no wonder that many students found themselves brought up before Patterson and the Faculty Disciplinary Committee for infraction of the rules.

Classical language and literature, morals, philosophy, ethics, and political economy formed a large part of the State College curriculum. However, in keeping with the needs of a growing agricultural and commercial economy, Patterson also emphasized the sciences. By the time that Morgan had arrived in the early 1880s, courses were offered in inorganic and organic chemistry, zoology, botany, veterinary science, geology, and mechanics. As an undergraduate, Morgan took courses in anatomy, physiology, hygiene, plant histology and physiology, microscopy, zoology, geology, and paleontology.[20] Particularly influential in Morgan's undergraduate career were A. R. Crandall, a geologist and former member of the State Geological Survey, and Joseph H. Kastle, an undergraduate who was two years ahead of Morgan. Many years later, in 1936, Morgan wrote: "Since my interests at that time were more in the direction of natural science, I came into closer relations with Professor A. R. Crandall than with other professors, and I should like to add that I have never met a better teacher."[21] Joseph Kastle went on to graduate school at Johns Hopkins, and it may well be because of Kastle's experience that Morgan resolved to attend Johns Hopkins after his graduation from the State College in 1886.

18 Hopkins, *The University of Kentucky*, pp. 154-161.

19 Ibid., pp. 166-167.

20 Stephenson, "Thomas Hunt Morgan," pp. 98-99.

21 Morgan to Dr. Frank McVey, president of the University of Kentucky, August 14, 1936; Hunt-Morgan Papers, Margaret I. King Library, University of Kentucky.

Although Morgan's career as a student was in many ways exemplary, he encountered difficulties during his undergraduate years. During his junior year he received demerits for "tardiness at chapel, disorder in the hall, and in section meetings."[22] Another incident was somewhat more serious. In Morgan's senior year, despite adequate classroom preparation, he found himself close to failing French. Morgan was unable to understand this, but soon enough an explanation became apparent. The professor of French and Germanic languages in the college, François M. Helveti, had been a Union soldier during the Civil War and had been captured by "Morgan's Raiders." As a hostage, Helveti had been forced to ride a mule about-face from Cincinnati to Lexington. Apparently Professor Helveti could never forgive a Morgan.[23]

Despite his problems with Professor Helveti, Morgan did pass French and graduated at the top of his class. His academic performance was outstanding, and he was voted highest honors by the faculty in April 1886.[24] This honor provided Morgan with the opportunity to present a valedictory address at the graduation ceremony. The manuscript of that address (or at least a rough version of it) has been preserved in the archives of the California Institute of Technology.[25]

It is difficult to estimate the influence of Morgan's college experiences on his developing intellectual and personal qualities. Many years later, from the perspective of old age and considerable success, Morgan wrote about his undergraduate days: "As I look back at those earlier days, I appreciate more and more the many kindnesses of the members of the teaching staff and also the really sound instruction that we received. When one recalls the simple conditions under which the A & M College carried on its work, it is surprising what an excellent group of teachers President Patterson brought together in the institute. . . . I realize the very great debt I owe to these earlier experiences."[26] Despite the fact that Patterson was a martinet and dominated every aspect of the operation of the college, he was a man of wide interests, a classicist and philosopher with a deep love of learning which he passed on to

[22] Stephenson, "Thomas Hunt Morgan," p. 98.

[23] Ibid., p. 99. [24] Ibid., p. 98.

[25] The address itself is undistinguished. It calls on Morgan's classmates to "go forth and do good deeds" in the formal style of commencement speeches.

[26] Morgan to Dr. Frank McVey, August 14, 1936; Hunt-Morgan Papers.

many of those around him. During his tenure as president, the State College grew into an important educational institution.[27] Yet, it was, in the last analysis, a small school oriented almost exclusively toward teaching. For Morgan, the contrast between his undergraduate institution and his graduate institution, the Johns Hopkins University, was considerable. It was a difference, in many ways, between the old style of university, typical of much of early nineteenth-century America, and the new style, typical of the post-Civil War period, influenced very much by the European (particularly German) model. Morgan may have appreciated his background at State College of Kentucky, but he dissociated himself from it and the type of institution it represented for the remainder of his life. All the other institutions with which he was later associated (Bryn Mawr College, Columbia University, and the California Institute of Technology) were private research-oriented institutions whose faculties were drawn from all corners of the globe.

MORGAN: THE MAN AND HIS PERSONALITY

According to his son, Morgan was over six feet tall, with black hair, and slightly stooped posture. His twinkling blue eyes were always engaged in active and keen observations, and he smiled easily.[28] Through most of his life he wore a beard and moustache, although on at least one occasion he shaved them off as a joke to see what his family's response would be. Predictably, his family (at least his wife and three daughters) were duly horrified, "one even bursting into tears."[29] Morgan paid relatively little attention to his personal appearance; he was often reported as looking a little unkempt and disheveled. Once when his collar became ripped away from his shirt, just before a lecture, he is reported to have said: "Oh, never mind, Bridges [Calvin B. Bridges, then one of his graduate assistants] will fix it with adhesive tape." Which is apparently just what happened.[30] William Bateson's description of Morgan during his stay with the Morgan family in 1921 is inter-

[27] Hopkins, *The University of Kentucky*, pp. 137-161.

[28] H. K. Morgan, "Notes on Thomas Hunt Morgan's Life," p. 1.

[29] Isabel Morgan Mountain, "Notes on T. H. Morgan," p. 21.

[30] Tove Mohr, "Personal Recollections of T. H. Morgan," address given at the centenary celebration of Morgan's birth, September 1966, Lexington, Kentucky; typescript, p. 2.

esting both for its graphic details of Morgan's appearance and for the insight it gives into Bateson's personality. In a letter to his wife, Bateson wrote: "Whether from poverty or neglect I don't know, but M's overcoat has an inch of bald collar where velvet was and gaping seams—his shoes are lacerated with long slits through which his socks appear,—real beggar's shoes."[31] Since Bateson often described people he met in a negative fashion, we might discount this description somewhat. However, Morgan's daughter reports that on at least one occasion her father, having forgotten to put on a belt, simply tied a piece of rope around his pants to hold them up for the day.[32] Yet his overall appearance was nonetheless commanding, as his tall figure, slightly stooped, would walk across the Columbia campus, coat blowing in the breeze.

All who knew him described Morgan's basic attitude as open and friendly—a very down-to-earth person. Despite his family and class background he was never aloof, frequently preferring the company of nonacademics and children to that of at least some of his scientific colleagues. When Otto L. Mohr and his wife Tove came from Norway to New York in 1918 to spend a postdoctoral year with Morgan, they were surprised at the informal atmosphere that Morgan created between himself and his students in the laboratory. Tove Mohr wrote, many years later: "How different everything was from the European Universities! We were used to a professor a little aloof, at a distance, who made a round in the lab now and then to guide and answer questions. But here in the New World, it was like a family having everything in common. We thought, 'Well, this is the way in the U.S.' Only later we learned that the atmosphere of the Morgan group was indeed unique."[33] A casual visitor at the Morgan household on any evening after supper might find Morgan romping on the floor with his children and their friends. In these circumstances, Tove Mohr wrote, "Morgan was not pretending—he *was* the children's age, enjoying them and their activities with everything else put out of his mind."[34] Sometimes his wife would have to rescue him from the overstimulated children.

[31] William Bateson to C. Beatrice Bateson, December 20, 1921; Bateson Papers, microfilm, American Philosophical Society.

[32] Mountain, "Notes on T. H. Morgan," p. 25.

[33] Tove Mohr, "Personal Recollections of T. H. Morgan," p. 1.

[34] Ibid., p. 3.

Although Tom and Lilian Morgan rarely entertained formally, they exhibited a constant hospitality to their friends, and particularly to Morgan's students and colleagues. While at Columbia, he used to hold weekly discussions with his graduate students in his home. On these occasions the group would discuss a new book or a new theory and have beer and other refreshments afterward. Similarly, he was very helpful to foreign visitors and constantly extended to them his hospitality in New York or at Woods Hole, Massachusetts, where the Morgans spent most of their summers.[35] Likewise, he did all he could for colleagues who needed help of any kind. His correspondence abounds in offers to help get materials, arrange lecture tours, or raise money for the publication of expensive monographs. On several occasions he was instrumental in negotiating new positions for colleagues who had been afflicted with poor health and needed a climatic change or who had become disaffected with their present posts.[36]

One of the most outstanding, if least visible, aspects of Morgan's personality was his financial generosity. On joint publication efforts, he rigorously made sure that each coauthor received his due share of funds and repeatedly refused to accept funds himself for translated publications of his own book.[37] Alfred Henry Sturtevant, one of Morgan's earliest associates in the study of heredity, told the following revealing story. He (Sturtevant) was about to graduate in the middle of the year, and thus would have to resign his undergraduate scholarship. At the time Sturtevant was working full time on research in the *Drosophila* lab, and thus wanted to stay on for the rest of the year. He went to Morgan, who sent him

[35] Morgan to Otto L. Mohr, May 19, 1919; T. H. Morgan Papers, American Philosophical Society; hereafter referred to as Morgan Papers, APS.

[36] For example, Morgan helped to secure a new position in 1926 for the geneticist, Dr. John Belling, who was at the time at the Carnegie Institution Laboratory at Cold Spring Harbor (Long Island, New York). Belling's wife had recently died, and he had suffered an acute case of depression which had led to hospitalization. In addition, a touchy controversy with his former collaborator, A. F. Blakeslee (1874-1954) had been taken very hard by Belling. Morgan wrote to E. B. Babcock, of the Department of Genetics of the University of California, Berkeley, suggesting that a complete change of location and climate might be desirable. Morgan offered also to make all the necessary negotiations with C. D. Davenport, then head of the Carnegie Institution Laboratory, if Babcock was favorable. The plan ultimately came to fruition, and Belling took up his new post and life in California in 1929.

[37] Morgan to Lippincott Company, 1920; Jacques Loeb Papers, Library of Congress.

over to the provost to see if he couldn't be fixed up with a scholarship renewal. When the provost said no, Sturtevant returned discouraged and explained the situation to Morgan. Several days later, Morgan told Sturtevant to go back and see the provost again; this time he was told that a scholarship was forthcoming. As Sturtevant wrote many years later: "I didn't have sense enough at the time, of course, to realize . . . that Dr. Morgan had put up the cash. . . . This is the way he would operate in those things if he could."[38] Similarly, a medical student who saved Morgan's life by preventing a severe hemorrhage after an automobile accident in Los Angeles in 1931, found himself the next year the recipient of a new medical scholarship. Later on, Morgan divided the Nobel Prize money equally among his childen and those of his two long-time collaborators, Calvin B. Bridges and A. H. Sturtevant.[39] When he did not actually give away money, he was always ready to lend it to friends and associates in an emergency.[40] However generous he was with his own financial resources, however, he was just the opposite with institutional funds. He seemed to take great pride in returning money from research grants at the end of the year and in finding new methods of saving money (as, for example, devising a new way of making *Drosophila* food). Although bookkeeping was an anathema to Morgan generally, he kept very accurate records of his departmental and research expenses. In several instances he carried meticulous financial honesty to the limit. In 1938, fifteen years after publication of *Laboratory Directions for an Elementary Course in Genetics* by Morgan, H. J. Muller, A. H. Sturtevant, and C. B. Bridges, the publisher (Henry Holt & Co.) decided to close out the final stock. They forwarded the accumulated royalties, amounting to $1.80, to Morgan. Morgan divided the amount four ways, sending 45¢ in stamps to each of the other three coauthors![41] Equally amazing, after returning from a trip to Berkeley in 1916, he wrote back to his host, Professor

[38] Interview with A. H. Sturtevant, July 24, 1965, Woods Hole, Massachusetts.

[39] Morgan to Sturtevant, January 17, 1934; Sturtevant Papers, California Institute of Technology Archives. The amount going to *each child* (three Sturtevants, three Bridges) was $4,700.

[40] Related by Dr. George MacGinitie, personal interview, February 14, 1973, Pasadena, California.

[41] See Morgan to H. J. Muller, May 16, 1938; from the Hermann J. Muller Collection, Manuscripts Department, Lilly Library, Indiana University, Bloomington, Indiana; hereafter referred to as Muller Papers.

J. C. Merriam, enclosing 41¢ in stamps for a telegram that he had sent from the Faculty Club. Morgan asked Merriam to make sure that this now settled his account!

Certainly one of Morgan's most outstanding personal traits, as judged by those who knew him, was his sense of humor. He seldom took himself seriously. One day in 1919 Tove Mohr was sitting in the lab at Columbia waiting for her husband. Morgan came in, stretched his two hands up in the air and yawned. Suddenly noticing Mrs. Mohr sitting there, he said, "Excuse my big yawn, but I have just come from one of my own lectures."[42] Similarly, Morgan described himself professionally in the following way: "I am professor of experimental zoology and I have three sorts of experiments: Fool experiments, Damned fool experiments, and Those that are still worse."[43] Likewise, his letters are filled with good-humored quips. In a letter to his good friend Hans Driesch in 1905, Morgan objected to Driesch's increasing excursions into the subjects of vitalism and mystical speculation. But at the conclusion of the letter, as if to soften the directness of his attack, Morgan wrote: "and believe me, always so long as there is a spark of *vitality* left in me, I remain your true and faithful colleague—T. H. Morgan."[44] Similarly, to his friend, Jacques Loeb, he wrote in 1907 about the problems involved in writing and publishing books: "Many thanks for your kind appreciation of my new book. It cost a lot of work and I only wish it were better. Bookmaking, as you know, is a perfect nuisance and leads only to indigestion."[45]

Just as Morgan did not take himself too seriously, he did not expect others to take him or themselves too seriously either. When he was with those who had a somewhat exaggerated sense of their own importance, he could not help but poke fun. Once, about 1932, following the International Genetics Congress at Ithaca, J.B.S. Haldane came to the Morgans' house at Woods Hole for

[42] Tove Mohr, "Personal Recollections of T. H. Morgan," p. 2.

[43] Ibid.

[44] Morgan to Driesch, October 23, 1905, p. 3; microfilm, Morgan-Driesch correspondence, origially obtained with the considerable help of Professor Reinhard Mocek, University of Leipzig, and through the courtesy of the Leipzig University Library. A copy of this microfilm material is now deposited in the library of the American Philosophical Society, Philadelphia. Hereafter, this material is cited as Morgan-Driesch correspondence. The frames of the microfilm are not numbered, but as the letters have been arranged chronologically, location of cited references is easy.

[45] Morgan to Jacques Loeb, March 13, 1907; Loeb Papers.

lunch. Haldane was telling the assembled group that he had signed a paper allowing a famous nerve specialist to get his brain when he was dead. Morgan pretended to find this a wonderful arrangement and asked Haldane a variety of more specific questions: Who would get to publish the results? Would Haldane get anything from that? Was the press to be properly notified? etc., etc. Although everyone else at the table realized what was happening, Haldane continued to answer every question very seriously and with a straight face. Tove Mohr, who was present, recalls: "How Morgan was amused. I will never forget the sparkle in his eyes, those very, very blue eyes. He never would tease people who could not defend themselves, but towards a man with Haldane's brain, he had no mercy."[46]

Despite his aptitude for humor and practical joking, Morgan always took both his work and his cultural interests very seriously. He worked continually, with what appeared to be a remarkable sense of concentration. Whenever he was sitting down, he was reading or writing. When he was walking, his mind was often engrossed in faraway thoughts. He would usually be in the laboratory by nine o'clock in the morning, where he worked steadily until noon. In New York he would have lunch and work through the afternoon until about four or five o'clock, at which point he would frequently play handball with friends. In Woods Hole during the summer, he would swim and then have lunch with his family. Occasionally they went on a picnic to the beach or out on E. B. Wilson's sailboat. Then he would work the rest of the afternoon, occasionally taking off an hour or so for tennis. In New York, Morgan would arrive home around 6:00 p.m., often stop in the pantry and take a small glass of bourbon and proceed to sip it while sitting at the old family player piano, pumping away at rolls of Beethoven's Kreutzer sonata or a Strauss waltz.[47] In the evenings, he would usually play with his children until eight o'clock or so, at which time he would help put them to bed by telling each one a different kind of story. Then, after lights out he would work—usually write—until midnight or one o'clock in the morning.

Such a schedule was more or less the routine pattern, but it found frequent variation. People were constantly dropping into his laboratory, both at the university during the winter and at Woods Hole during the summer. Foreign visitors, invited lecturers, and

[46] Tove Mohr, "Personal Recollections of T. H. Morgan," pp. 2-3.
[47] Mountain, "Notes on T. H. Morgan," p. 14.

guests had to be entertained, and the Morgans did their share. Not a rigid person, Morgan never seemed to object to spending time talking with students or colleagues when they had a genuine question or problem to raise. In fact, his students often came to him for personal as well as scientific advice, and he was well acquainted with many aspects of their personal lives. When interrupted, he would often appear brusque, but if the matter were serious he would continue the conversation. His friend E. G. Conklin once remarked much later in his life, "I never saw a person who wasted so little time."[48]

As a young man Morgan made frequent trips to Europe, where he became acquainted with classical painting, sculpture, and European (especially Italian) literature.[49] Through his European associations, he became particularly interested in the philosophy of Schopenhauer, but wrote that he had to give it up because his scientific work did not leave him enough time to wade through such heavy philosophical tracts.[50] Particularly exciting to Morgan had been his introduction to Dutch painting through a trip to Holland in 1898. As he wrote to Driesch:

> Thanks for your valuable directions. I had a most delightful trip down the Rhine and in Holland and in Belgium. . . . I should regret very much that I did not return by way of Italy except for the pictures in Holland and Belgium. It was an entire revelation to me and I was quite carried off my feet for I had never before realized what Dutch art was. From what you told me I did not expect to find very much, and it was a great surprise to me. It is true that *some* of the best pictures are out of Holland, but I did not appreciate them until I had seen what remained in Holland.[51]

Aside from Flemish painters, Morgan was especially fond of the Italian Renaissance painters Botticelli, Crivelli, and Bellini.[52] However, strong as it was in his youth, Morgan's interest in visual art seemed to pass away as he grew older, no doubt a result of the increased demands for time that his work placed on him.

[48] E. G. Conklin, "Some Personal Recollections of Thomas Hunt Morgan," unpublished manuscript, Morgan Papers, Caltech.

[49] Morgan to Driesch, February 24, 1896, p. 4; Morgan-Driesch Correspondence.

[50] Ibid., December 6, 1896, p. 3.

[51] Ibid., September 18, 1898, pp. 3-4.

[52] Ibid., p. 6.

The one cultural interest he retained from early until late in his life was music, particularly opera. Later on, in New York City, whenever he heard a street singer from behind his house on 117th Street, Morgan would exclaim, "What a glorious voice, a former Italian grand opera basso, I'm sure." While at Bryn Mawr College, where he taught from 1891 to 1904, his letters to Driesch are filled with references to the various performances he was able to see in Philadelphia. Particularly fine, to him, was a Carmen that he saw in 1896 with Emma Calvé, Nellie Melba, and Jean and Eduard DeReszke. As Morgan wrote to Driesch: "You see, even in the *back woods* of America it is not so dull as you might think."[53] In New York, one of the few activities the Morgans attended with any regularity were concerts. Mrs. Morgan played the violin in the Columbia University Orchestra, and Morgan's friend E. B. Wilson was a cellist of professional caliber (Wilson's daughter Nancy became a professional cellist). Music figured prominently in the Morgans' lives.

Like most scientists of his time and social class, Morgan paid relatively little attention to contemporary social or political issues. He operated largely on the principle that scientists serve humanity best by concentrating their efforts on their scientific work rather than by making public pronouncements and being socially active in one or another cause. He apparently looked with disfavor on the political and social activism of his students H. J. Muller, Alexander Weinstein, and Edgar Altenburg, and he is reported to have told J.B.S. Haldane (a noted political activist in the 1920s and 1930s) that he (Haldane) would do best to stay in the laboratory and leave politics to others. In those days, as today, the prevalent view was that political or any other kind of activism decreases the quality of scientific output.

Morgan did not oppose, and in fact, encouraged, participation in the general duties of citizenship, such as voting. He also recognized that in periods of national emergency scientists would be in some ways obligated to participate in practical research. For this reason he supported formation of the National Research Council in 1918 as a governmental advisory arm of the National Academy of Sciences. All these activities were, in his mind, necessary but peripheral to the normal functioning of science in a society. Morgan's view of the relationship between science and society was

[53] Ibid., February 24, 1896, p. 3.

basically one of separateness: he held that science functioned largely apart from society and that those who practiced science should focus on solving scientfic problems, leaving social and political problems to others. None of this means that Morgan was callous or insensitive to large-scale problems around him. The letters he wrote to Driesch between 1914 and 1919 testify to the horror with which he viewed the events of World War I, and he is reported to have once claimed that he could never really appreciate the great medieval cathedrals of Europe because of the suffering and poverty that had gone into producing them.

Evidence that Morgan had relatively little interest in conventional politics comes from the fact that his family can recall very few occasions on which he discussed political issues or even for whom he voted in presidential elections. In general, his outlook might be described as conservative, middle-of-the-road, though with an open-minded and humanitarian turn. He was distinctly opposed to extremism on either the right or the left, and thus appears to have found distasteful some of the socialist movements in the United States during and after World War I. When one of his graduate students was seeking employment in 1922, he wrote a favorable recommendation to Herbert Spencer Jennings at Johns Hopkins. Apparently in response to a question from Jennings about the man's political persuasion, Morgan wrote: "There is not the slightest possibility of his going into any 'radical social movements' so long as he is allowed to earn a modest living doing scientific work. . . . He has never joined any radical clubs or associations of any kind and in my opinion is much too broad-minded to get involved in any set of fanatics or fakers. He has a liberal outlook, but adjusts himself quietly and peacefully to the ordinary conventional life that engrosses us all."[54] Morgan did not object to the student's political views per se. What he might have found objectionable was a political conviction that competed with science for the student's time in any serious way. As he remarked later in the same letter, the saving grace of this student's political and social interests was that they were "largely theoretical."

Morgan was a person whose feelings were not always easy to determine. In his vast correspondence he seldom discussed his emotional reactions. He appears to have operated in a similar way in day-to-day contact with both his colleagues and his famliy. Sel-

[54] Morgan to H. S. Jennings, March 22, 1922; from the Ross G. Harrison Papers, Yale University Library; quoted with permission.

dom were harsh words spoken in the Morgan household.[55] Feelings, both positive and negative, were difficult to elicit from Morgan. One of his daughters writes that when as a young woman she sought some insight into his relationship with their mother, she asked, "Mother is a wonderful woman, isn't she?" "Of course she is," Morgan replied, and that ended the conversation.[56]

Yet Morgan was neither an unfeeling nor an uncommunicative person. There was never any doubt among those close to him about how he felt about them. But it apparently took time to learn to read these feelings, and, of course, not everyone who came into daily contact with him learned to read his unexpressed feelings accurately. Some thought him aloof and detached and occasionally felt that he disliked them. But the bulk of the evidence suggests that Morgan was congenial, friendly, and a personal as well as scientific inspiration to all who knew him well, and even to many who knew him only casually.

As with most people, Morgan was more complex than any description could adequately communicate. He was himself a curious mixture of the romantic and classic. As a romantic he sought new ideas and pushed continually toward new projects and new methods. As a classic, he often came back again and again to the same problems in biology, of which he never seemed to tire. Like a romantic, he had a warm and friendly feeling toward most people, but like the classic he often kept those feelings hidden from public view. The impish twinkle in his eye over a new idea or joke was set against a rather quiet outward demeanor.

[55] H. K. Morgan, "Notes on Thomas Hunt Morgan's Life," p. 1; also, Mountain, "Notes on T. H. Morgan," pp. 8-10.

[56] Mountain, "Notes on T. H. Morgan," p. 8.

CHAPTER II

Graduate School, Morphology, and Sea Spiders (1886-1891)

After graduation from the State College of Kentucky in 1886, Morgan spent the summer at the Marine Biology School at Annisquam, Massachusetts. Originally organized by Alphaeus Hyatt (1838-1902) under the auspices of the Boston Society of Natural History, the small station offered courses in marine biology and the natural history of marine organisms.[1] The Annisquam experience gave Morgan his first systematic contact with marine organisms, thus providing an opportunity to learn important collecting and handling techniques for live specimens. To Morgan the experience was exhilarating, and he wrote back to his friend Ballard Thurston of Louisville:

> And now about the summer school. Annisquam is about thirty miles North (and a little East) of Boston. It is a beautiful place, the country and bay I mean, not the village. The coast is rocky (like all of New England I have seen) but between the rocks some very good beaches are found and to these we repair very often. The laboratory is situated on an inlet of the bay. A wharf runs out one side, with a float at its end to which the boats are tied. These are furnished by the laboratory as are also dip nets, skimming nets, etc. The water is fairly alive with animals and a great variety of them are found.
>
> I continually congratulate myself that I am here and not on the survey [i.e. the U.S. Geological Survey on which Morgan had worked several previous summers in Maryland and Kentucky]. Besides what I shall learn, the weather is so cool and invigorating, not to say anything about the food which is somewhat better than in H [Hancock, Maryland where Morgan had worked on the Geological Survey] County. Alto-

[1] W. K. Brooks, "Alphaeus Hyatt (1838-1902)," *Biographical Memoirs, National Academy of Sciences* 6 (1908): 313-325.

gether I am delighted with myself for being here and without doubt the work will be of the greatest assistance to me next winter.[2]

In his first experience with marine organisms, Morgan displayed a basic love of living things in their natural habitat which was always central to his study of biology.

This was apparently Morgan's first trip to New England. He traveled from Norfolk, Virginia to Providence, Rhode Island by steamer, a trip that he described as "very cold and rather rough."[3] The journey from Providence to Boston and north to Annisquam was by rail. With his typical humor and zest Morgan wrote back: "I have been in Boston two or three times and strange to say did not get lost which I believe is 'quite the thing' to do. I account for this by the fact that I always took the streetcars. I went to Harvard, but did not remain long enough to see much. Perhaps you will say this was not the case at Wellesley."[4]

In the fall of 1886, Morgan entered Johns Hopkins University to begin work for a Ph.D. degree in zoology. It is not quite clear how Morgan chose to go to Hopkins, then still a fairly young school (it was founded in 1876). Later in his life he recounted what may have influenced him: "My days at Johns Hopkins were probably not very different from those of other students who were attracted there by the rather vague rumor that reached us as undergraduate students in far distant colleges. In my own case it was through Joseph Castle [*sic*], who had preceded me by a couple of years. . . . Perhaps the fact that my mother's family were Baltimoreans had some effect; but little did I know then how little I appreciated that a great university had started in their midst, and I think this was typical of most of the old families of that delightful city."[5]

[2] Morgan to Ballard Thurston, July 9, 1886; from the archives of the Filson Club, Louisville, Kentucky.

[3] Ibid., p. 1.

[4] Ibid., pp. 1-2.

[5] Undated letter from T. H. Morgan to A. H. Sturtevant; Sturtevant gives the date of the letter as about 1943; Sturtevant Papers, California Institute of Technology Archives; quoted with permission. Joseph Kastle (Morgan misspelled it) had graduated from the State College of Kentucky in 1884 and had entered the graduate school of Johns Hopkins in the fall of that year. It may well be, as James F. Hopkins has suggested, that Morgan was inspired to attend Hopkins by his friendship with Kastle when the two were students at the State College of Kentucky; see James F. Hopkins, *The*

Prior to the Civil War, most Americans seeking doctoral level training in scholarly and scientific (and even in many cases, medical) fields went to European universities (especially to Germany for the sciences). In choosing to seek his graduate education at Johns Hopkins, Morgan was among the first generation of Americans to break with the European tradition. Although Johns Hopkins was young in 1886, its reputation had spread far and wide. Morgan was energetic and ambitious, and the new challenge of America's first truly graduate university appealed strongly to him.

JOHNS HOPKINS UNIVERSITY: THE FOUNDING AND GUIDING PRINCIPLE

The post-Civil War period was one of educational renaissance in the United States. The Morrill Act of 1862 made possible the granting of public lands to individual states for the founding of state colleges and universities. In 1865 the Massachusetts Institute of Technology was founded, in 1868 Cornell. In 1869 with the election of Charles William Eliot to the presidency of Harvard a new era was initiated in curriculum reform at the college level with a particular emphasis on the sciences.[6] On the heels of this increased activity came the founding of Johns Hopkins University in 1876.

Johns Hopkins (1795-1873) was a financier who made his fortune by establishing the Baltimore and Ohio (B & O) Railroad. A bachelor, Hopkins left half of his $7,000,000 estate in trust for the founding of a university and hospital. The twelve trustees of this estate were drawn from prominent Baltimore citizens, the majority being businessmen, including Hopkins's trusted friend John Work Garrett, a leading mogul of the B & O Railroad.[7] The trus-

University of Kentucky: Origins and Early Years (Lexington, Ky.: University of Kentucky Press, 1951), p. 289.

[6] For a sketch of these developments, see the following: Richard J. Storr, *The Beginnings of Graduate Education in America* (Chicago: University of Chicago Press, 1953), pp. 9-14; Francesco Cordasco, *Daniel Coit Gilman and the Protean Ph.D.* (Leiden: E. J. Brill, 1960), pp. 6-10; and Hugh Hawkins, *Pioneer: A History of the Johns Hopkins University, 1874-1889* (Ithaca, N.Y.: Cornell University Press, 1960), pp. 7-8.

[7] Hawkins, *Pioneer*, pp. 3-4. Four of the trustees were lawyers and one, James Carey Thomas, a physician and the father of Martha Carey Thomas, later president of Bryn Mawr College. Two were judges of the Supreme Court in Baltimore, and one of these, George William Brown (1812-1890),

tees invited Daniel Coit Gilman (1831-1908) to become the first president of Johns Hopkins University. In many ways Gilman was an ideal choice for the type of university the Hopkins trustees envisioned. Trained at Yale's Sheffield Scientific School (1848-1852), he had studied for a year at the University of Berlin (1854-1855) before returning to the Sheffield Scientific School in 1855, where he served, successively, as librarian, secretary, and eventually professor of physics and political geography. He had left the Sheffield in 1872 to assume the leadership of the University of California.[8]

Gilman maintained that the university of the future (i.e., Johns Hopkins) should be research oriented and nonsectarian, and bring together the best minds available at the time. While classical education was to be preserved, Gilman was strongly committed, by his own background as well as by his reading of the times, to emphasis on the sciences. To build his faculty, he toured the United States and Europe seeking out research-oriented scholars. The first faculty was not large. Significantly, of the initial seven faculty positions, four went to science. In addition, however, there were approximately eleven junior faculty or assistants, and several invited faculty from other institutions who presented series of lectures. It is interesting to note that when T. H. Morgan undertook the task of organizing the Division of Biological Sciences at the California Institute of Technology in 1928, he followed many of Gilman's procedures. For instance, he toured Europe seeking research-oriented, younger biologists to make up his first faculty. And, like Gilman, he strongly maintained that the university should remain nonsectarian.

Along with establishing a good faculty, it was Gilman's aim to

---

had in earlier years been active in quelling riots in the city of Baltimore. In 1835 Brown had joined a body of citizens organized by Samuel Smith to suppress the riots attending the failure of the Bank of Maryland (ibid., pp. 5-6). As mayor of the city, he had played a considerable role in quelling the Baltimore riot of April 19, 1861; the riot grew out of the response of angry citizens of the city to the passage of Union troops through the city on the Northern Central Railroad. See Faye Thomas Scharf, *The Chronicles of Baltimore* (Baltimore: Turnbull Brothers, 1874), pp. 587-601; see also George William Brown's *Baltimore and the 19th of April, 1861* (Baltimore: Johns Hopkins University, 1887).

[8] Fabian Franklin, et al., *The Life of Daniel Coit Gilman* (New York: Dodd Mead & Co., 1910), pp. 1-14; also Cordasco, *Daniel Coit Gilman*, pp. 2-6; 17-32.

attract to Hopkins good students, especially graduate students. He sought to do this by designing a series of appointments that capable graduate students could hold and that could, if followed in a stepwise fashion, lead ultimately to a Hopkins professorship. The initial appointment was as a "scholar," which could lead on to an appointment as a fellow, an assistant, an adjunct, and finally a university professor. Of course, it was not envisioned that every graduate student would follow the complete sequence. But most capable students traversed at least the first two or three stages. Gilman's use of one stage, the fellowship, was particularly important in attracting to Hopkins in its early years some of the best students in the United States. The granting of fellowships was not based primarily on need, but rather on academic merit.

Gilman created an exciting academic community by focusing on the essential ingredients: the people. By gathering highly motivated and first-rate faculty and students, he emphasized that the core of any educational process is mutual stimulation and exchange of ideas. Morgan has left an interesting picture of the excitement he felt as a student at Hopkins and the central role that Gilman played in fostering that atmosphere:

> My first year was one of great enthusiasm that did not diminish during the subsequent four years that I was there as a graduate student.
>
> Only slowly and little by little did I come to realize that it was President Daniel Gilman who was the kingpin of all that was taking place. Like all graduate students I saw him only occasionally, usually at the beginning of the fall term when we assembled to hear the "professors" tell of their summer experiences. . . . On these occasions, Dr. Gilman was the life of the party, smiling, friendly, hearty, and without a trace of pomposity. The only other occasion when I caught a glimpse of him was when I went to the post office to get my mail. The President was often seen flitting between his office door to and from Mr. Ball's desk in that half of the post office when he carried on the business of treasurer.
>
> As was probably the case with most of the students at Johns Hopkins, I came to realize later what Gilman had accomplished not only for this university, but for higher education throughout the country. Of hundreds of colleges both great

and small, all of them carry on today the policy that Gilman inaugurated, emphasizing the importance of research not only in science, but in all branches of scholarship. I can think of not one man that has been as influential in the growth of our institutions.[9]

At the time Morgan was a student at Johns Hopkins, the zoology department was under the leadership of H. Newell Martin, a former student of Michael Foster (1826-1907) and, before coming to Johns Hopkins, an assistant to T. H. Huxley. Martin was an animal physiologist of considerable note, strongly influenced by Foster's progressive views of the importance of laboratory and experimental work in both teaching and research. Foster had, in fact, set up the first teaching laboratory for physiology in England, and Martin expected to follow his lead. To Martin, physiology, with its emphasis on the experimental method, was to be the dominant science in the new university, with the more descriptive areas such as anatomy and embryology in important, but subservient, positions. The predominance of physiology was made overtly clear some years after the founding of Hopkins, when the new biological laboratory was dedicated in 1884. At the dedication ceremony, Martin stated: "So many distinct branches of biological sciences are pursued in it [the new laboratory], we call it in general the biological laboratory; but it is a biological laboratory deliberately planned that physiology in it shall be queen, and the rest her handmaidens."[10] Martin went on to say that some day a sister building for the descriptive areas of biology would be built. But rather than do two things poorly, it was necessary to concentrate on one at a time.

The close ties that Johns Hopkins established from the beginning with medical practice contributed significantly to the initial prominence being given to physiology. But there was more to it than that. Behind the emphasis on physiology lay a historical struggle within the biological community which was coming to a head in the last decades of the nineteenth century. This struggle developed between two distinct traditions: the naturalist and the experimentalist.

[9] T. H. Morgan to Abraham Flexner (no date, but in a folder labeled 1943); Morgan Papers, Caltech; quoted with permission, California Institute of Technology Archives.

[10] *Johns Hopkins University Circular* 30 (April 1884): 87.

THE NATURALIST-EXPERIMENTALIST DICHOTOMY

Neither the naturalist nor the experimentalist traditions can be regarded as specific fields or disciplines within the biological sciences, nor as a set of well-defined principles or practices. Rather, each represents a different viewpoint about organisms, the methods for studying them, the kinds of questions that are most meaningful to ask, and the kinds of answers that are most acceptable. In some sense these two traditions represent, at least among biologists, two different world views. Both traditions have existed in one form or another during most of the history of biology. The naturalist tradition, however, was dominant from ancient times through most of the nineteenth century.

The naturalist tradition is characterized by a number of different but interconnected concepts and practices. In general, it involves an interest in the whole organism as opposed to its isolated parts and is concerned largely with form or structure as opposed to function. The naturalist tradition took its cue from the field naturalist, who was concerned with organisms in their natural state, and in the post-Darwinian period, at least, with the problem of adaptation and evolution. At the same time the naturalist tradition encompassed museum studies of preserved specimens, with a great deal of emphasis on taxonomy and classification. In the post-Darwinian period natural history became especially concerned with historical questions: how did the variety of living forms, particular adaptations, or particular ecological interrelationships come to exist? Methodologically the naturalist tradition has been largely descriptive. Whatever the subject, its data have been qualitative, and largely observational, from the growth of embryos to description of fossil forms. The naturalist tradition received its greatest stimulus, of course, from the work of Darwin in the last third of the nineteenth century.

Ever since its deliberate introduction into biology in the late seventeenth century, the experimentalist tradition has been characterized by a strong interest in function as opposed to structure. It thus found the study of animal and plant physiology particularly congenial. The experimental approach was tied much more closely to the laboratory than to the field and often saw organisms wholly out of the context of their natural surroundings. With its emphasis on controlled conditions, the experimental approach tried to reduce the number of variables operating in a particular situation at any one time. One consequence of this approach was that experimentalists often investigated only a single process in isolation from others. A cardinal assumption of physiologists was that the principles of function determined in experiments with isolated parts (e.g., organs) could be applied to an understanding of the organ's function within the intact animal.

As the name implies, the chief method by which experimentalists studied biological problems was through experiments. By definition, experimentation means the imposition of unnatural conditions upon a system. Experimentalists persisted in the assertion that the response of the system under artificial conditions reveals a great deal about its function under normal conditions. They further maintained that the data obtained from laboratory experiments can be validly extrapolated to the field situation. The experimentalist tradition also emphasized a rigorous approach to scientific problems. This meant, among other things: (1) the posing of hypotheses in such a way that rival views could be distinguished from one another by an appropriate experiment; and (2) the collection of quantitative, rather than qualitative, data. From the very earliest days, experimentalists in biology often attempted to model their methodology on the physical sciences. Many experimentalists had as their stated aim to make biology as rigorous and "scientific" as physics and chemistry.

One prominent aspect of the experimental tradition is "reductionism." Reductionism is a complex philosophical and methodological approach, not easy to characterize. However, the central element of reductionism is the view that complex processes can be reduced to already familiar component systems often, but not necessarily, at a lower or simpler level of organization. Physicochemical reductionism is thus one, and perhaps the most prominent, form of the reductionist view that prevailed during the late nineteenth and early twentieth centuries. Historically, however, reductionism has not been a necessary component of the experimentalist tradition. In fact, it represents only one trend within the experimentalist tradition, especially evident in Germany from the 1840s on. Claude Bernard, for example, the great French physician and physiologist, was an avid experimentalist who opposed the strong reductionist tradition in Germany in the 1860s and 1870s.

From the viewpoint of contemporary philosophy there is no necessary contradiction between the experimentalist and naturalist approaches to biological problems. Each deals with different levels of organization or with different kinds of problems. However, through much of the history of nineteenth- and early twentieth-century biology, the two views were often very much in conflict. The conflict was open and conscious in the minds of many workers, and in many cases they came to view themselves as belonging to one or the other camp. Naturalists accused experimentalists of seeing only a part of a system and generalizing from it to the whole; experimentalists accused naturalists of drawing from merely descriptive observations, speculative conclusions which could never be tested. Naturalists saw in experimental work an artificial distortion of the living state which they felt could have no relation to normal life processes; experimentalists saw in naturalists' endless descriptions no information that led beyond trivial, individual problems.

The naturalist-experimentalist dichotomy was amply characterized by a number of writers in the late nineteenth century, among them William Bateson and E. B. Wilson. In his 1894 book, *Materials for the Study of Variation,* Bateson wrote:

> [The data of variation] attract men of two classes, in tastes and temperament distinct, each having little sympathy or even acquaintance with the work of the other. Those of the one class have felt the attraction of the problem. It is the challenge of nature that calls them to work. But disgusted with the superficiality of "naturalists" they sit down in the laboratory to the solution of the problem, hoping that the closer they look the more truly will they see. For the living things out of doors, they care little. Such work to them is all vague. With the other class, it is the living thing that attracts, not the problem. To them the methods of the first class are frigid and narrow. Ignorant of the skill and of the accurate final knowledge the method has bit by bit achieved, achievements that are the real glory of the method, the "naturalists" hear only those theoretical conclusions which the laboratories from time to time ask them to accept. With senses quickened by the range and fresh air of their own work, they feel keenly how crude and inadequate are these poor generalities and for what a small and conventional world they are devised. Disappointed with the results, they condemn the methods of the others, knowing nothing of their real strength. So it happens that for them the

study of the problems of life and the species become associated with crudeness and meanness of scope. Beginning as naturalists, they end as collectors, despairing of the problem, turning for relief to the tangible business of classification, accounting themselves happy if they can keep their species apart, caring little how they became so, and rarely telling us how they may be brought together. Thus each class misses that which in the other is good.[11]

[11] William Bateson, *Materials for the Study of Variation* (London: Macmillan & Co., 1894), pp. 574-575. That Bateson was not alone in recognizing the existence of a dichotomy between the naturalist and experimentalist traditions is readily discernible by reading through biological journals during the period from 1890 to 1910. C. S. Minot at the Harvard Medical School pointed out its pervasive influence on medical education (C. S. Minot, "Certain Ideals of Medical Education," *Journal of the American Medical Association* 53 [1909]: 502-508). And Henry Fairfield Osborn, a paleontologist at the American Museum of Natural History and adjunct professor of zoology at Columbia, complained that by the early 1920s natural history was in such disfavor that it was difficult to obtain money for research and teaching positions in paleontology and other related areas (H. F. Osborn to E. G. Conklin, March 29, 1921; Conklin Collection, Firestone Library, Princeton University). In the field of neurology a similar dichotomy over the question of the origin of nerve fibers has been pointed out and described recently by Levi-Montalcini. The issue centered on whether nerve fibers originated from bridges left between cells in the dividing embryo, or grew outward from the central nervous system. As Levi-Montalcini summarized the fundamental problem: "The significance of this controversy . . . did not, perhaps, lie only in the decision as to the basic mechanism by which connections in the nervous system are formed, fundamentally important as that is. As so often with such hotly contested disputes, it seems to have symbolized more. It was a struggle between those who felt that the mystery of the organism as a whole is utterly lost to sight when it is subjected to analysis, and those whose approach is atomistic, who try to resolve the whole into parts from the interrelations of which the whole can be derived." (See Rita Levi-Montalcini, "Problems and Perspectives in Neurobiology: A Brief Survey of Old and New Concepts," *Archivio Zoologico Italiano* 51 [1966]: 629-666.)

Bateson's description cited above further indicates the deep-seated hostility and even bitterness attending the dichotomy between the naturalist and experimentalist traditions. This is due in part at least to the fact that the naturalist tradition, predominant in the pre-1880 period had begun to lose ground to the experimental by the 1870s. A quick survey of articles published in the journal *The American Naturalist* in three years (1900, 1906, 1912) shows a significant change in the percentage of articles dealing with natural-history subjects on the one hand, and experimental on the other. Somewhere between 1906 and 1912 a dramatic drop occurred in the number of papers devoted to natural history, while a corresponding upsurge is noted

In a very interesting and revealing article written in 1901, "Aims and Methods of Study in Natural History," E. B. Wilson echoed Bateson's observations. Wilson captured particularly well the psychological milieu in which the dichotomy was perpetuated:

> I shall never forget the impression made on me many years ago, shortly after returning from a year of study in European laboratories [including the Naples Zoological Station], by a remark made to me in the friendliest spirit by a much older naturalist, who was one of the foremost systematic and field naturalists of his day, and enjoyed a world wide reputation. "I fear," he said, "that you have been spoiled as a naturalist by this biological craze [experimentalism] that seems to be running riot among the younger men. I do not approve of it at all." I was hardly in a position to deny the allegations; but candor compelled me to own to having had a suspicion that while there might have been a moat in the biological eye, a microscope of sufficient power might possibly have revealed something very like a beam in that of the systematist of the time. However that may have been, it is undeniable that at that period, or a little later, a lack of mutual understanding existed between the field naturalist and the laboratory workers which found expression in a somewhat picturesque exchange of compliments, the former receiving the flattering

---

in the number of experimentally oriented works. The data (in percentages) are as follows:

Year	*Natural History Papers*	*Experimental Papers*
1900	89	11
1906	92	8
1912	47	53

For the purposes of this survey natural history included articles dealing with paleontology, invertebrate and vertebrate zoology (but not physiology), botany, descriptive embryology, and general field natural history; experimental included general physiology, heredity (where experimental breeding was involved), experimental embryology (including regeneration), and cytology. *The American Naturalist* is a journal with a natural-history bias and therefore was chosen specifically to illustrate an upsurge in interest in experimental topics. Other journals have not been surveyed; it would be interesting to know whether a similar shift can be observed in others at the same, or possibly at a different, time.

> appellation of the "bug hunters," the latter the ignominious title of the "section-cutters," which on some irreverent lips was even degraded to that of the "worm-slicers." . . . I dare-say there was on both sides justification for these delicate innuendos. Let us for the sake of argument admit that the section-cutter was not always sure whether he was cutting an *ornithorhynchus* [duck-billed platypus] or a Pearly Nautilus and at times perhaps he did lose sight of out-of-doors natural history and the living organism as he wandered among what Michael Foster called the "pit-falls of carmine and Canada balsam" [a stain and preservative material, respectively, used in preparing and staining sections of organs and tissues for microscopic observation]; but let us in justice mildly suggest that the bug hunter, too, . . . was sometimes a trifle hazy as to whether the cerebellum was inside or outside the skull, and did not sufficiently examine that hoary problem as to whether the hen came from the egg or the egg from the hen, and by what kind of process.[12]

Wilson also pointed out that experimentalists, buoyed by their successes, particularly in neurophysiological work such as that of Hermann von Helmholtz (1821-1894) in the 1840s and 1850s, often adopted an arrogant attitude. Hailing experimentalism as the *only* method of any possible value in scientific work, they seemed to forget that observation and description had any role in the gaining of knowledge either in the past or present.

The descriptive tradition in biology in the post-Darwinian period was, of course, composed of numerous fields and subdisciplines. Because Morgan was trained largely in the descriptive tradition, it will be necessary to turn attention to the particular areas that influenced him.

THE MORPHOLOGICAL TRADITION AND W. K. BROOKS

At the time Morgan was a graduate student at Hopkins, the dominant area of the naturalist tradition was known as "morphology." By definition "the study of form," morphology encompassed a number of disciplines that today would be considered

[12] E. B. Wilson, "Aims and Methods of Study in Natural History," *Science* 13 (1901): 14-23; quotation, p. 19.

independent fields of biology: comparative anatomy, embryology, paleontology, and cytology.[13] Although morphology as an area of biology has had a long history, with changing goals and methods over the years, in the post-Darwinian period it was understood to encompass three major aims:

1. to determine the basic unity of plan underlying the diversity of living forms
2. to discover the common ancestor, the so-called "archetypal" form that related two (or more) divergent groups of organisms, or served as an ancient progenitor of a single modern line
3. to reconstruct family trees, or phylogenies as they were called (the evolutionary course of phyla over time), of various animal and plant groups based on evidence from comparative anatomy, the fossil record, and careful study of embryonic development.[14]

Morgan was trained at Johns Hopkins under William Keith Brooks (1848-1908), a student of Louis and Alexander Agassiz at Harvard, and a specialist on the embryonic development of marine invertebrates. Brooks was a morphologist par excellence. His standards and practice of morphological study provide a specific example of the naturalist tradition in the late nineteenth century. His influence on Morgan and other members of the younger generation was striking—as much for what he taught them not to do as for what he taught them to do. No understanding of Morgan's growth as a biologist can be complete without noting the influence that the morphological tradition in general, and Brooks in particular, had on his thinking.

Born in Cleveland, Ohio, Brooks was educated at Williams College, receiving the bachelor's degree in 1870. In 1873 he enrolled in the first session of the Anderson School of Natural History on Pennikese Island, off the coast of Cape Cod near Woods Hole, Massachusetts. This summer program, a brainchild of Louis Agassiz (1807-1873), was designed to train advanced students and teachers in the elements of natural history and marine biology.

[13] Today, the term "morphology" is used almost synonymously with "anatomy." Its present-day meaning also encompasses the concept of developmental anatomy, that is, the developmental pattern of various structures. It is important to keep in mind that the term has a much more limited use today than it did in the 1880s and 1890s.

[14] One of the most complete and concise accounts of the whole range of activities encompassed by morphology in the nineteenth century can be found in Patrick Geddes's article, "Morphology," *Encyclopedia Britannica*, 9th ed. (1883), vol. 16, pp. 837-846.

As with so many biologists in the latter part of the nineteenth century, Brooks's first contact with marine organisms began a fascination he retained for life. As a result of his experience in the Anderson School, in the fall of 1873, Brooks went to study as a graduate student with Agassiz at Harvard's Museum of Comparative Zoology. Agassiz died that fall, but Brooks continued his work in descriptive morphology under Louis's son, Alexander Agassiz (1835-1910). After receiving his Ph.D. (the third granted by Harvard University) in 1875, Brooks accepted Gilman's offer of a position at Hopkins, working under Martin to train medical and graduate students in areas of descriptive natural history. Brooks remained at Hopkins throughout his academic career, rising to the position of associate professor in 1883, professor in 1891, and head of the Department of Biology in 1893 (a position he maintained until his death). During his first seventeen years at Hopkins (1876-1893), Brooks published sixty-five scientific papers in the fields of morphology and embryology. His studies included a detailed and comprehensive history of the Chesapeake Bay oyster (work that, by the way, was instrumental in helping to save the oyster from near extinction, thus greatly helping the Maryland oyster industry),[15] and detailed morphological studies on a number of marine species, including the genus *Salpa.* Brooks's work on this organism is representative of the morphological tradition.

A confirmed Darwinian, Brooks felt the morphologist could most accurately come to understand the evolutionary significance of structures by studying their process of formation. Following the general tenets of the biogenetic law, developed in the 1860s by Ernst Haeckel (1834-1919)[16] Brooks studied the comparative natural history of adult and larval forms (larvae are immature,

[15] Brooks's work on the oyster was one of the most immediate demonstrations during the early years of the university of the practical benefits of pure research within institutions of higher learning. Gilman did not allow this work to go unpublicized and continually pointed out to his trustees and potential benefactors that higher education could, in fact, strongly serve the needs of industry (Hawkins, *Pioneer*, p. 144).

[16] The biogenetic "law" is not a law at all but a generalization stating that "ontogeny recapitulates phylogeny," i.e., the history of every individual's embryonic development repeats, in telescoped fashion, the phylogenetic history of the species. This law provided the rationale for studying embryonic development in great detail, for in that process was supposed to lie the history of the whole species. Biologists today do not believe that the correspondence between embryonic and evolutionary stages is significant enough to provide any real clues about the course of a species' evolutionary history.

hence embryonic, stages in development) of *Salpa* to gain an insight into the adaptive pressures at work on the species during its evolutionary history.[17] He felt it was possible to deduce from studying larval or other embryonic stages of an organism not only the evolution of structure, but more importantly the evolution of the interaction of structure and function, i.e., the evolution of adaptations.

To Brooks, marine organisms provided the most valuable materials for morphological studies. Marine forms, he believed, often retained many more of their primitive ancestral characteristics than terrestrial or fresh-water forms. Because he felt that competition and selective pressure had always been less in the oceans than in other ecological situations, he felt that marine forms had undergone less transformation in the course of their evolutionary history. Furthermore, to Brooks, marine life offered such an immense diversity that evolutionary relationships could be seen more clearly than in land forms, where the diversity in any one area was often more restricted. To this end, Brooks was instrumental in founding the Chesapeake Zoological Laboratory of Johns Hopkins in 1878. In addition, he conducted summer courses at the biological stations at Beaufort, North Carolina and Bermuda.

However, to suggest that Brooks was concerned only with describing the life cycles of marine vertebrates or invertebrates, and speculating about evolutionary relationships, is to greatly underplay the breadth of his thinking and interests. All his life he was especially concerned with the larger problems of contemporary biology, particularly the mechanisms of heredity and embryonic differentiation. His book, *The Law of Heredity* (1883) combined Darwinian ideas of pangenesis (a particulate theory of heredity which Darwin used partly to account for what he felt to be the inheritability of acquired characteristics) with elements of August Weismann's germ-plasm theory.[18] Among the larger issues, evolu-

[17] W. K. Brooks, *The Genus Salpa* (Baltimore: Johns Hopkins Press, 1893).

[18] W. K. Brooks, *The Law of Heredity. A Study of the Cause of Variation, and the Origin of Living Organisms* (New York: John Murphy, 1883). Weismann's germ-plasm theory was formulated from observations on early stages of cell division in certain roundworms (*Ascaris*). Weismann, an outstanding microscopist until poor eyesight forced him to abandon observational for speculative work, noted that very early in embryonic development a single cell gave rise to all the adult germ tissue (ovary or testes). This observation suggested to Weismann that during embryonic

tion and adaptation figure prominently in Brooks's writings. His views on heredity were developed, as were those of many of his generation, in response to the questions about the origin of variation, which Darwin had raised, but not satisfactorily answered, between 1859 and 1883. Brooks was no mere taxonomist or collector. He possessed a strong interest in large, general problems and an ability to focus on the study of specific details.

Like most morphologists, Brooks did not refrain from speculation about the great biological issues of the day. He maintained that speculation was valid and even essential, especially in areas such as evolutionary history where direct verification was almost never possible. A complete series of fossil forms was almost the only definitive evidence of a particular line of development. Such a series was seldom, if ever, obtained. Thus, Brooks once wrote that "the evidence [for constructing phylogenies] is circumstantial, but still, with a modicum of restraint it could be carried out in a rational way. It did, however, require the educated imagination:

> The morphologist unhesitatingly projects his imagination, held in check only by the laws of scientific thought [Brooks does not say exactly what these "laws" are], into the dark period before the times of the oldest fossils and feels absolutely certain of the past existence of stem forms [primitive ancestral groups from which several phyla subsequently developed] from which the classes of echinoderms have inherited the fundamental plan of their structure, and he affirms with equal confidence that the structural changes which have

---

development, the sperm- and egg-producing tissues, the germ plasm, remained separate and were distinct from the body tissues, the somatoplasm. This implied further that changes in the body tissue (mutilation, atrophy, increase in size through use, etc.) could not alter the germ plasm, and thus could not be passed on to the next generation. Weismann became a strong and persistent opponent of the idea that acquired characteristics could be inherited. He later performed a series of experiments in which he cut off the tails of mice for many generations to demonstrate that such repeated alterations of body tissue did not produce an inherited change in the germ tissue (i.e., shortening of the tails in future generations). On Weismann, see: Helmut Risler, "August Weismann (1834-1914)," *Berichte Naturforschung Gesellschaft Freiburg im Breisgau* 58 (1968): 77-93; also, Frederick Churchill, "August Weismann and a Break from Tradition," *Journal of the History of Biology* 1 (1968): 91-112.

separated this ancient type from the classes which we know were very much more profound and extensive than all the changes which each class has undergone from earliest palaeozoic times to the present day.[19]

In Brooks's work, emphasis on description and license for speculation were combined with a skepticism about the value of experimentation. As Brooks saw it, experimentation tended to isolate parts from wholes and thus not see the broad actions that were the key to life.[20] Similarly, employing the tools of physics and chemistry appeared to Brooks inappropriate for investigating the fundamental processes of living organisms. Toward the end of his life, Brooks wrote that one could achieve important discoveries and unravel the mysteries of life without grants of money, foreign travels, the latest equipment, and "above all, without undertaking to resolve biology into physics or chemistry."[21] Brooks was, indeed, a naturalist and morphologist of the old school, which dominated so much of biology in the later nineteenth century.

It is important to note that Brooks, like many of his contemporaries, often confused the concepts of experimentalism and reductionism. Experimentalism, the manipulation of variables to produce some observable outcome is not synonymous with reductionism, the explanation of one theory in terms drawn from another theory, usually in a different branch of science. While it is true that many advocates of experimental method in biology have also championed one or another form of reductionism, the two methods are philosophically quite distinct. The development of reductionist ideas has comprised a long and often confusing chapter in the history of philosophy. More frequently than not, it has been understood by scientists, lay people, and some philosophers, to be the process of explaining phenomena at a higher level of organization in terms of phenomena at a lower level of organization. Describing the physiology of the nervous system in terms of the ionic properties of its individual units, the neurons, or cellular respiration in terms of biochemical pathways, are both common examples of what most writers mean when they speak

[19] Brooks, *The Genus Salpa*, p. 164.

[20] W. K. Brooks, "The Lesson on the Life of Huxley," *Smithsonian Institution Annual Report, 1900* (Washington, D.C.: U.S. Government Printing Office, 1902, p. 710.

[21] Ibid.

of reduction. As some contemporary philosophers of science point out, however, reductionism does not philosophically require that an explanation be couched in terms of lower levels of organization.[22] For example, to describe interneuronal connections in the nervous system in terms of electrical circuitry does not necessarily reduce the central nervous system to lower-level components. Yet it does explain one theory (neurophysiology) in terms of another (electricity). Philosophically it involves the same sort of explanatory process as the more classical reductionism which might be couched in terms of atoms or molecules. The important point here is to emphasize that Brooks was characteristic of most biologists, and especially naturalists, of his era, in thinking of reductionism as a necessary philosophical component of experimentalism.

Personally, Brooks was an enigmatic and often withdrawn person who baffled his friends and students alike. A corpulent man, he was described as possessing a dignified, but somewhat eccentric, appearance.[23] Lacking, it is said, many of the more traditional social graces, he chewed tobacco, was fidgety and ill at ease in most social gatherings, rambled in his discourse, and appeared to be most at ease stretched comfortably in his steamer chair reflecting on broad philosophical issues.[24] Morgan provides one picture of Brooks, in a recollection written about 1943. Early in the fall of every year at Johns Hopkins, it was customary for faculty and graduate students to come together for an evening session in which the faculty described its summer's research. "Professor Brooks was asked to speak of his summer experience as a member of a

[22] For a fuller discussion of the problem of reductionism, see Kenneth Schaffner, "Antireductionism and Molecular Biology," *Science* 157 (1967): 644-657; "The Peripherality of Reductionism in the Development of Molecular Biology," *Journal of the History of Biology* 7 (1974): 111-139; and "Approaches to Reduction," *Philosophy of Science* 34, no. 2 (1967): 137-147.

[23] See E. G. Conklin, "William Keith Brooks," *Biographical Memoirs, National Academy of Sciences* 7 (1910): 25-88.

[24] G. Lefevre, "William Keith Brooks, A Sketch by Some of His Associates," *Journal of Experimental Zoology* 9 (1910): 17. In addition to his biological interests, Brooks was also a devotee of philosophy, being an especial admirer of the English phenomenologist, George Berkeley (1685-1753). Brooks continually admonished his students to study philosophy as well as biology, advice which few of them took literally, though he may have impressed upon them the importance of relating detailed observations to broad, general issues.

commission to study 'the Maryland oyster.' He always prepared his talk very carefully. In the course of the evening his turn was overlooked. After adjournment we had a general pow-wow. One of our group approached Dr. Brooks who was silent and glum. 'Good evening, Dr. Brooks,' he said. Dr. Brooks suddenly awakened, looked at him vaguely, and said oratorically, 'The American oyster is about to be exterminated.' Obviously the opening sentence of his prepared address."[25] Yet Brooks was a kind man, not unconcerned about the welfare or attainments of his students. More often than not he seemed to have been lost in his own world of biological problems and philosophy.

Brooks's role as a teacher at Hopkins has been the subject of several historical papers.[26] There is little doubt that his students included some of the most original and influential biologists in the twentieth century. In addition to T. H. Morgan, he taught Edmund Beecher Wilson (cytologist), Ross G. Harrison and E. G. Conklin (embryologists), and William Bateson (geneticist). At the time of Brooks's death in 1908, his students held commanding positions around the world according to E. A. Andrews of Johns Hopkins. Andrews wrote that as a friend and teacher Brooks was "an inspiration and example."[27] A whole generation of young biologists, of which Morgan was a member, owed their biological heritage to Brooks.[28]

One reason for this may have been Brooks's remarkable ability to relate the smallest observation to problems of broad and general biological interest. Bateson, who spent two months in the summer of 1883 and one month in 1884 studying with Brooks at the seaside laboratory in Beaufort, wrote: "For myself I know that it was through Brooks that I first came to realize the problem which for years has been my chief interest and concern. . . . For me this whole province was new. Variation and heredity with us had stood as axioms. For Brooks they were problems. As he

[25] Morgan to Abraham Flexner; Morgan Papers, Caltech; quoted with permission.

[26] Most notably, Dennis M. McCullough, "W. K. Brooks' Role in the History of American Biology," *Journal of the History of Biology* 2 (1969): 411-438; and E. G. Conklin, "The Life and Work of William Keith Brooks," *Anatomical Record* 3 (1909): 6.

[27] E. A. Andrews, "William Keith Brooks," *Science* 28 (1908): 784.

[28] J. Walter Wilson to D. H. McCullough, personal communication; cited in McCullough, "W. K. Brooks' Role in the History of American Biology," p. 411.

talked of them, the existence of these problems became imminent and oppressive."[29]

Another reason for Brooks's apparent success as a teacher may have been his tendency to let students go their own ways, with very little direct interference on his own part. Like his own teacher, Louis Agassiz, Brooks allowed his graduate students to work largely on their own initiative.[30] W. H. Howell, H. N. Martin's successor in physiology at Hopkins, corroborated this statement: "In many cases the routine work for the Ph.D. degree was accomplished with but little direct supervision on his [Brooks's] part."[31] Those students who were capable of independent research stayed on and developed the skills necessary to pursue their own courses. Those who needed continual guidance usually did not make it. As E. G. Conklin commented: "He [Brooks] believed so thoroughly in the law of natural selection, as he once said, that he thought it best for a student to find out for himself as soon as possible whether he was fitted for independent investigation or not, and by this rigid discipline the unfit were weeded out from the fit."[32]

Morgan said very little about his own experiences with Brooks. Thus it is difficult to determine what feelings he may have carried away from his graduate experience. From the few remarks that are preserved, it would seem that Morgan was equivocal about Brooks's influence. On leaving Hopkins in 1891 for a post at Bryn Mawr, Morgan wrote to President Gilman of Hopkins to resign the Bruce Fellowship he had held for a year, stating: "I will always look back with pleasure to the time spent with Dr. Brooks in his laboratory."[33] Morgan evidently thought enough of Brooks to dedicate his first book (*Evolution and Adaptation*, 1903) to him. However, even the circumstances of that dedication are enigmatic. In a letter to E. B. Babcock at the University of California in 1917, Morgan wrote: "Did I ever tell you what Brooks said to me when I asked him if I could dedicate my book on

[29] Quoted in McCullough, "W. K. Brooks' Role in the History of American Biology," p. 443.

[30] R. T. Bigelowe, "Memoirs—W. K. Brooks," *Proceedings of the American Academy of Arts and Sciences* 71 (1937): 491; cited in McCullough, "W. K. Brooks' Role in the History of American Biology," p. 429.

[31] McCullough, "W. K. Brooks' Role in the History of American Biology, p. 429.

[32] Conklin, "The Life and Work of William Keith Brooks," p. 6.

[33] Morgan to D. C. Gilman, 1881; from Johns Hopkins University Archives, letter supplied by D. M. McCullough; quoted with permission.

'Evolution and Adaptation' to him? He thought a while and then delivered the ponderous statement: 'Do just as you like, Morgan.' I confess that I felt rather dashed but had gone too far to back down."[34] Morgan appears to have liked the freedom to pursue his own research undisturbed. But he also appears to have disliked the aloofness and peculiar habits Brooks sometimes displayed.

On at least one occasion Morgan directly confronted Brooks in an incident that may have produced a lasting pique in the professor. Even as a graduate student Morgan was not one to venerate authority. At a biological seminar one day, Morgan confronted Brooks about the latter's views on the question of the inheritance of acquired characteristics. Morgan's fellow student and longtime friend E. G. Conklin told the story as follows:

> I recall in a session of the biological seminar at Johns Hopkins when Professor Brooks, a staunch supporter of Weismannism in opposing the inheritance of acquired character, pointed out the improbability if not impossibility of finding a mechanism for the inheritance of acquired characters. Morgan interjected, "But, Dr. Brooks, you have approved pangenesis [a theory proposed by, among others, Darwin, to account for the heredity of acquired variations] for this purpose in your book on *The Law of Heredity* [1883]. Brooks insisted that he had not done so. Morgan proposed to get the book and prove it, when Brooks got up and left the room and as he passed out of the door he looked back and said, "I think I ought to be given credit for knowing my own mind."[35]

There was also a negative side to Brooks's influence which went much deeper than personal idiosyncrasy. A number of his students

[34] Morgan to Babcock, December 29, 1917; from Papers, Department of Genetics, University of California, Berkeley; housed at the American Philosophical Society.

[35] E. G. Conklin, "Thomas Hunt Morgan," unpublished manuscript (ca. 1945); E. G. Conklin Papers, Firestone Library, Princeton University. A modified version, omitting the above anecdote, was published by Conklin in *Biological Bulletin* 93 (1946): 14-18. Morgan was evidently correct in his challenge to Brooks. As Conklin pointed out, in *The Law of Heredity* Brooks wrote: "The mode of origin and transmission of the gemmules is essentially like Darwin's conception, and we must acknowledge that Buffon's view of the part played by his organic molecules was very near the truth."

have documented their exasperation at the aims and methods of morphology that Brooks championed. To many of the younger biologists, phylogeny seemed a fruitless quest. The data were too circumspect and the conclusions too fanciful to ever be rigorously tested. One phylogenetic scheme seemed about as good as another, with often no way to distinguish between them. In a letter to his friend Hans Driesch in 1899, Morgan remarked about Brooks's new book, *The Foundations of Zoology*: "My old professor, Dr. Brooks, of Balto [*sic*] has published a book that you may have seen. A strange production in which are trotted out all the worn out themes of the metaphysicians. . . . Perhaps you may be interested in his theme that life is a response to nature. But his thought is vague and he fails to handle his subject or to tell clearly what he means."[36] Morgan and many of his fellow students at Hopkins felt that in the long run Brooks's style of biology was not likely to lead to the solution of crucial biological problems. As Dennis McCullough has shown, most of Brooks's students ended by working in areas quite different from that of their teacher. Almost all became strong experimentalists and renounced both the methods and aims of morphology. Thus, perhaps Brooks's greatest service was allowing each student to follow his own leads.

The conflict between naturalist and experimentalist which had become so prominent in biology from the 1850s on was present at Hopkins from its inception. Martin was undisputed head of the biological sciences; Brooks was hired explicitly as his younger assistant, to handle the morphological subjects. Any hopes Brooks may have had of parity between physiology and morphology were dashed soon after his arrival in Baltimore. As we saw in a preceding section, Martin was unequivocal about the dominant role of physiology in the new university. Morphology was definitely secondary. Also, because of the medical orientation of physiology, most of the graduate students in the first decade or so were M.D.'s interested in physiological research. In dedicating the new biological laboratory in 1884, Martin pointed out that morphology had had its day—it was time for the experimental side of biology (i.e., physiology) to come to the fore: "I think even the morphologists will admit that hitherto, and especially in the United States, they have had rather more than their fair share; numerous museums

[36] Morgan to Driesch, February 12, 1899, p. 5; Morgan-Driesch correspondence, APS.

and laboratories have been built for their use; while physiology, if she got anything, had been usually allotted some out of the way room in an entirely unsuitable building, if no one else wanted it; and been very glad to get even that."[37] Despite being younger than Martin, Brooks was a representative of the older and declining tradition of the naturalists. The experimental tradition was beginning to win out in its long struggle against descriptive natural history. The relative positions of Brooks and Martin at Johns Hopkins clearly symbolized this development. Morgan entered biology at a crucial transition period, and this transition greatly influenced the subsequent course of his biological work.

MORGAN'S WORK UNDER BROOKS

Morgan's doctoral dissertation under Brooks concerned the embryology and phylogeny of a group of invertebrates known as Pycnogonids (sea spiders). Morgan's sketch of the adults of one of these forms is shown in figure 3. This group of curious organisms had been a taxonomic enigma for over a century, ever since the Swedish systematist Karl von Linné (Linnaeus) described them in the twelfth edition of his *Systema Naturae* (1767). Most naturalists prior to Morgan agreed that Pycnogonids were Arthropods (animals with jointed legs, including insects, crabs, lobsters, and true spiders). What was not clear was the relation of the Pycnogonids to other Arthropods: in particular, were they members of the class Arachnida (which includes the terrestrial spiders), or the class Crustacea (the lobsters, crabs, crayfish, etc.)? Pycnogonids seemed to have characteristics of both groups. In his thesis of 1891, Morgan questioned the conclusions on the evolution of Pycnogonids that Anton Dohrn, then director of the Stazione Zoologica in Naples, had reached in his recent studies. Dohrn worked in great detail on the comparative anatomy of the adult and late embryonic stages of the Pycnogonids and concluded that they descended neither from the Arachnids nor the Crustaceans, but represented an independent, parallel evolutionary line from the Annelids (segmented worms such as the common earthworm). Morgan felt that before this conclusion could be accepted it was necessary to study the early embryological stages: "These stages should be known and take equal rank with com-

[37] *Johns Hopkins University Circular* 30 (April 1884): 87.

parative anatomy of the adult in disentangling the relationships of the group."[38] In his studies on early development, Morgan ultimately came to quite different conclusions from Dohrn's.

Morgan collected the specimens for his work in the summers of 1889 and 1890 at Woods Hole, Massachusetts.[39] His procedure was to collect fertilized eggs, prepare them for microscopic observation by fixing and embedding them in paraffin, thin-sectioning them, and then staining the sections to bring out certain cell structures in sharper detail. The fertilized eggs would be in various stages of cleavage (the cell divisions of early embryonic development), so that from a single batch Morgan could observe nearly a complete series of embryos. It was much like looking at a randomly selected set of frames from a motion picture. In the pattern of early cell divisions, Morgan hoped to find clues about the ancient phylogenetic relationships of the sea spiders. This procedure could be meaningful, of course, only when the details of early embryonic development in Pycnogonids were compared to those in other groups to which it was supposedly related (Annelids or Arachnids). Morgan chose to compare the early, rather than later embryonic or adult stages because he felt, like Brooks, that the earliest stages would have shown the least evolutionary changes over time. They would thus represent the most ancient traits of the species. At this point in his career Morgan, like Brooks, followed more or less the basic tenets of Haeckel's biogenetic law, which saw embryonic development as a telescoped glimpse of phylogenetic history.

In his thesis, Morgan concluded that the Pycnogonids descended from Arachnids many eons after the whole Arthropod group had split off from the Annelids, thus openly opposing the Annelid theory of Dohrn. The evidence on which Morgan based his conclusions was rigorously stated, and consisted of several different observations: (1) The pattern of endoderm formation (figure 3) is very similar to that found in the Arachnids, but quite different from other Arthropods or the Annelids. (2) The Pycnogonids show signs of vestigial organs which appear to resemble Arachnid

[38] T. H. Morgan, "A Contribution to the Embryology and Phylogeny of the Pycnogonids," *Studies from the Biological Laboratory, Johns Hopkins University*, vol. 5, no. 1 (Baltimore: Isaac Friedenwald, 1891), pp. 1-76.

[39] In the first summer (1889) he worked in the laboratory of the U.S. Fish Commission and in the second (1890) at the Marine Biological Laboratory.

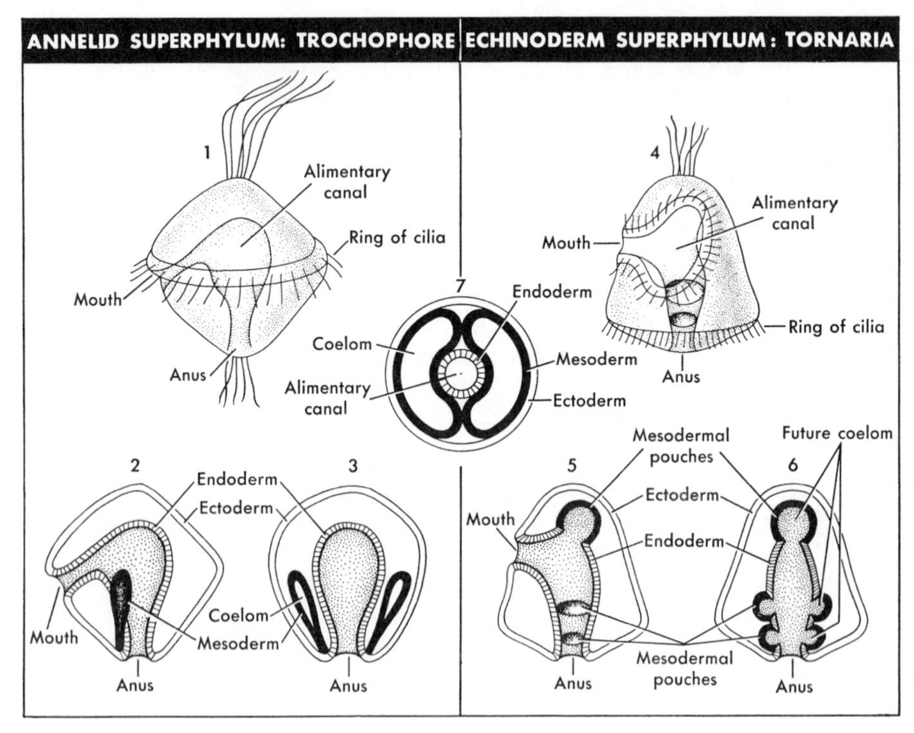

Figure 3. Comparison of larval forms representing two major groups of the invertebrates: the Annelid and the Echinoderm superphyla. The trochophore larva, shown at left, characteristic of the Annelid superphylum, is seen in a side and front view; the Tornaria larva shown at right, characteristic of the Echinoderm superphylum, is seen in similar views. A fundamental difference between the two superphyla occurs in the formation of the mesoderm (a germ layer) during embryonic development of the larva. In members of the Annelid superphylum, the mesodermal layer originates from special cells near the point of transition between ectoderm and endoderm near the primitive anus. In members of the Echinoderm superphylum the mesoderm arises from pouches which fold in from the wall of the developing alimentary canal (gut). By detailed study of the development of the mesoderm in Pycnogonids, Morgan was able to establish their phylogenetic relationship to the Echinoderm superphylum, a group that includes both the Arthropods (crustaceans, insects) and the Arachnids (spiders, scorpions). Further study showed Morgan that within the Echinoderm superphylum, the Pycnogonids more closely resembled the Arachnids than the Arthropods.

rather than Annelid or other Arthropod organs. (3) The first areas of the young Pycnogonid embryo to differentiate are the same as those in the young Arachnid embryo, but quite different from the Annelid or other Arthropod groups. (4) In both Pycnogonids and Arachnids, the first pair of appendages have been modified into feeding organs, and are innervated directly from the brain. In most other Arthropods the feeding organs are modified from second, third, and fourth appendages, and are innervated from a postoral ganglion; Annelids, being segmented worms, have no appendages at all. (5) Early studies in the nineteenth century had revealed that Pycnogonids have one more pair of appendages than Arachnids—an observation that had figured strongly in Dohrn's contention that the two groups were not phylogenetically related.[40] In studying the early embryonic development of Arachnids, however, Morgan noted that these forms possess rudimentary appendages on the abdomen; as the embryo develops, these appendages degenerate and are not visible in the adult. Thus, the Arachnids and Pycnogonids appear to have the same basic number of appendages after all.[41]

From these and other observations, Morgan tried to show how the Arachnid theory was the only view that made sense of all the data about Pycnogonid development. In drawing his conclusions, Morgan emphasized that study of the early stages in embryonic development was crucial for drawing phylogenetic relationships. Later embryonic or adult stages were often so highly modified or degenerated that they were not reliable for comparative studies between groups. He also emphasized that the availability of fresh material was crucial for carrying out such studies. Previous investigators (such as Dohrn) had worked with specimens collected and preserved (in alcohol or formalin) and observed some time later. In materials stored for long periods, many of the subtler details of early development are obscured. As a result of this experience, Morgan came to feel strongly that seaside laboratories, where fresh material was always available, were of crucial importance in carrying out morphological studies in depth.

[40] This observation was described in detail by Francis Maitland Balfour, "Notes on the Development of the Araneina," *Quarterly Journal of Microscopital Science* 20 (1880): 167-189; cited in Morgan, "A Contribution to the Embryology and Phylogeny of the Pycnogonids," pp. 31-32.

[41] All the above summarized from Morgan, "A Contribution to the Embryology and Phylogeny of the Pycnogonids," pp. 29-33.

Although the morphological nature of Morgan's early research is obvious from the outset, there are some distinct and important differences between Morgan's work and that of old-style morphologists such as Brooks. First, Morgan laid out neatly and clearly the conflicting opinions about the phylogenetic history of the Pycnogonids. With several hypotheses from which to draw, Morgan's aim was to provide further observations on which one or another of the hypotheses could be rejected. It is often assumed that a distinction between rival hypotheses can be made only by devising a "crucial" or critical experiment. Morgan's work showed, however, that new observations can also be a valid means for distinguishing between differing views. Brooks seldom presented his evidence in such a clear-cut, rigorous way. He did not so often attempt to eliminate alternative hypotheses as simply to champion overtly the one (or ones) he thought most likely. Second, there is in Morgan's work almost a complete lack of the kind of speculative suggestions about phylogenetic relationships that is so often found in Brooks's writings. Morgan's thesis sticks closely to the problem of the Pycnogonids' relation to the Crustacea, Arachnida, and Annelids. He limited himself to disproving the hypotheses proposing a Crustacean and Annelid origin for the Pycnogonids and to substantiating a hypothesis proposing an Arachnid origin.

This ability to focus clearly on a central issue remained characteristic of Morgan throughout his life. Although he occasionally indulged in speculation, the vast bulk of his work remains close to the available facts. Although trained as a morphologist, Morgan's thought processes demonstrated a critical quality that was often lacking in morphologists of the older generation.

With his thesis behind him and his credentials obtained, Morgan was prepared to leave Brooks and Hopkins and begin his independent career in the academic world. In the fall of 1891, at the age of twenty-five, he accepted a position at Bryn Mawr College.

CHAPTER III

Entwicklungsmechanik and The Revolt from Morphology (1891-1904)

At Bryn Mawr College Thomas Hunt Morgan was an associate professor of biology. Thirteen years later, in 1904, he left Bryn Mawr to accept a post as professor of experimental zoology at Columbia University in New York City. In that period his research interests switched almost completely from the morphological and descriptive to the experimental and analytical. He would later become one of the most outspoken proponents of the experimental approach to biology in the twentieth century.

EVENTS, 1891-1904

In August of 1891, when Morgan returned to Woods Hole from a collecting trip to Jamaica, he found awaiting him a telegram from President James Rhoades of Bryn Mawr College. The telegram offered him a position not only as associate professor of biology but also as head of the department. Bryn Mawr was a young and vigorous college for women, modeled in many respects on Johns Hopkins. The decision was not an easy one. For the second time, Morgan had been awarded a prestigious Bruce Fellowship at Johns Hopkins, for the year 1891-1892. Similar to today's postdoctoral grants, the Bruce Fellowship would have given Morgan another year of full-time research, uninterrupted by teaching responsibilities. After some deliberation, however, Morgan accepted the Bryn Mawr position. He succeeded his friend E. B. Wilson, who was leaving Bryn Mawr for a position in the zoology department at Columbia University. Morgan's respect for Wilson had convinced him that Bryn Mawr was a college of distinguished academic standing. Shortly after making his decision, Morgan wrote a note to President Gilman at Hopkins to resign the Bruce Fellowship, and made plans to take up his new position as quickly as possible.[1]

The struggle of women for equality in education began to have its effects in the post–Civil War period. Vassar was opened in 1865, Smith and Wellesley in 1875, and Bryn Mawr in 1885. The last was founded by a bequest from Joseph W. Taylor, a philanthropist whose fortune had been made in the tannery business in Ohio. Like its predecessors, Bryn Mawr was intended to give young women a chance for a bona fide liberal arts education on a plane equal to that of the men's colleges. But more than the others, Bryn Mawr was committed to providing full-scale graduate opportunities as well.[2] From the beginning, Bryn Mawr had ties with Johns Hopkins much closer than mere geographic proximity. Taylor was a close friend of Francis T. King, one of the original trustees of Johns Hopkins, and a man of considerable wealth. In addition, a number of other Johns Hopkins trustees were also members of the original board of Bryn Mawr College.[3] Furthermore, a later dean and president of the college, Martha Carey Thomas (1857-1935) was the daughter of one of the original trustees of Johns Hopkins. Thus, the same upper-middle class, eastern financial and business interests that guided the founding of Hopkins were also responsible for establishing Bryn Mawr. So much was Bryn Mawr, in fact, parallel in its principles and designs to its Baltimore model that it was often referred to as the "Janes Hopkins."[4]

The biology department at Bryn Mawr (of which Morgan became chairman) consisted of two faculty members, both of them new to the college and to each other. Morgan was to teach all the morphological subjects, and the German-born physiologist, Jacques Loeb (1858-1924), the physiological ones. Loeb had just come to the United States to accept the Bryn Mawr position. Although their teaching duties appeared to perpetuate the experimentalist-naturalist split, Loeb and Morgan became quite close friends dur-

[1] Morgan to D. C. Gilman, August 1891; from Archives, the Milton Eisenhower Library, Johns Hopkins University. Obtained with the help of D. M. McCullough.

[2] Hugh Hawkins, *Pioneer: A History of the Johns Hopkins University, 1874-1889* (Ithaca, N.Y.: Cornell University Press, 1960), pp. 265 ff.

[3] Merle and Roderick Nash, *Philanthropy in the Shaping of American Higher Education* (New Brunswick, N.J.: Rutgers University Press, 1965), p. 99; Hawkins, *Pioneer*, pp. 265-266.

[4] Hawkins, *Pioneer*, p. 266.

ing their year together. Although Loeb subsequently took positions at the University of Chicago, Berkeley, and finally the Rockefeller Institute, he and Morgan retained a lifelong association that was rich and productive for both (see chapter VIII).

With only two faculty members responsible for all biological instruction, Morgan had a considerable amount of teaching to do during his first few years at Bryn Mawr, including courses in vertebrate embryology, theories of metamerism,[5] structure of protoplasm and the cell, criticisms of Darwinism, invertebrate embryology, regeneration, graduate seminars each semester dealing with specific biological problems, and laboratory instruction for both the embryology and regeneration courses. Normally, Morgan taught two courses each semester, as well as advising a number of graduate and undergraduate students in research work. Morgan always liked teaching, and although he felt sometimes that his duties at Bryn Mawr overtaxed him, he appears to have done an outstanding job as a teacher. In accepting Morgan's resignation in 1903, M. Carey Thomas, then president of Bryn Mawr, wrote: "You have been a wonderful instance to me of equal success in undergraduate and graduate teaching, and your interest in your students has in no way seemed to interfere with your investigations."[6] Throughout his career Morgan retained a personal interest in and continual association with his students.

President Thomas was correct in claiming that Morgan did not let his teaching responsibilities curtail his research work. His thirteen years at Bryn Mawr were highly productive. He was continually at work on his investigations, and all but a few summers he spent in full-time research, either abroad or at the Marine Biological Laboratory in Woods Hole. During that time he published 103 articles and three books. From the beginning, both in the laboratory and in his writing, Morgan displayed the enormous industry that characterized his scientific work at the peak of his career. Although he enjoyed contact with students, it was research that fired Morgan's imagination and captured his keenest interest.

[5] Metamerism is commonly found in animals such as Annelids and Arthropods. It involves the repetition of a pattern of elements belonging to each of the main organ systems of the body. It is commonly exhibited along the antero-posterior axis of the body, but is also applied to repetition of parts along appendages. A theory to account for the origin of metamerism was of particular interest to morphologists.

[6] M. Carey Thomas to T. H. Morgan, November 9, 1903; from the Archives, Bryn Mawr College.

There were always so many questions to be asked, so many answers to be pried from the enigmatic organism. Each problem solved revealed a multitude of new questions. There was no end to biological research for Morgan. He was still working on problems of regeneration and early development up until a few months before his death. Like many of his contemporaries at Johns Hopkins, Morgan had been imbued with the investigator's spirit. In a scientific career that lasted over fifty-five years, it was never to leave him.

While Morgan enjoyed his teaching post at Bryn Mawr, he continually made humorous references to working at a "ladies college." Especially in his letters to his close friend Hans Driesch, Morgan frequently poked fun at his "plight" surrounded by so many young women. Once, after returning from Europe, Morgan wrote to Driesch: "Write me in America and pity me in the wilds of my country surrounded by ladies and wild Indians." And, "You can think of me there giving eloquent lectures to my nice girls and wishing heartily that I were with you in Naples."[7] In a letter the following winter Morgan wrote to Driesch that he was having trouble finishing a paper that had some difficult parts to it; he then added: "But what can you expect with 75 girls to attend to!"[8] That Morgan was not as aloof from his "ladies" as these remarks playfully suggest is evident by the fact that in 1904 he married Lilian Vaughan Sampson, one of his former graduate students at Bryn Mawr, and a biological investigator of considerable merit in her own right.

As interesting and challenging as Bryn Mawr undoubtedly was, it was nevertheless a limited environment for a young man of Morgan's research interests and intellectual drive. He missed especially other colleagues with whom to discuss his work on a day-to-day basis. Also, Morgan had begun to hear that exciting new kinds of biology were being initiated in Europe. From the journals he read about revolutionary experiments being performed in Italy and Germany on living embryos! From his colleague Jacques Loeb, he learned of the new school of physicochemical biologists which had grown up in Berlin and Würzburg. He was most excited on hearing about the wonders of the Stazione Zoologica, a marine laboratory in Naples established in 1872. Increasing numbers of

[7] Morgan to Driesch, August 29, 1895, pp. 3-4; Morgan-Driesch correspondence, APS.

[8] Ibid., February 24, 1896.

biologists from all over the world had done research in recent years at the Naples Station, as it was called. The station had become a gathering place for investigators who came for a month, a summer, a year or more, to carry out studies on a variety of topics, from cell structure and function to embryology and natural history. The Bay of Naples was particularly rich in marine fauna; and it was marine organisms (almost exclusively invertebrates) that formed the common denominator of all the research carried out at the Naples Station.

Morgan heard most directly of the wonders of the Naples Station from several of his personal friends. E. B. Wilson had been the first American to work at the station, as far back as 1883. He had returned there again in 1892, sending back glowing reports of the exciting research possibilities that Naples offered. Meanwhile, two other friends, George H. Parker (1864-1955) and William Morton Wheeler (1865-1937), had both been to Naples in 1893. Morgan thus determined to spend an academic year at the Naples Station. Use of the station was obtained by renting "tables" (research benches, equipped with running sea-water tanks). In the 1890s there were three American tables under subscription at Naples: one supported by the Smithsonian Institution, one by Alexander Agassiz in Harvard's name, and one by a private philanthropist, Major Davis. Morgan applied to the Smithsonian for permission to work at its table. With permission granted, and a year's leave of absence (1894-1895) secured from Bryn Mawr, Morgan left for Europe in June of 1894. Part of the summer he worked at the Biologische Anstalt at Helgoland, studying fertilization in fragments of sea-urchin eggs. Toward the end of the summer he went to the Zoologisches Institut in Berlin, and thence to Naples, where he arrived on October 29.[9]

What Morgan found at Naples fulfilled more than his fondest expectations. The physical plant as well as the personnel were enormously exciting to a home-trained American biologist. In the mid-1890s the United States was still relatively isolated from the great research centers of the world, and unlike many of his older colleagues, Morgan had not had the advantage of taking his doctorate in a foreign university. Europe in general, and the Naples Station in particular, captured Morgan's heart.

The Naples Station had been founded in 1872 through the per-

[9] Obtained from the records of the Dohrn Archives, Zoological Station, Naples, courtesy of the Archives.

sistent efforts of Anton Dohrn (1840-1909), who became its first director (from 1872 to 1907). A student of the great Darwinian popularizer and morphologist Ernst Haeckel, Dohrn was a zoologist of considerable skill and erudition who had pioneered a number of studies of invertebrate morphology and phylogeny. By the early 1890s, his reputation as an outstanding biologist as well as his abilities as an organizer and administrator had made the Naples Station the foremost marine laboratory in the world. In an era that saw a great increase in the establishment and growth of such biological stations, Naples stood foremost and supreme, "The Mecca of the biological world."[10]

Dohrn's foresight had secured a particularly favorable location for the laboratory. The facilities at that time were housed in the Villa Nazionale, with an added wing, facing the waterfront of the Bay of Naples. The water of the bay was exceptionally clear and unpolluted, abounding in marine life almost unequalled in any other locality. The efficient organization of the laboratory, with sea water piped in to each table, made it possible to work for extended periods on fresh, living specimens. As one observer wrote:

> Every morning at 10 o'clock ten or more fishermen of the station with buckets or baskets full of glass jars poised gracefully on their heads, march into the court and receiving room of the station with their prizes from the sea—scarlet orange feather stars, sea cucumbers, sea urchins, squirming serpent stars, and bristling purple sea-urchins; or it may be a bit of red coral or a wriggling creeping octopus. . . . Equally prized by the naturalists is the great array of seemingly less interesting worms, crabs, and amorphous sponges, or even the brown slime that the nets of finest silk sift from the blue waters of the bay. Along with the fishermen of the station come others silently offering their wares—some rare prize, or merely the inedible waste, the few types of marine life, without value at the mercato, which his nets have yielded, for the Neapolitan fisherman has long since learned that such spoils have a value and a sure market at the station. The varied material thus assembled is speedily distributed to the

[10] Charles A. Kofoid, *Biological Stations of Europe* (Washington, D.C.: United States Bureau of Education, bulletin #4, 1910), pp. 1, 9.

specialists for investigation, sent to the exhibition tanks, or preserved for shipment to museums and biological laboratories.[11]

Morgan found the atmosphere in Italy both friendly and stimulating. He enjoyed the Italian people and felt much at home with them and in his later life often joked about his ideal wife being an Italian princess. The warmth and friendliness of the Italian people stayed with him always. It can safely be said he lost his heart to Italy during his stay there in 1894.

However, of all the friends that Morgan made among the regular workers at the Naples Station, the closest was Hans Driesch (1867-1941), who profoundly influenced the direction of Morgan's work in subsequent years. Morgan also came to know and like Driesch's special friend, Curt Herbst (1866-1946) who in 1894-1895 was carrying out experiments on the effect of the ionic concentration of sea water on the development of embryos. Morgan also came to have enormous respect for the director of the station, Dr. Anton Dohrn, and his assistant, Professor Dr. Hugo Eisig. Driesch and Herbst, working at this time year round at Naples, introduced Morgan to the ways of the laboratory and the countryside. A favorite pastime was eating Italian food and drinking wine at the local tavern, Vincenzo's.

Morgan stayed at Naples through July of 1895, working on problems of experimental embryology, including a long study of

[11] Ibid., p. 11. For a contrasting view, see William Morton Wheeler's reaction: "You have heard the song Santa Lucia and have probably had pleasant dreams when you heard it, of sunny Italy and its charming (??) women, etc., etc., but you should see the real harbor or street of Santa Lucia at Naples! Of all the filthy places! It is here where they sell the oysters with their gills full of cholera bacilli—and sea urchins—think of eating a sea urchin! and nasty little squids that have been out of water so long that they have deliquesced. It is interesting to study the faces of these people in Santa Lucia, or, in fact, anywhere in Naples. Not a line that tells of any better feeling or a trace of intellectuality. The men look as if they were capable of stabbing anybody for a few liras. The women are no women at all. And then watch them when a few priests carry the consecrated host through the street! What a kneeling and bowing and taking off of hats!" Wheeler's description may tell more of him than of Naples, but it is revealing in its opposition to Morgan's more sunny, romanticized reaction. From Mary Alice Evans and Howard Ensign Evans, *William Morton Wheeler, Biologist* (Cambridge, Mass.: Harvard University Press, 1970), p. 91.

the development of ctenophores (called by the popular name "comb jellies") in collaboration with Driesch. In August, Morgan went with Herbst and Driesch to Zurich. Morgan then departed for Paris, London, and his eventual return to the United States.

After the exciting and stimulating year in Naples, Morgan felt something of a letdown on returning to the hectic but more humdrum atmosphere of the United States. No sooner had he arrived than he wrote to Anton Dohrn: "My greatest wish is to return again to Naples. When I compare the freedom there with the artificial and busy life here I am filled with a desire to go back."[12] The United States appeared too confining, and there was much less chance for fruitful scientific collaboration and exchange of ideas at Bryn Mawr. Teaching seemed more onorous than before, and he voiced strongly his sense of isolation. But, as always, Morgan adjusted to what had to be done, and quickly resumed a pattern of work as strenuous as before he had left.

The effect of Morgan's year at the Naples Station appears in two areas. Personally, it had introduced him to the culture of Europe: its art, history, music, and cosmopolitan customs. Scientifically, through Driesch and others, he had been introduced to experimental embryology; through the station itself, he had become convinced of the importance of a well-organized, equipped, and staffed marine laboratory in facilitating modern biological research.

As a result of his Naples experience, Morgan returned to Europe on a number of occasions, usually in the summers, during his Bryn Mawr period.[13] He had become enormously attracted by the excitement of Europe, and wanted to have his fill of it. On all these trips he spent some time sight seeing. He enjoyed a lovely voyage through the Rhine Valley and Switzerland in the summer of 1898. He particularly enjoyed the art and architecture he observed along the way. Music was also a great passion, and he planned to meet Driesch in Bayreuth during the summer of 1902 to see a Wagner opera.[14] Of particular interest to Morgan during his European sojourns was painting: Italian, of course, but also Flemish, with which he had been relatively unfamiliar up to that

[12] Morgan to Dohrn, October 5, 1895, p. 3; Dohrn Archives.

[13] The best evidence indicates that Morgan went to Europe in the summers of 1896, 1898, 1900, and 1902.

[14] Morgan to Driesch, March 27, 1902; Morgan-Driesch correspondence. There is no indication from future letters whether the proposed Bayreuth trip ever occurred.

time. British painting he found drab and uninteresting, like British cooking and British science. It was the Continent, particularly the southern part, that always attracted him most.

During the summers of 1896 and 1898, which Morgan devoted mostly to seeing Europe rather than to biological research, he took long pleasure trips through Spain, northern Italy, Venice, southern Germany, Switzerland and the eastern Mediterranean. He generally found large cities much less interesting than small, provincial ones. After a brief stay in Paris, for instance, he wrote back to Driesch that he had "seen all the sights and was willing to go away."[15] Paris was, in his words, too nervous a city with too many people and too much noise.

In England in early September of 1895, Morgan took the occasion to visit a meeting of the British Association for the Advancement of Science held that year at Ipswich. He found the meeting a "beastly performance."[16] Morgan felt that British biological thinking was "simple-minded and out of date," and much of their work old fashioned. Most of the papers presented were descriptive and morphological and seemed to take no account of the newer work in experimental embryology which Morgan had seen so actively pursued at Naples. Many years later when Morgan was visiting England to give a series of lectures, his host, R. C. Punnett, wanted to know what sights Morgan wanted to see. "Why don't we just sit down and talk?" Morgan asked. "No more castles and museums for me. I only want to hear the skylark which we don't have in the United States." When Punnett, a little offended, protested and said, "Is that really all you want to see in England?" Morgan is reported to have answered, "Yes, the skylark."[17]

Morgan's feeling for Europe did not stem from its cathedrals, monuments or castles; not even from its lovely scenery which he always appreciated. It was the people; people who could talk of so many subjects, who had enthusiasms they did not hide. Morgan found much intellectual and personal kinship with his European colleagues. To this end, for instance, he always made a point of setting aside at least a week in each of his trips to Europe to spend with Hans Driesch. These meetings usually took place at Naples, but, when possible, elsewhere such as Munich or Zurich. Both men enjoyed good meals and pleasant surroundings, and their days and evenings were spent in constant talk, scientific and non-

[15] Ibid., August 29, 1895. [16] Ibid., September 25, 1895.

[17] Related by Tove Mohr, personal communication, August 12, 1974.

scientific. Morgan always felt that no European trip would be complete without a lengthy visit with Driesch.

After returning from his long stay at Naples in 1894-1895, Morgan continued to work diligently and enthusiastically among American biologists for further support of the station's activities. Convinced of the important benefits of an international center for the study of marine organisms, Morgan devoted much effort to publicizing the work of the Naples Station, singing its praises on every possible occasion. He constantly encouraged young American workers to take advantage of its rich and varied resources. When the station fell into some financial difficulty shortly after Morgan's return, he did all he could to rally support.

Morgan's response is indicative of his strong attachment to the station and the high scientific esteem in which he held it. During the year Morgan had been in residence at Naples, Alexander Agassiz at Harvard had undertaken a campaign to discourage various American institutions and professional societies such as the Smithsonian or the American Society of Naturalists from continuing their table subscriptions at the station. Such subscription fees were vital to the financial solvency of the station. The exact reason for Agassiz's antagonism remains unknown, as he circulated his reports in private.[18] However, Morgan was particularly incensed to hear that Agassiz had been campaigning against the Naples Station—especially in a behind-the-scenes-way—with very influential people. "I have heard of the contemptible means that Agassiz has employed to harm the Naples Station, and am ready to do all that I can towards preventing further infection from him. I should further be very glad to see him nailed sharply up for the wretched stories that he has already set afloat. I shall be in Baltimore and New York before long and *quietly* find out if Agassiz is still circulating his reports. I shall also go to Washington to the Smithsonian if necessary.[19]

Dohrn greatly appreciated Morgan's warm show of support and the help he was ready to offer to counteract Agassiz's efforts:

[18] No clue is obtained from Agassiz's correspondence books at the Museum of Comparative Zoology, Harvard University. In that collection are several letters between Agassiz and Dohrn, relative to the discontinuation of Harvard's subscription. But no substantive reasons are given other than Agassiz's claim of lack of funds. However, such a reason would not account for Agassiz's campaign to encourage other institutions to follow his lead.

[19] Morgan to Dohrn, October 5, 1895; Dohrn Archives.

> My thanks for your answers to my Berlin letter, by which you confirm the rumours which have reached me about Agassiz' miserable and mean action. Let us take him by the throat and make him speak out publicly what he had the enviable (?) [*sic*] courage to circulate secretly. He will be very much astonished, if he is taken unawares on his cowardly proceedings. I wrote to you that I would send him an answer to his last letter, by which he told me he could not go on taking a table for Harvard College: as yet he has not answered that letter, though I put it into such terms, that if he felt anything like loyalty he ought to have told at once all the possible criticisms he might have had to urge against the Naple Station. If he continues his silence, he commits himself; if he answers without giving his censure, then he is disloyal, telling what he tells everywhere [i.e., privately but not publicly], and if he speaks out,—well then I can publicly contradict him. In every way he is placed in a dilemma, and most likely things, [*sic*] that I am a harmless man, with whom he may deal at liberty. We'll make him awake to the truth of the situation ere long.[20]

Immediately on his return to the United States, at the suggestion of Henry Fairfield Osborn (1857-1935), then of Columbia University and the American Museum of Natural History, Morgan wrote an account of his year-long stay at the Naples Station. This account was published the next year in *Science*.[21] In this article Morgan praised the exciting research possibilities and the international atmosphere that pervaded the Naples laboratory. For Morgan it was a labor of love, as he wrote to Dohrn in October of 1895:

> I shall rejoice if I can be of any service whatsoever to the Naples Station. I appreciated so fully the splendid opportunities which you have given the world for original work that I am eager for others of my countrymen to have the same opportunity that I have had there. . . . Meanwhile I shall do what I can for a full endorsement of the Smithsonian table by the American Society of Naturalists and by the morphologists. . . . Whitman, E. B. Wilson, Osborn, and all of the

[20] Dohrn to Morgan, October 19, 1895; Dohrn Archives.

[21] T. H. Morgan, "Impressions of the Naples Zoological Station," *Science* 3 (1896): 16-18; see also Morgan to Dohrn, November 1895, p. 1; Dohrn Archives.

> younger men are enthusiastic admirers of the Naples Station and there will be, I am confident, a unanimous and warm expression of their views.[22]

Morgan further took upon himself the task of encouraging other American groups to take subscriptions for laboratory tables at the station. By 1898 Morgan's efforts had paid off. He wrote to Dohrn that not only had Agassiz remained quiet and done no further harm, but that Morgan and a group of zoologists had raised enough money for an additional table (to replace Agassiz's), thus keeping the number at three:

> We have good news to send you from America for the new year. At the meeting of the Naturalists a few days ago we got $100 toward another table at Naples. The next day at a meeting of the new trustees of the Marine Laboratory [MBL] another $150 was promised, mainly through the efforts of Bumpus [H. C. Bumpus, a long time trustee of the MBL, an invertebrate zoologist, and an enthusiastic advocate of marine biology].
>
> Columbia has $250 from Mr. Dodge, and you see, that makes the entire amount [$500, the normal subscription fee]. I saw Miss Hyde [Ida Hyde, a physiologist and perennial summer worker at the MBL], and she tells me that the women have also enough for another table and the Smithsonian will make the third table of which you spoke last summer.
>
> You must congratulate us on our great good fortune in obtaining these tables. We have always looked on the Stazione Zoologica as the "Mecca of Zoologists" and now we have at least an opening into the holy city.[23]

To Morgan it was essential that Americans continue to support this important enterprise, not only so that the station could survive, but so that American workers could be exposed to its many benefits. To that end, for example, he encouraged some of his most promising graduate students at Bryn Mawr to take advantage

[22] Morgan to Dohrn, October 5, 1895, pp. 1-2; Dohrn Archives.

[23] Ibid., January 2, 1898, pp. 1-2. The remainder of this letter is particularly interesting as it describes in detail the exact negotiations that Morgan, Osborn, and Wilson went through with various friends and officers in the professional societies to ensure the necessary financial support for the third table.

of the "women's table." In 1901 when one such outstanding woman, Nettie M. Stevens (1861-1912), wished to work at the station, Morgan wrote a letter of introduction for her. Stevens had just been appointed a research assistant at Bryn Mawr in that year (at the age of forty!). Morgan wrote of her outstanding abilities to Hugo Eisig, Dohrn's administrative assistant at Naples:[24]

> Dear Professor Eisig:
>
> Miss Nettie M. Stevens is leaving here for Naples and I want to make certain that she will know you by giving her this letter of introduction.
>
> Miss Stevens has been appointed to one of the Bryn Mawr College Fellowships that were invented to make trouble for European Professors, and as there is no escape, I hope you will be kind to her. Indeed I know so well how hospitable the Station is to all newcomers that I can have no doubts but that Miss Stevens will have, as we say over here, "the time of her life."
>
> With best wishes for yourself and greetings to all
>
> Sincerely,
> T. H. Morgan

Although Morgan wanted to encourage as many Americans as possible to work at the Naples Station, it was obvious that no more than a handful (perhaps two or three) could conveniently do so in any one year. There was a need for a Naples Station in the United States, one that combined the availability of a variety of marine organisms with the benefits of a national (if not international) center for biological research. To Morgan, the most likely place to fill these demands was the already existing, but small, Marine Biological Laboratory (MBL) at Woods Hole. Before his year abroad Morgan had spent several summers doing research at Woods Hole (spelled then, Woods Holl). He was aware of its potential and was enthusiastic about the prospects of helping to build an American Stazione Zoologica.

The Marine Biological Laboratory at Woods Hole grew out of several earlier, primarily summer, seaside stations on the northeast coast of the United States. One was Louis Agassiz's short-

[24] Morgan to Eisig, September 17, 1901; Dohrn Archives.

lived Anderson School of Natural History on Penikese Island (outermost in a chain of islands stretching away from the coast of Cape Cod, just off Woods Hole), founded in 1873.[25] Another was the small station at Annisquam, Massachusetts (which Morgan had attended in the summer of 1886), supported by the Women's Education Association under the auspices of the Boston Society of Naturalists. To a number of naturalists who felt strongly about the value of seaside laboratories, there appeared to be value in consolidating these various enterprises into one strong and prosperous institution. That viewpoint spearheaded the formation of the MBL in 1888.

Woods Hole was chosen as a site partly because the United States Bureau of Commercial Fisheries Biological Laboratory was already there[26] and was amenable to cooperation with another research institution. Also, many of the organizers of the MBL had worked as students at the Fisheries Laboratory (as Morgan had in the summer of 1888), and knew of the favorable collecting opportunities in the Woods Hole, Buzzards Bay area. The possibility of collaborating with the Fisheries Laboratory in expeditions, collecting, and library facilities made Woods Hole seem like the ideal location. The MBL opened its doors for its first summer session in July 1888; Morgan was there the next summer as a student in the invertebrate zoology course.

The MBL was organized with the explicit aim of developing the best ideals of the Naples Station in conjunction with the goals originally set by the Anderson School and the Annisquam laboratory. Naples was a wholly research-oriented station; the Anderson and Annisquam labs were schools providing instruction in marine biology for secondary and college teachers. As the first director of the MBL, Charles Otis Whitman (1843-1910), stated at the founding of the laboratory in 1888: "Other things being equal, the investigator is always the best instructor. The highest grade of instruction in any science can only be furnished by one who is thoroughly imbued with the scientific spirit, and who is actually engaged in original work."[27] Organized on the table system, as at

[25] Agassiz died the following winter; the school operated for a second summer (1874), and then was closed. Its buildings burned in 1891.

[26] That laboratory had been established in 1885 by Spencer Fullerton Baird under the aegis of the United States Fish Commission.

[27] F. R. Lillie, *The Woods Hole Marine Biological Laboratory* (Chicago: University of Chicago Press, 1944), p. 38. Lillie's book remains the only history of the founding and historical development of the MBL.

Naples, the MBL thus attracted two groups of people: teachers seeking further instruction in modern (and especially marine) biology, and original investigators taking advantage of the abundance of organisms available in the Woods Hole area. The institution prospered, and by the mid-1890s had grown from its original one to over seven buildings (see figure 4).[28]

After his return from Naples, Morgan became an even firmer supporter of the MBL than he had been before. Although in its early days the MBL was populated largely by morphologically oriented workers, Morgan saw clearly after 1895 that it could be a center for the new experimental biology he had seen at Naples. The eggs of marine organisms such as the sea urchin offered ideal material for experimental studies of early embryonic development, and Morgan was more than anxious to develop this line of work at MBL. He also saw that in the scientific growth of the institution lay the prospect of establishing an international center as well known as Naples. Consequently, he was far more interested in the research programs at the MBL than he was in the teaching programs. In those summers after 1895 when he did not return to Europe (1897, 1899, 1901, and 1903) Morgan carried out research on experimental embryology, particularly artificial parthenogenesis[29] and regeneration, at the MBL. He became a trustee of the MBL Corporation in 1897, a position he held until his death in 1945. Morgan used his experience from Naples, along with his conviction that the United States should have its own *Stazione*, to help formulate and direct the policy of the MBL during the formative years of its growth.[30]

As a trustee, Morgan was convinced that the strength of the MBL lay in its attaining a size where personal discourse and exchange of ideas could be encouraged among a variety of different workers. Most important, however, was the problem of securing proper funding so that the laboratory did not have to operate on a year-to-year basis. In 1897 the board of trustees was compelled

[28] Ibid., pp. 39-46.

[29] Parthenogenesis is the process whereby an unfertilized egg undergoes normal embryonic development. In a number of invertebrate species this process occurs normally. It can be induced by special physical or chemical processes in a number of other species, producing adult organisms derived only from the female hereditary material.

[30] MBL was supported wholly by private donations, student fees, and subscriptions for tables. The ownership of the laboratory was in the name of the MBL Corporation, which consisted wholly of scientists who worked at the lab.

Figure 4. Marine Biological Laboratory, approximately 1893. The Old Lecture Hall, which contained both a lecture room and laboratories.

The interior of a typical laboratory room. Nearly all work was microscopical, and thus investigators were clustered around windows. A salt-water table was located in the center of the room and contained live specimens collected from nearby waters. The buildings were frame covered by clapboard and were not insulated. In this period electricity was still not in widespread use; thus most night work was carried out with kerosene lamps (bottom, right). (Courtesy of the Library, the Marine Biological Laboratory, Woods Hole, Massachusetts.)

to overcome a financial crisis. One set of proposals was that some large, relatively well-endowed institution, such as the University of Chicago or the Carnegie Institution of Washington, should take over operation of the lab. Another was that the MBL retain its independent financial base by organizing more extensive fund-raising activities. C. O. Whitman, as director, favored the latter course. Morgan and the younger trustees, such as E. B. Wilson, preferred the former. Morgan had all the sympathy in the world for keeping the MBL independent and relatively small. However, he realized it took money to keep the laboratory going. He felt the aims of modern scientific investigation would be better served in the long run by giving the MBL a permanent financial base. Morgan felt that if independence had to be bought at the price of good science, the MBL would have gained a Pyrrhic victory. In the long run, Whitman's persistence won out, and the MBL retained its corporate ownership.[31] In confronting the issue, however, Morgan revealed that his love of promoting scientific research went above the dictates of any specific institution. As much as he cared for the MBL, it was primarily as a place to pursue biology, and not for the sentiment of its past or its high, though perhaps unrealistic, ideals.

The period between Morgan's return from Naples and his leaving Bryn Mawr for Columbia (1904) was extremely busy and productive. It was during this time that he collected the results of many of his studies on regeneration in planaria and the earthworm, presenting them first in two sets of lectures at Columbia in December of 1899 and in Chicago in March of 1900. The material composing these lectures was reworked into Morgan's first major book, *Regeneration*, which appeared in 1901.[32] On occasion Morgan took on more projects than he could easily finish, and at times he complained (principally to Driesch) that he was overworking. So, the summer of 1900 was spent on a travel vacation to Europe, principally to Naples, Greece, and Constantinople. He visited friends along the way, spent some time with Driesch, and returned fresh to Bryn Mawr by September.

By the early 1900s Morgan was beginning to feel the desire

[31] See Lillie, *The Woods Hole Laboratory*, pp. 42-46. Throughout subsequent economic crises the MBL managed to maintain a similar autonomy from governmental or other private institutional control.

[32] T. H. Morgan, *Regeneration*, Columbia University Biological Series, vol. 7 (New York: Macmillan Co., 1901).

for a family of his own. The decade since taking the Bryn Mawr job had been busy and exciting. His mode of research had changed, his horizons within the biological community were greatly enlarged, and his personal and cultural appetites to see and be a part of the world had been largely satisfied. In addition, he was rapidly becoming respected as one of the most authoritative of the younger generation of embryologists. His reputation had spread, as the invitation to give the lecture series in both New York and Chicago indicated. But he had lived the life of a bachelor thus far, and had begun to find it a limitation.

In 1904 his wish was realized by marriage to his former graduate student, Lilian Vaughan Sampson (1870-1952). Like Morgan, Lilian Sampson was from a well-to-do upper-middle-class family (from Philadelphia). On her father's side she was a descendant of John and Priscilla Alden. Her maternal grandfather was Samuel Vaughan Merrick (1801-1870), an engineer, industrial entrepreneur, first president of the Pennsylvania Railroad, and cofounder of the Franklin Institute. She and her older sister were orphaned quite young and were raised largely by their maternal grandmother in Philadelphia. Lilian attended Bryn Mawr College as an undergraduate (1887-1891) where she became interested in biology and took courses from E. B. Wilson. The Sampson family was politically and socially progressive, and Lilian was much concerned with the new movement for women's rights and women's education. This remained part of her perspective throughout her life, though she appears to have expressed herself more by example than by public activism.[33]

In the fall of 1891, Lilian entered the graduate school of Bryn Mawr College just as T. H. Morgan took up his teaching position there. Because of her interest in biology in general, and descriptive morphology in particular, Morgan was her advisor, and she took all the courses, including laboratories and seminars, that he offered between 1891 and 1899. Her major area of interest was embryology and morphology, with a minor in cytology.[34] Her work as a

[33] Isabel Morgan Mountain, personal communication, September 1972.

[34] She published a number of papers in this area between 1895 and 1906, including "The Musculature of Chiton," *Journal of Morphology* 2 (1895): 595-628; "Unusual Modes of Breeding and Development among Anura," *American Naturalist* 34 (1900): 687-715; "A Contribution to the Embryology of *Hylodes martinicensis*," *American Journal of Anatomy* 3 (1904): 478-504; "Regeneration of Grafted Pieces of Planarians," *Journal of Experimental Zoology* 3 (1906): 269-294.

student was apparently of good quality. However, despite her exposure to the women's equality movements of the latter part of the nineteenth century, Lilian placed her own career second to that of her husband and remained out of active scientific work during the early years of raising her four children between 1906 and 1920. Later, after her children were well started in school, she returned to the laboratory and made a number of important contributions to the *Drosophila* work in which her husband was engaged at that time.

It is difficult to determine how well Morgan knew Lilian Sampson during much of the period of her graduate study at Bryn Mawr or at what point a personal relationship between them began to develop. At any rate, early in 1904 Morgan announced his engagement to his friends in his typical ebullient style. As he wrote to E. G. Conklin in February: "Here is news for you out of a clear sky. I am engaged to be married to Lilian Sampson. So it's all up and I shall have to settle down with the rest of the married community and be 'good'. . . . Before long I shall come in to see you both [Conklin and his wife] and let you tease me as much as you like. I shall only be flattered. . . . Send me your blessing and best wishes."[35] And to Driesch he wrote in a more enthusiastic and playful vein: "This time I am really engaged to be married. You will be delighted to hear it, I know. In fact, the unholy delight of all the married men begins to frighten me. But I have forgotten to tell you the really important part of my news. I am engaged to Miss Lilian Sampson of whom you may have heard since she studied one year with Lang in Zürich. I need not tell you that I am the luckiest fellow in Christiandom."[36] The marriage took place on June 4, 1904.

Immediately after their wedding, Tom and Lilian Morgan set off for California, to combine a wedding trip through the West with embryological work in marine labs on the West Coast (those of Berkeley and Stanford). Morgan's studies that summer included regeneration in the coelenterate *Tubularia* and in lizards; studies on the external factors that determine the switch from nonsexual to sexual reproduction in aphids;[37] and finally on the process of

[35] Morgan to Conklin, February 29, 1904; Morgan Papers, Caltech; quoted with permission.

[36] Morgan to Driesch, March 14, 1904; Morgan-Driesch correspondence.

[37] Aphids, like various other animals, are capable of reproducing by parthenogenesis. Almost all forms that reproduce by parthenogenesis also re-

self-fertilization in certain of the lower chordates.[38] The Morgans spent the entire summer in California, including among their activities sightseeing, field-collecting trips, and laboratory work.

It was understood from the beginning that Morgan's biological work was the primary focus of his life. Always interested in his family, he appeared increasingly as the years went on to have gotten great joy out of being with them. Nonetheless, management of household and practical matters was left to Lilian while Morgan spent considerable amounts of his time absorbed in his work.[39] His relationship with his wife appears to have been one of conscious but unspoken communication and understanding. It is certainly clear that much of his ability to carry out his scientific work in an atmosphere of peace and freedom from worry was due to Lilian's ability to take the major burden of raising a family and running a household. Yet Morgan did not take her for granted, for he was always conscious of her contribution. The easy outward manner he displayed throughout his life was certainly in part attributable to their relationship.

In 1904 when E. B. Wilson invited Morgan to join his faculty in the zoology department at Columbia University, Morgan recognized the chance to join scientific colleagues, not only those on the university's staff, but also those who passed through this cosmopolitan city. Furthermore, the fact that the appointment was as professor of experimental zoology offered a rare opportunity to pursue his investigations free from a burdensome teaching load. Thus, on returning from the West Coast at the end of the summer of 1904, Morgan and his wife prepared to move to New York City.

FROM MORPHOLOGY TO EXPERIMENTAL EMBRYOLOGY

In the period between 1891 and 1904 Morgan's research interests metamorphosed from a dominant concern with morphology

---

produce at least one generation per season by sexual means. One problem of great concern to many biologists around the turn of the century was the mechanism by which the sexual reproduction phase was triggered after several generations of parthenogenesis.

[38] Self-fertilization is the process whereby organisms capable of producing both sperm and egg in the same individual (so-called hermaphroditic forms) fertilize their own eggs.

[39] Isabel Morgan Mountain, personal communication, September 1972.

to strong involvement with the new experimental methods being introduced into embryology. Several influences contributed to Morgan's move away from morphology while also stimulating his attraction toward experimental work. There was the negative influence of Brooks and the corresponding positive influence of H. Newell Martin at Hopkins. At Bryn Mawr there was the exciting presence of Jacques Loeb, fresh from the stimulating atmosphere of Würzburg, where he had worked with two prominent experimentalists, Adolf Fick (1829-1901) and Julius Sachs (1832-1897). And at the Naples Station there was the personal influence of Hans Driesch and the widespread and growing interest in experimental embryology.

As Morgan and others at Hopkins tired of speculative, morphological work, they found in Martin a representative of a refreshing tradition. Through Michael Foster, Martin had become an advocate of laboratory work, and especially experimentation, in teaching. Martin tried to implement this approach at Hopkins. There is no evidence to document what Morgan learned directly from Martin, or even how well he knew him. The only clue is given by Sturtevant, who wrote in his sketch of Morgan's life for the *Biographical Memoirs of the National Academy of Sciences*: "He [Morgan] was also influenced, in his student days, by H. Newell Martin and by W. H. Howell [physiologist and M.D. at Hopkins]. From them he learned to appreciate the value of physiological approaches to biology; and I think he was inclined to turn to them rather than to Brooks at times because he felt the latter was somewhat too metaphysical for his tastes."[40] It is possible to surmise only that the naturalist-experimentalist dichotomy, so obvious in the persons of Brooks and Martin, could not have failed to attract Morgan's attention. Because he was apparently sometimes dissatisfied with Brooks's concerns and methods, it is likely that he was receptive to an alternative—the more exact and rigorous experimental approach represented by Martin.

More certainly an influence on Morgan's awareness of the value of physiological and experimental work was his association with Jacques Loeb. Loeb had been trained in the best tradition of German materialistic physiology.[41] When he arrived at Bryn Mawr in

[40] A. H. Sturtevant, "Thomas Hunt Morgan," *Biographical Memoirs, National Academy of Sciences* 33 (1959): 285.

[41] Loeb had been a student of Fick who was a student of Emil Ludwig and a close associate of Hermann von Helmholtz. Loeb was thus a second-generation physiologist of the Berlin "medical materialist" school, as it was

the fall of 1891, he brought with him both a physiologist's concern with function (as opposed to structure), and a strong bias toward materialistic (to Loeb this meant especially physicochemical) explanations in biology. Morgan and Loeb became friends, and their relationship lasted until Loeb's death in 1924. As we will show in more detail in chapter VIII, Loeb had a distinct impact on Morgan's adherence to a form of mechanistic philosophy. But Loeb's direct and explicit influence came later, after 1910, when the *Drosophila* work was well underway. In this earlier period, we can only hypothesize about Loeb's role in Morgan's move toward experimental problems. Because they worked together at Bryn Mawr so closely for a year, there is little doubt that Loeb's views that biology should become experimental like physics and chemistry made a strong impression on Morgan in the period between 1891 and 1894. Loeb was eight years older than Morgan, and a man of forceful and intense views. Morgan could not have escaped being influenced toward the virtues of experimental work.

By the time he was teaching at Bryn Mawr, Morgan was generally aware of the new experimental approach to embryology passing under the name of *Entwicklungsmechanik*. *Entwicklungsmechanik*, as a method of studying embryos, represented a significant departure from earlier nineteenth-century studies in embryology, which had been largely descriptive in nature. Wilhelm Roux (1850-1924) and others among the new experimentalists sought to learn from the embryo not what its phylogenetic history might have been, but rather what factors *cause* embryos to differentiate during development. To find answers to that question, Roux had resorted to modifying embryos to observe the resulting changes in their developmental form. He had dared to interfere with development. In a word, he had *experimented*.

---

called, openly espousing a materialistic and mechanistic philosophy of physiology. For more details on Loeb, see Donald Fleming's introduction to a reprint of Loeb's *The Mechanistic Conception of Life* (1912; reprint ed., Cambridge, Mass.: Harvard University Press, 1964). On the medical materialists, see Paul Cranefield, "The Organic Physics of 1847 and the Biophysics of Today," *Journal of the History of Medicine and Allied Sciences* 12 (1957): 407-423; "The Philosophical and Cultural Interests of the Biophysics Movement of 1847," *Journal of the History of Medicine and Allied Sciences* 21 (1966): 1-7; and Oswei Temkin's splendid "Materialism in French and German Physiology of the Early 19th Century," *Bulletin of the History of Medicine* 20 (1946): 322-327.

Differentiation—the process by which the similar appearing cells of a young embryo (as in the blastula or gastrula) gradually become specialized (differentiate) into nerve or skeletal or skin or gut tissue—is one of the great enigmas of embryological development that has attracted the interest of investigators from Aristotle to the present. In the mid-nineteenth century (and earlier) the questions posed by differentiation had been the focus of considerable speculation, but it still had taken a back seat to the overriding concern for phylogeny. It was only with the work of Roux in the 1880s that the question of differentiation, and the methods of experimentation applied to investigating it, came into prominence.

Roux had been a student of Ernst Haeckel's at Jena. Partly in reaction to Haeckel, he had developed a skepticism toward speculating about phylogenies and a dislike of his teacher's concentration on evolutionary problems. On the positive side, however, he had learned from Haeckel the importance of seeking the *material causes* (as opposed to idealistic, nonmaterial causes such as essences or vital spirits) for biological phenomena.[42] Through Haeckel, Roux became interested in embryology and heredity, in how differentiation is triggered and regulated. By 1885 Roux had formulated in his mind a concept of heredity and development that provided a mechanism for embryonic differentiation: the mosaic theory. Most of his experimental work was designed to test, ultimately, one or another aspect of this theory. The mosaic theory held that hereditary particles in the cell were divided in a qualitatively uneven fashion during the cell divisions (called cleavage) that formed the multicell embryo from the single-celled egg. At each division, the two daughter cells would contain different hereditary potential. According to the mosaic theory, as cleavage continued, the potential of individual cells became more and more restricted; ultimately, a cell would express only one major hereditary trait, being recognized as belonging to one particular tissue type. The beauty of this formulation was that it led to predictions that could be checked. If the hypothesis were true, then destruction

[42] A detailed discussion of the many influences leading to the establishment of Roux's experimental embryology can be found in Frederick Churchill, "Wilhelm Roux and a Program for Embryology" (Ph.D. Dissertation, Harvard University, 1968), especially chap. 11. Some of this information is also contained in Churchill's "Chabry, Roux, and the Experimental Method in Nineteenth Century Embryology," in *Foundations of Scientific Method: The Nineteenth Century*, ed. R. N. Giere and R. S. Westfall (Bloomington, Ind.: Indiana University Press, 1973).

of one cell (blastomere) at the two- or four-cell stage, for example, ought to produce a deformed embryo. If the hypothesis were not true, then destroying one blastomere ought to produce little or no effect.

Roux set out to perform the appropriate experiment. With a hot, sterilized needle he punctured one of the blastomeres in the two-cell stage of a frog embryo. The punctured cell was killed, but the other blastomere continued to develop. Roux's results showed that all such embryos developed abnormally. They invariably lacked some particular set of embryonic parts and usually failed to develop beyond the late gastrula stage. In essence, Roux got half embryos which, when studied microscopically, showed that cells on one side were well developed and even partially differentiated, while those on the other side were highly disorganized and undifferentiated. Roux interpreted these results as support for the mosaic theory. When its sister cell was killed, the remaining blastomere was unable to develop into a whole embryo because, according to the mosaic theory, it contained hereditary particles for only one-half of the adult organism.[43]

However, a few years later, in 1891, Morgan's friend Hans Driesch, already working at the Naples Station, published a set of contradictory experiments. Driesch had tested Roux's theory with eggs of another organism, the sea urchin. Instead of killing one of the first two blastomeres with a hot needle, as Roux had done, Driesch shook sea water containing two-celled embryos so that the blastomeres were separated from their partners, but none was killed. Driesch then allowed the isolated blastomeres to develop and found, to his astonishment, that each produced a normal, though somewhat small larva. Driesch's results were in direct opposition to the prediction of the mosaic theory. Rather than seeing differentiation as a result of parceling out of hereditary units, Driesch concluded that the embryo was a series of cells bound together as a self-adjusting whole. If each blastomere, even when separted from other cells in the embryo, could still develop into a full adult, then there

[43] Wilhelm Roux, "Beiträge zur Entwickelungsmechanik des Embryo. Ueber die künstliche Hervorbringung halber Embryonen durch Zerstörung einer der beiden ersten Furchungskugeln sowie über die Nachentwickelung (Postgeneration) der fehlenden Körperhälfte," *Virchow's Archiv für Pathologische Anatomie und Physiologie und für Klinische Medizin* 114 (1888): 113-153; Resultate 289-291. This classic article is translated and reprinted in B. H. Willier and J. Oppenheimer, *Foundations of Experimental Embryology*, 2d ed. (Englewood Cliffs, N.J.; Prentice-Hall, 1974).

must not have been any qualitative separating out of hereditary material as Roux had hypothesized. Driesch observed, however, that if the two blastomeres were left together, and the fate of their daughter cells traced, each blastomere would give rise to a different part of the embryo. In other words, when a two-celled stage is allowed to develop normally, each cell gives rise to qualitatively different sets of tissues in the adult organism.[44]

During the later nineteenth and early twentieth centuries, Roux became the most vociferous advocate of the experimental and mechanistic approach to biology. From the 1890s until his death, he devoted enormous energy to expounding his program for research in embryology. Given the impressive name *Entwicklungsmechanik*, Roux's method sought to find, through physical and chemical means, an explanation for how developmental changes are brought about. *Entwicklungsmechanik* is roughly translated as "developmental mechanics," although the English terms are not quite equivalent to the German. Roux himself took considerable care in selecting the name for his research program, rejecting such terms as *Entwicklungsphysiologie* because they did not adequately reflect the strong mechanistic bias he wished to give to the new school of thought.

The mechanistic aspect of Roux's program was emphasized when he repeatedly pointed out that biological research should seek the same kind of explanation as the physical sciences:

> Since, moreover, physics and chemistry reduce all phenomena, even those which appear to be most divers, e.g., magnetic, electrical, optical, and chemical phenomena, to movements of parts, or attempt such a reduction, the older more restrictive concept of mechanics in the physicist's sense as the causal doctrine of the movement of masses, has been extended to coincide with the philosophical concept of mechanism, comprising as it does all causally conditioned phenomena, so that

[44] The apparent contradiction between the Roux and Driesch experiments was the result of differences both in the experimental techniques used and in the species of organism involved. Several years later other embryologists showed that if the one punctured blastomere in Roux's frog embryo was removed from physical contact with its partner, the latter would develop into a full-fledged, viable embryo. The physical contact between the injured and the normal blastomere appeared to alter the developmental process. In addition, the different results seem also to have been to some extent the result of differences between the sea-urchin and frog embryos.

> the words "developmental mechanics" agree with the more recent concepts of physics and chemistry, and may be taken to designate the doctrine of all formative phenomena.[45]

Thus, Roux tried to reduce the complex problem of differentiation, with its myriad component parts, into mechanical causes such as the qualitative separation of determinants by cleavage. The important feature of Roux's analytic procedure was that it led to the formulation of experimental tests.[46] *Entwicklungsmechanik* introduced the experimental method into embryology in a highly formal way. In a very real sense, Roux created the field of experimental embryology by separating the study of the causes of development from the more complex field of general morphology.

Although Morgan was familiar with the literature of the Roux-Driesch controversy, and with the growing school of *Entwicklungsmechanik*, he had had little direct contact with experimental embryology prior to 1894. One paper, published in 1893, dealt with experimental studies of fish eggs.[47] The preface to that paper indicates that Morgan's knowledge of, and interest in, experimental embryology was already keenly developed before he went to Naples. What was lacking was a close, firsthand involvement with the technical and conceptual problems that the experimental approach to embryology encompassed. Since *Entwicklungsmechanik* had not yet reached the United States in any force, Morgan knew he could only gain the necessary experience abroad. Thus, his immersion in this concept at Naples, especially the close contact with Driesch, was crucial in catalyzing Morgan's progress in the direction of experimental embryology.

The Naples Station was not only a mecca for the best biologists, it also provided an atmosphere in which some of the most challenging biological ideas of the early years of the century were spawned. By the early 1890s it had became a center for research in the *Entwicklungsmechanik* school. Morgan has left an account of the general atmosphere at Naples in the article, "Impressions of the Naples Zoological Station," published in 1896.[48] The con-

[45] Roux, "Beiträge zur Entwickelungsmechanik des Embryo," p. 150.

[46] See also Frederick Churchill, "Chabry, Roux, and the Experimental Method," pp. 161-205.

[47] T. H. Morgan, "Experimental Studies on Teleost Eggs," *Anatomischer Anzeiger* 8 (1893): 803-814.

[48] *Science* 3 (1896): 16-18. This was the article meant to counteract Alexander Agassiz's campaign to cut off American subscriptions to Naples. See the first section of this chapter.

stant turnover of investigators brought people of all shades of thought and background to the station. Visitors came to work for short or long periods, or sometimes just to visit and exchange ideas with whoever was there. Morgan pictured the Naples Station as a giant kaleidoscope in which people of all nationalities and diversity of ideas constantly interacted:

> At the Naples Station are found men of all nationalities. Investigators, professors, privatdocents, assistants and students come from Russia, Germany, Austria, Italy, Holland, England, Belgium, Switzerland and "America"—men of all shades of thought and all sorts of training. The scene shifts from month to month like the turning of a kaleidoscope. No one can fail to be impressed and to learn much in the clash of thought and criticism that must be present where such divers elements come together. And through all the changes of life and thought Professor Dohrn and his staff remain always open-minded, courteous, helpful and generous.[49]

This intellectual atmosphere was exciting and stimulating to Morgan. It was unlike anything he had ever experienced before, even at Hopkins. Dohrn's open-mindedness seemed to set the stage for free exchange of ideas that was unique.[50] The effect of this open communication was that visitors at Naples could get in touch with the most modern trends in biology. Here, unlike America, Morgan found himself face to face with the most advanced workers and the newest ideas: "Isolated, as we are in America, from much of the newer, current feeling, we are able at Naples, as in no other laboratory in the world, to get in touch with the best modern work."[51]

The affection Morgan felt for his Naples experience is echoed again and again in his letters to Driesch after 1895. Hardly had he left Italy for England and the return trip home in September 1895, than he wrote back to Driesch who was then on holiday: "How I wish I were going with you back to Naples! You will find the teapot and cups in the closet in my room and I wish you many a pleasant cup of tea out of them. Give my best regards to Eisig

[49] Ibid., p. 18.

[50] The relationship between Morgan and Dohrn was always cordial, earmarked by that openness Morgan mentions. This is especially interesting in view of the fact that Morgan had strongly attacked Dohrn's theory of the Annelid origin of the Pycnogonids in his doctoral dissertation.

[51] Morgan, "Impressions," p. 18.

when you get to Naples and let me hear from you as often as you can. I shall always remember the pleasant winter we had in Naples together and not forget the many good things I have learnt from you during that time."[52]

The most specific influence operating on Morgan during his nine-month stay at the Naples Station was undoubtedly Driesch. Virtually the same age (Driesch was one year younger), Morgan and Driesch from the start hit off a lasting friendship which survived their later differences of opinion on both World War I and Driesch's vitalistic philosophy. Both men were exceedingly quick and perceptive intellectually and had reciprocal senses of humor that made their association a continual delight. Professionally, Morgan found in Driesch an enthusiastic experimentalist who had in his own research thrown over the old style of descriptive morphology and adopted the principles of Roux's *Entwicklungsmechanik* school. Although Driesch had disagreed profoundly with Roux's interpretation of the half-embryo experiments, he nonetheless considered himself one of the most enthusiastic exponents of the new experimental methods in embryology.

It was largely through Driesch that Morgan's conversion from a morphologist to an experimentalist was completed. Like Roux, Driesch was a student of Haeckel's at Jena, where he received his doctorate in 1889. He was inspired to study embryology by Haeckel's popular books and by his emphasis on the importance of embryology for phylogeny. However, for his doctoral dissertation, Driesch deemphasized phylogenetic speculation in favor of a study of the laws governing the growth of colonial hydrazoans.[53] Because his family was wealthy and provided him with an independent income, Driesch did not find it necessary to proceed through the arduous ranks of the German academic hierarchy. For over ten years after receiving his degree, he traveled extensively, usually with his close friend Curt Herbst whom he had met at Jena in 1887. His travels took approximately one-third of the year, the remaining time being spent principally, though not exclusively, at the Naples Station. It was here in 1887 that Driesch carried out his most important experiments on the development of sea-urchin

[52] Morgan to Driesch, September 25, 1895; Morgan-Driesch correspondence.

[53] The best short biographical sketch of Driesch is that by Jane Oppenheimer in *Dictionary of Scientific Biography* (New York: Charles Scribner's Sons, 1971), vol. 4, pp. 186-189.

embryos produced by separating the first two blastomeres. Although settling in Heidelberg in 1900, he did not receive an appointment at the university until 1909. He became extraordinary professor in 1912 and in 1919 accepted the ordinary professorship in philosophy at the University of Cologne. Two years later (1921) Driesch became professor of philosophy at Leipzig, where he remained until his "early retirement" by the National Socialist regime in 1933.

In his early years, Driesch had been convinced that by applying the experimental approach, especially by the use of physical and chemical procedures, it would be possible to understand completely such complex biological processes as growth or differentiation. By the early 1900s, however, he became convinced that a mechanistic and materialistic explanation of the differentiation process, for example, was totally impossible. His own experiments had shown that the whole embryo (at the two-cell stage), when divided in half, could produce two complete new embryos. No machine, Driesch reasoned, could produce complete wholes from separate halves. These observations finally led Driesch to envision a vitalistic principle operative in living systems, which he called by Aristotle's word "entelechy."[54] Driesch's vitalism was a source of disagreement between him and Morgan. The latter had no use for nonphysical forces or principles, or anything that was not subject to investigation. Eventually, Driesch's vitalism led him to abandon biological research altogether. His position at Leipzig was as professor of philosophy, not natural science.

During his most productive experimental years, between 1890 and 1900, Driesch profoundly extended the methods of the *Entwicklungsmechanik* school. Unlike Roux, who remained fixed on the mosaic theory, Driesch developed his ideas continually and was always seeking new experiments and testing new hypotheses. Oppenheimer has summarized these succinctly.[55] They were in one way or another all concerned with the question of the *causes* of differentiation, and more specifically, with the degree to which nuclei in the cells of developing embryos always remained capable of forming whole, new embryos. From a variety of experiments, Driesch concluded that the nuclei of embryonic cells remain totipotent (i.e., capable of developing into whole embryos) until well

[54] Ibid., p. 188. In Aristotelian terminology, "entelechy" meant guiding principle, teleological in nature.

[55] Ibid., pp. 187-188.

through the blastula stage. He also concluded, in 1896, that the mesodermal tissue of sea-urchin embryos differentiated as a result of tactile stimulus from the overlying ectoderm. This concept was very close to Hans Spemann's later idea of embryonic induction.[56] Two years earlier (1894) Driesch had proposed the possibility of embryonic induction as a mechanism for differentiation; at the same time he proposed that such induction is the result of chemical influences on the nuclei of cells, causing those cells to produce substances that affect the cytoplasm. He even inferred that nuclear influence was mediated through "ferments" (the older name for enzymes). These ideas were contained in Driesch's widely disseminated and influential monograph, *Analytische Theorie der organischen Entwicklung*.[57] They were far-reaching in their significance, and way ahead of their time. They suggested a way of viewing embryonic induction that was neither as purely mechanical as some of the older models (which postulated stress forces as the cause of differentiation), nor as rigid and nonexplanatory as the mosaic theory.

But it was in its search for causes that the *Entwicklungsmechanik* school had its most profound influence on young morphologists such as Morgan. To Morgan, the kind of causal questions Driesch and Roux were asking were fundamentally different from the historical questions that Brooks and other morphologists had asked. The morphologists' questions were not susceptible to rigorous tests. The best morphological answer was still largely speculation. However, Roux's and Driesch's questions could be put to the test. For example, Morgan was convinced that the mosaic theory, at least as championed by Roux, was not acceptable as an explanation for differentiation. His reasoning was concise: if embryonic cell division qualitatively parceled out hereditary materials to daughter cells, then no cell beyond the egg should be able to give rise to a complete embryo. Since this prediction was contradicted by Driesch's experiments, Morgan concluded that the mosaic theory could be effectively rejected. On the other hand, few phylogenetic hypotheses could be put to a test, and hence it was seldom possible to distinguish the correct from the incorrect.

Morgan was also attracted to the idea of emphasizing the inter-

[56] Ibid., p. 187.

[57] Driesch, *Analytische Theorie der organischen Entwicklung* (Leipzig, 1894); Oppenheimer, *Dictionary of Scientific Biography*, vol. 4, p. 187.

action of embryo and environment during development. Embryonic cells could be viewed as differentiating in a specific direction as a result of both internal (hereditary) and external (position, chemical factors, etc.) influences. Morgan's already deep appreciation of the delicacy and balance of living systems, their tremendous range of adaptability, made Driesch's approach far more attuned to what he thought organisms were like, than Roux's overly simplistic mosaic model. Besides, Roux's mosaic idea postulated the existence of hereditary particles that were parceled out during embryonic cell division, which for Morgan sounded too much like the speculations of older morphologists when they talked about "pangenes," "biophors," "living molecules," and the like. To Morgan's way of thinking, it was sheer folly to invent particles to explain a process about which very little was known. As we shall see later, it was this view that made Morgan remain skeptical about the Mendelian theory for a number of years after his first contact with it in 1900.

While at Naples, Morgan and Driesch collaborated on some experiments dealing with embryonic development of Ctenophore eggs. Morgan and Driesch attempted to test again the validity of the mosaic theory of development as it had been applied to this group of animals. Earlier investigators, working under the influence of Roux's theory, claimed that isolated blastomeres of Ctenophores could produce only half, quarter, or at best three-quarter embryos.[58] When Morgan and Driesch carried out their own experiments with Ctenophore eggs, they noted that while most isolated blastomeres did not produce full-grown larvae (the number was significantly less than in sea urchins, for example), some did in fact develop completely. Morgan and Driesch studied the Ctenophore egg more carefully and noted that the cytoplasm was organized quite differently from that in sea-urchin eggs. It was this peculiar cytoplasmic structure, they concluded, that accounted for the greater number of partial embryos, and not a mosaic parceling out of nuclear material. The production of *any* whole larvae, Morgan and Driesch emphasized, showed that the mosaic theory could not be correct. Since in principle the Ctenophore egg followed the same lines of development as that of other animals, it was not necessary, Morgan and Driesch argued, "to make any

[58] See, for example, C. Chun, "Die Bedeutung der direkten Zelltheilung," *Sitzbr. Physik-Okon Gesellschaft Königsberg* 31 (1890): 1-3.

compromise with the theory of the qualitatively unequal nuclear division."[59]

From Morgan's correspondence, it is apparent that he and Driesch, along with Herbst, spent long hours at Vincenzo's, a local bar and restaurant near the Naples Station, drinking wine and discussing problems in embryology. Morgan felt in Driesch a kindred spirit, one who was not only fed up with the old-style descriptive morphology, but was also striking out in a bold and new direction. His letters indicate the close bond between the two men and an honesty and good-humored playfulness that seems all too rare between scientific workers. For example, in 1897 when Driesch had been somewhat remiss in writing to Morgan, the latter wrote back reprovingly:

> Dear Driesch. You are a lazy fellow not to write to me—I who am always thinking of writing to you and finding so little time to do it. But you! All your time your own, and living in the middle of the world and yet not helping a poor, isolated down-trodden devil like me. Fu! Fu! Allora. . . . Now tell me what you are doing. If I lived in Naples I would "sweetly do nothing" but here I keep fairly well occupied. You, however, are a philosopher and oblivious to your surroundings and even Naples does not seem to subdue the activity of your mind, or soften the temper of your criticism of our friend Roux.[60]

And, when Driesch wrote in 1897 that he was engaged to be married, Morgan responded enthusiastically:

> How shall I tell you of my surprise! . . . Your postal came like a thunderclap from a clear sky and of course being of skeptical disposition I shall not believe a word of it until you demonstrate to me your sincerity. At least you might send me something better than a short postal. How did it happen and who is your Gretchen? Now we can no longer have pleasant

[59] Hans Driesch and T. H. Morgan, "Zur Analysis der ersten Entwickelungsstadien des Ctenophoreneies. I. Von der Entwickelung einzelner Ctenophorenblastomeren. II. Von der Entwickelung ungefurchter Eier mit Protoplasmadefekten," *Roux' Archiv für Entwickelungsmechanik der Organismen* 2 (1895): 204-215; 216-224; quotation, p. 216. The original German reads: "und glaubten nicht genöthigt zu sein, der Lehre von der qualitativ-ungleichen Kerntheilung irgend welche Zugeständnisse zu machen."

[60] Morgan to Driesch, January 8, 1897, p. 1; Morgan-Driesch correspondence.

> summers in Munich together—or rather, you cannot and Herbst and I shall have to do without you. Give Herbst my warmest condolences—you have certainly deserted the Dreibund. . . . But I congratulate you from the bottom of my heart. Here is great good luck and a world of happiness to you! In regard to your earlier expressions about womankind I always knew that they were wrong, but I also knew that it would be hopeless to convince you of your error. And now you have acknowledged it yourself. Va bene! There is something else in the world than science and art. Nicht wahr? But it took you a long time to find it out—about 30 years! Far from thinking that you are lost, I think you are saved to humanity. Only it is a little hard for your friends.[61]

And in another letter a few months later, Morgan wrote: "I do not believe that you are half impressed with your good fortune. Yet while I envy you, I do not begrudge a bit of it to you. It is one of those rare cases where the best things come to the right man! Ecco!"[62]

The good spirit and candor carried over to biological matters as well. In 1897 Morgan wrote to Driesch explaining that he had been using the latter's book *Analytische Theorie der organischen Entwicklung* in his graduate biology course dealing with invertebrate embryology. Morgan wrote: "I have been lecturing to graduate students in experimental work all this semester. Think of it, I have boiled down the *Analytische Theorie* and gave it all in two lectures. There were some places I did not understand, but that was your fault. I put on a bold face and went right through. But believe me it is good stuff for 'ladies.' "[63] Similarly, when Driesch upbraided Morgan for misstating some of his (Driesch's) conclusions, Morgan wrote back:

> I am very sorry to have *misrepresented your views* on the action of the red pigment of Tubularia, but really if you will look back over your Studien . . . you will see how an unwary foreigner might make such a mistake. You speak of a "hydranthbildenden Stoff" and account for the different kind of structures, probosces, etc. as a result of its greater or less amount. That is all that I meant in my paper, though I may have implied more than I ought to have done. But I think you

[61] Ibid., October 9, 1898.

[62] Ibid., February 12, 1899, pp. 7-8.

[63] Ibid., January 8, 1897, pp. 1-3.

were also not entirely clear in the place I have referred to. . . . It is mighty hard to get things exactly right and I may quote a statement in your *Ergebnisse* article by way of illustration.

You say that "Morgan had a fish embryo to develop without a yolk sac very far." Never so far as I know, have I made such a statement or anything like it. At best I found a good deal of the yolk could be removed and an embryo be formed, but there was always a yolk sac present.[64]

Morgan's association with Driesch and the Naples Station and his own work in the *Entwicklungsmechanik* tradition completed his metamorphosis to a staunch experimentalist. For the remainder of his life Morgan remained a champion of the experimental approach to biology.

MORGAN AND EXPERIMENTAL EMBRYOLOGY

In the years after his return from Naples, Morgan's biological interests ran in a number of directions. He collected information on the problem of evolutionary adaptation; he initiated numerous studies in experimental embryology, especially regeneration; and shortly after 1900 began, in addition, a new line of work dealing with the problem of sex determination. In all of these studies Morgan shows clearly the influence of the *Entwicklungsmechanik* tradition and of Driesch. This is amply illustrated by two examples of Morgan's experimental embryology. The first is taken from the general area of regeneration which figured so prominently in Morgan's thinking at this time. The second comes from Morgan's study of "partial" larvae, a problem derived directly from his association with Driesch and European experimental embryology.

Morgan had carried out descriptive studies on earthworm regeneration as a graduate student, but had not pursued that line of work until after his return from Naples.[65] It is well known that most plants have a high degree of regenerative ability. Among animals, the ability to regenerate is common among lower forms, but it decreases as one moves higher up the phylogenetic tree. Thus, for example, a sponge can, if macerated, regenerate several new sponges; a hermit crab can replace an eye if that organ is

[64] Ibid., March 4, 1901, pp. 1-4.

[65] Sturtevant, "Thomas Hunt Morgan," p. 289. Morgan's first paper on earthworm regeneration was published in 1897: "Regeneration in Oligochaete Worms," *Science* 6 (1897): 692-693.

damaged, a salamander a whole tail or limb. Human beings or other mammals, on the other hand, can generally regenerate only structures such as skin, hair, or nails; they cannot regenerate limbs or other complex parts. For Morgan, the problem of regeneration was an obvious link between his older phylogenetic interests and his newer interests in experimental embryology. The obvious question of evolutionary interest was to determine what makes regeneration possible in some lower animal forms, and impossible in other, "higher" forms. From the embryological side, regeneration involves the regrowth and redifferentiation of tissues. When a salamander limb regrows, the tissue of the injured stump reorganizes and goes through a process of differentiation to form new bone, muscle, and nerve endings in a manner similar to that of the formation of the limb during early embryonic development. Morgan saw regeneration as a natural aspect of embryology, and in 1895 he set about to study regeneration from an experimental point of view.[66]

The problem of regeneration developed further significance in light of the debate between Roux and Driesch. The question was essentially one of the potency (to use Driesch's term) of the cells in adult organisms. If the potency of adult cells were restricted, as Roux's theory maintained, then regeneration would have to be explained by some process other than regrowth and development of the already differentiated cells at the site of the injury. Some workers had, in fact, proposed that regeneration was the result of nonspecialized cells located at certain key positions throughout the adult organism which moved to the site of a wound and were responsible for regrowth. Morgan tended to reject this view, not only on the grounds that it presupposed a restriction in total potency of adult cells (and thus opposed Driesch's views), but also that it could not be supported by any observational or experimental evidence.

In studying regeneration, Morgan worked with the earthworm partly because it could be easily obtained and maintained in the laboratory throughout the year.[67] From surveying the literature as

[66] Morgan's summary of the current as well as some past views on regeneration are given in "Some Problems of Regeneration," *Biological Lectures Delivered at the Marine Biological Laboratory of Woods Hole in the Summer Session of 1897 and 1898* (Boston: Ginn & Co., 1899), pp. 193-207.

[67] He also did a number of regeneration experiments on marine forms that he obtained at Woods Hole during the summers or on one trip in 1901 to the Marine Lab at Beaufort, North Carolina.

well as from his own laboratory work, Morgan established a number of observations about regeneration in the earthworm which seem to be valid. (1) If a few anterior segments are cut off (between one and five), the same number of segments always regrow anteriorally. (2) If more than five segments are cut off, the process of regeneration anteriorally takes longer, and at most only four or five segments grow back. (3) If the anterior cut involves more than half the worm, then the time before regeneration begins is still longer, and fewer worms succeed in regenerating at all. As the cut is made further and further from the anterior end, the remaining posterior segments show no regeneration of anterior segments. (4) If a small piece (eight to ten segments) is cut off at the anterior end, that piece will not regenerate a posterior region. However, if eight to ten anterior segments are cut off and then a few of the anterior segments are cut off of that piece, the remaining portion *will* regenerate anterior segments.[68] With these observations or "facts" in mind, Morgan posed the current problems biologists face in trying to develop any understanding of how the process of regeneration is regulated. The facts show that the rate or time of regeneration is not controlled simply by the size of the pieces that are cut off (or conversely of those that remain). The facts also show that there are no cellular distinctions between the anterior and posterior ends of the earthworm. That is, there are no observable cellular or other structural differences that might account for the differences in regenerative ability between the two ends.

Like the study of differentiation itself, two theories of the *cause* of regeneration had been put forward in the later years of the nineteenth century. One was that regeneration was governed by external stimuli acting on the injured part (such as gravity, chemical agents, or light), and the other that regeneration was guided by internal factors (such as organization of cells, differential cell growth, etc.). When he first became interested in the problem of regeneration, Morgan meticulously surveyed the older as well as the more recent literature on the subject. A forced stay in London in September of 1898, while waiting for his boat to leave for the United States, provided the opportunity to read the older literature (particularly the work of Abraham Trembley and Charles Bonnet) at the British Museum. At that time Morgan found himself impressed particularly with Bonnet, whose observations and

[68] Morgan, "Some Problems of Regeneration," p. 198.

ideas he said were more stimulating and valuable than anything such contemporaries as Weismann or Haeckel had produced.[69] It was characteristic of Morgan to read voluminously and to systematically cover all previous accounts on a subject to which he was devoting experimental effort. As he wrote to Driesch, "I find one thinks a good deal while one reads."[70]

After surveying the literature, Morgan noted that the distinction between the role of external and internal factors was in many ways artificial. Those who proposed external factors, for instance, often tried to pinpoint one specific type of influence, such as light, or a specific chemical, i.e., sought *one* external factor that was *the* determinative factor in regeneration. Those who proposed internal factors, on the other hand, tended to see regeneration as Roux saw embryonic differentiation: as a rigid process, somehow inherent in cells, unresponsive to environmental changes. As with the process of differentiation itself, regeneration seemed to Morgan to result from the constant interaction of internal and external factors. The ability to regenerate was certainly inherent in the cells of certain species, or certain organs within a species. Yet, the nature and form of regeneration was very much governed by certain external factors (such as number of segments cut off of the earthworm; or, in higher animals, the site of the damage). The organism was neither a biological automaton, nor a completely plastic object buffeted this way and that by external conditions.

Yet it was profoundly interesting to Morgan to try to understand what affected the inherent capacity of certain cells, in certain species, to carry out regeneration. Why did not all cells have that capacity? Is there something else (other than regenerative ability) that distinguishes cells that can regenerate from those that cannot? In a letter to Driesch of January 5, 1897 (published in *Roux's Archiv* later that year), Morgan outlined a series of experiments designed to investigate the problem of regenerative capacity. He suggested cutting two earthworms at approximately their midpoints; with a very fine needle and thread he would then sew the two anterior ends together. He proposed the same experiment for the two posterior ends. He would thus have two normal-length worms, one consisting of two heads, the other of two tails (see figure 5). He would then cut off a certain number of segments from

[69] Morgan to Driesch, September 18, 1898; Morgan-Driesch correspondence.

[70] Ibid.

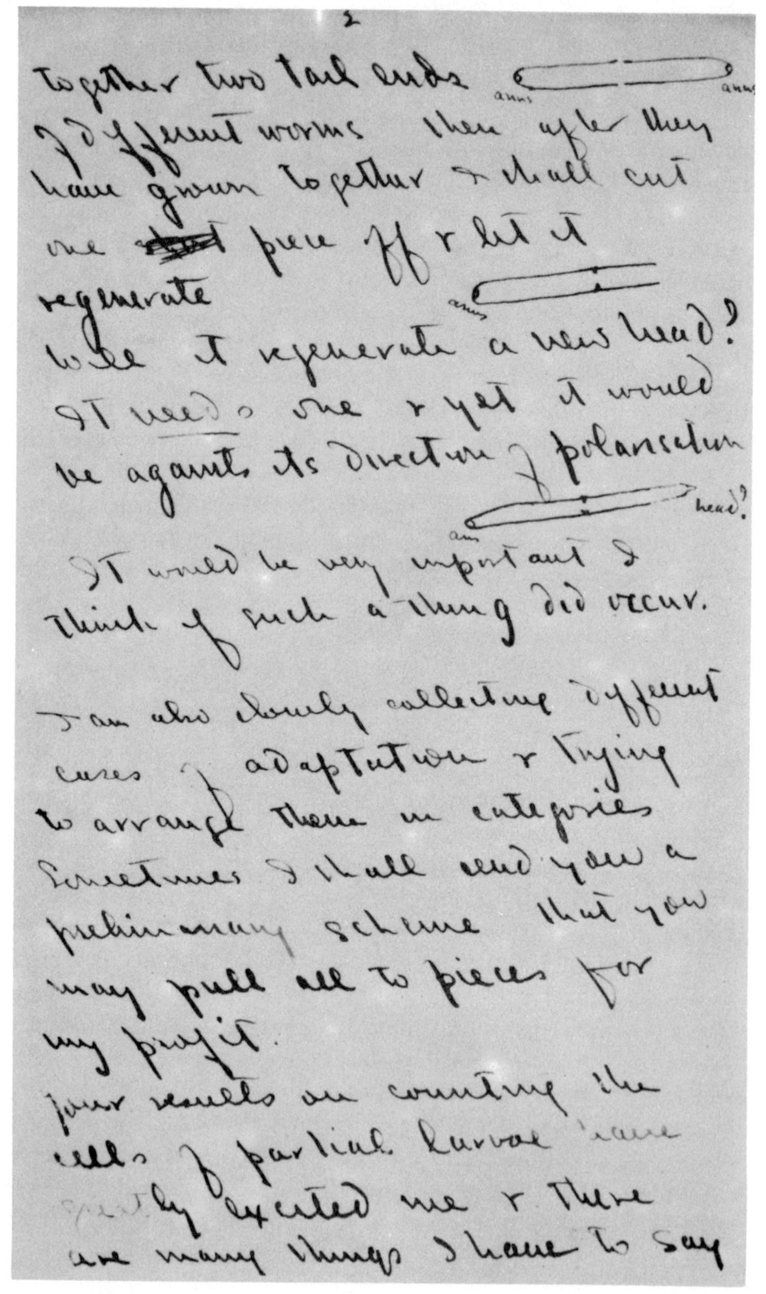

2

together two tail ends
of different worms then after they
have grown together I shall cut
one piece off & let it
regenerate
will it regenerate a new head?
It needs one & yet it would
be against its direction of polarization

It would be very important I
think if such a thing did occur.

I am also slowly collecting different
cases of adaptation & trying
to arrange them in categories
Sometimes I shall send you a
preliminary scheme that you
may pull all to pieces for
my profit.
Your results on counting the
cells of partial larvae have
greatly excited me & there
are many things I have to say

Figure 5. Manuscript pages from Morgan's letter to Hans Driesch of January 5, 1897. Here Morgan outlines one of his characteristically simple experiments, in this case dealing with regeneration in earthworms where two

3

BIOLOGICAL LABORATORY,
BRYN MAWR COLLEGE BRYN MAWR, PA., ... 189

on the subject. In general I think my observations were correct although I am free to admit that the particular interpretation may have been wrong. Further, and this is perhaps more to the point, the comparisons between particular stages of partial & whole larvae may be misleading because as you point out, and as I also said (page 84) it is difficult to get exactly the same stage of the two kinds of larvae. It would have certainly been better to have taken organs like the mesenchyme that have a definite ~~[illegible]~~ period of formation & ending. (With regard however to this mesenchyme I shall have more to say later. Here I should like to ask, however, did you use partial larvae from ~~the~~ broken eggs or from isolated blastomeres?)

anterior or two posterior sections are sewn together. See text for a more detailed description of the experiment and its results. (Courtesy of Professor Reinhard Mocek, Martin-Luther-Universität, Halle-Wittenberg.)

one end of each specimen. The experiment was simple in conception and, characteristically, was stated in precise terms in his published paper: "If, for instance, two posterior ends are united, and then later a portion of one of these ends be cut off, will a new head or new tail regenerate? Under ordinary circumstances a new tail would appear from such a piece, but since the double piece has already one tail, and obviously needs a new head, it seemed worth finding out whether a new head would appear. Conversely, if two anterior ends are united and then a portion of one of these be cut off, will a new head or a new tail appear?"[71] Morgan's results were not wholly conclusive, but nonetheless pointed in the direction of further research. First of all, he encountered some experimental difficulties: for example, the anterior segments would not stay sewn together, so none of those grew back together or could be further operated upon. Of the eleven posterior pairs that were sewn together, only five ultimately grew together. It was on these five that he performed the second part of the operation by cutting off a small section from one or the other end. All five regenerated new posterior ends only—that is, no heads appeared. As Morgan wrote: "This is, perhaps, what we should have anticipated, but I was not without hope, that since the pair had already one posterior end, an anus, a new head might appear, as it alone was needed to make a complete (?) worm."[72]

Morgan's results supported his basic idea that regenerative capacity is not caused primarily by external factors, but is, as he phrased it, "connected with the organization of the worm itself."[73] Morgan went on to formulate a hypothesis to explain the results. The first part of Morgan's hypothesis was that earthworms contain "stuffs" that are more or less abundant in different regions of the organism.[74] Morgan further assumed that the stuffs were

[71] T. H. Morgan, "Regeneration in *Allolobophora foetida*," *Roux' Archiv für Entwickelungsmechanik der Organismen* 5 (1897): 570-586; quotation, p. 578.

[72] Ibid., pp. 578-579.

[73] Ibid., p. 582.

[74] The concept of "stuffs" was not Morgan's invention, but was persistent among various embryologists around the turn of the century. The German term "*Stoff*" had been invoked to explain embryonic differentiation itself. Different cells were thought to contain different material, originating from cytoplasmic differentiation of the egg cell. The different *Stoff* present in these different regions were thus parceled out to the daughter cells during cleavage. The qualitatively different substances thus came to act in different ways upon their respective cell nuclei, producing differentiation. See also, T. H. Morgan, "The Hypothesis of Formative Stuffs," *Torrey Botanical Club Bulletin* 30 (1903): 206-213.

distributed throughout the earthworm in a gradient: "The head stuff would gradually diminish as we pass posteriorly, and the tail stuff increase in the same direction. We should also think of this stuff in the cells as becoming active during regeneration. Where there is much of the head stuff, the cells can start sooner to regenerate anteriorally: where there is less it must increase first to a certain amount or strength before the part can begin to regenerate."[75] By this hypothesis, Morgan purported to account for both the fact that the more posterior the region, the less regenerative ability (anteriorally) and the longer the time required between the injury and the onset of regeneration.

Morgan's experimental methodology in approaching the problem of regeneration was very much in the *Entwicklungsmechanik* tradition. This means, primarily, that he paid strict attention to good experimental design. First and foremost, Morgan set up his experiment, and in fact asked his initial question, in a precise and rigorous way. He was interested in determining whether the intact worm was more or less programmed to regenerate in certain directions, or whether that regeneration could be modified according to special circumstances in which the need to regenerate one part (for example a head when two posterior segments were united) could override the more customary response. The experiment was designed to provide a "yes" or "no" answer. Whichever way the experiment turned out, Morgan would have some answer to his original question. This was what was meant in the *Entwicklungsmechanik* school by a rigorous experiment. Morgan could not always work this way, of course, as no worker is able to design only experiments where the answer will come out yes or no. In fact, Morgan even poked fun at himself for the experiments he often carried out with no apparent plan whatsoever. As he wrote to Driesch about his regeneration experiments: "Then I have some planarians and some lumbriculus [*sic*] lots of earthworms and a *Necturus* [a salamander] with a finger and a part of a tail cut off. It is great fun! I laugh when I think of what I said about a *well planned* experiment."[76] Nonetheless, the rigorous experiment was the ideal of the *Entwicklungsmechanik* school, and Morgan was particularly gifted in designing such experiments.

Another characteristic of the *Entwicklungsmechanik* school emphasized in Morgan's work on earthworms is the strong skepti-

[75] Morgan, "Regeneration in *Allolobophora foetida*," p. 582.

[76] Morgan to Driesch, January 8, 1897, pp. 3-4; Morgan-Driesch correspondence.

cism toward hypothetical explanations. As he wrote: "I do not pretend that this, the concept of a gradient of stuffs, explains anything at all, but the statement covers the results as they stand."[77] In later years Morgan found himself opposing the concept of gradients and stuffs to explain embryological and regenerative phenomena. However, in the period between 1894 and 1906, Morgan repeatedly invoked a gradient or polarity concept of substances to account for the regenerative phenomena he studied in fish tails, salamander limbs, hermit crabs, and a variety of other species.[78] However, the important point is that Morgan approached the use of such hypotheses quite differently from the older morphologists. He considered that hypotheses were of value only if they could suggest further experiments to test their validity. To take a hypothesis more seriously, as if it were true because it was suggested, was a danger of which Morgan was well aware. It was one of the influences *Entwicklungsmechanik* had on him: to build a healthy skepticism of all hypotheses.

A second example illustrative of Morgan's approach to biological problems appears in the period prior to 1904: the study of partial larvae. Driesch's experiments concerning the regulation of development in embryos that originated from isolated blastomeres indicated that the isolated blastomeres could regulate their development to compensate for their separation from the whole. This regulatory ability was, according to Driesch, part of the "harmonious equipotential system," self-adjusting and self-regulating, as he thought of the developing embryo. The immediate question arising from this concept of self-regulation was: how does the embryo do it? Does the isolated blastomere proceed exactly as if it were a fertilized egg? Driesch's own experiments

[77] Morgan, "Regeneration in *Allolobophora foetida*," p. 582. The concept of a "gradient" became increasingly popular during the early years of the twentieth century and was invoked by many workers to explain all types of physiological and regenerative phenomena. See, for example, the work of C. M. Child, "Susceptibility Gradients in Animals," *Science* 39 (1914): 73-76; and the work of Sven Hörstadius, "The Mechanics of Sea Urchin Development, Studied by Operative Methods," *Biological Review* 14 (1939): 132-179.

[78] Morgan continually mentions gradients of substances in his letters to Driesch between 1904 and 1906; also, these ideas were published in "An Analysis of the Phenomena of Organic 'Polarities,'" *Science* 20 (1904): 742-748; and "'Polarity' Considered as a Phenomenon of Gradation of Material," *Journal of Experimental Zoology* 2 (1905): 495-506.

had shown that it did not, for the partial larvae that developed, while complete in all their parts, were smaller than embryos directly developed from the fertilized egg.

Morgan asked the logical next question: why were the experimental embryos smaller? Was the difference in size the result of a smaller *number* of cells making up the embryo, or was each individual cell smaller? Here was a problem that could be investigated by experimental and quantitative means. Furthermore, the data obtained would help elucidate the underlying mechanism by which the embryo regulated its own growth and development. Driesch had claimed in the early 1890s that partial embryos contained proportionately less cells than their normal counterparts. That is, a one-half larva would contain half as many cells as the normal larva (of the same developmental stage), a one-fourth larva one-fourth the number of cells, etc. Driesch claimed that this proportionality held for both total cell number and cell number in individual organs. Morgan took up the study of this problem with the sea-urchin egg initially, because preliminary evidence from his own studies had indicated that the proportionality was not as fixed as Driesch would have it. Morgan showed at first that he and Driesch agreed on proportionality in terms of total cell number in the partial larvae. In contrast to Driesch, however, Morgan maintained that for some organs, at least, a proportionately greater number of cells was involved in the partial larvae than in the normal larvae.[79]

Morgan set out to study in sea urchins the development of the archenteron, the "primitive gut," an invaginated region of the gastrula which will eventually form the digestive tract of the embryo. The archenteron is a long tube, lined with cells some of which have migrated in from the outside. Morgan chose to determine what variations, if any, could be produced under different experimental conditions in the number of cells lining this cavity.[80] Morgan studied the archenteron because it was the first organ to develop out of the blastula, and its cells could be counted early and easily. By observing numerous embryos under the microscope, and count-

[79] T. H. Morgan, "Studies on the 'Partial' Larvae of *Sphaerechinus*," *Roux' Archiv für Entwickelungsmechanik der Organismen* 2 (1895): 81-126.

[80] Morgan summarized these experiments in: "The Proportionate Development of Partial Embryos," *Roux' Archiv für Entwickelungsmechanik der Organismen* 13 (1901): 416-435, especially pp. 417-419.

ing their archenteron cells, Morgan concluded that in half embryos the number of cells involved in forming the gut was greater than one-half the number normally involved for whole embryos, and that the same lack of proportionality held for one-fourth embryos, etc.[81] Morgan and Driesch exchanged their contradictory evidence both in letters and in print, and for some time maintained opposing positions on this issue.

However, Morgan continued to try to resolve the problem by studying the development of the archenteron in other species than *Sphaerechinus*. He studied bony fishes such as *Fundulus*, and another type of sea urchin, *Toxopnustes*. In the last species, Morgan noted something that helped explain the discrepancy between his own and Driesch's views. The problem, it appeared, arose out of not taking into account the difference in time of invagination in one-half, or one-fourth larvae compared to normal larvae, or the time differences when invagination occurs in different species. The longer the invagination process is delayed, the more cell divisions that can occur prior to the formation of the gut. Thus, when invagination occurs at a later, rather than an earlier time, more cells are present to be pushed inward. Thus, counts of cell numbers mean nothing without a reference point in time of development in the embryo's life. Morgan studied hundreds of specimens before finally relaying his findings to Driesch in a letter. Both of them were at fault, he claimed, and freely acknowledged his own contribution to the misunderstanding:

> You will be much interested to hear that the results in regard to the endoderm [which lines the archenteron] show that we were *both* right. It is like this. The one-half blastulae, that gastrulate *at the same time* as (or a little later than) the whole eggs, invaginate only one-half the total number of cells. Later the number continues to increase as in the normal. Those one-half blastulae that *gastrulate later*, i.e., after four or five hours, invaginate a larger number of cells than one-half the whole number, but one-half the same number that the whole larvae possesses *at this time* [of invagination]. In other words the number of potential endoderm cells also increases in the one-half blastulae and therefore more are invaginated than one-half the whole number. The same facts hold for the one-fourth embryos, but in a larger degree because their gastrulation period is longer delayed—also for

[81] Ibid., p. 418.

> the one-eighth embryos. My former estimates were correct, but I failed to take into account the time at which gastrulation takes place. . . . The facts seem so positive and the explanation is so simple, that I think we must accept them.[82]

Because he had specifically wanted to test the conclusions on another species than *Sphaerechinus*, Morgan had spent several weeks of the summer of 1901 at Beaufort, North Carolina, where he had been able to collect specimens of *Toxopnustes*, a fact that nicely illustrates the thoroughness and persistence Morgan directed toward solving problems that confronted him.

This work on cell number in partial larvae did not lead to any revolutionary progress in understanding how embryos regulate growth and differentiation. It did, however, show that at one stage of development there is a regular pattern to the formation of organs. The results showed that one factor involved in organ formation is the number of cells available in the embryo at the particular time and place the organ is formed. Once the time factor (reflected in the number of cells) was taken into account, Morgan saw that there was no fundamental difference in the process of organ formation between partial and whole embryos. The archenteron was formed from the same proportional number of cells available at the time of invagination in both partial and whole larvae. Thus, disturbance of the larvae by separating the blastomeres had not altered the fundamental internal patterns of development. Had the number of cells in partial larvae been significantly different from that in whole embryos, or had the relationship proved to be a random one varying from one specimen to another, Morgan and Driesch would both have found that their concept of internal regulatory mechanisms for differentiation might have needed serious revision.

Like all scientific workers, Morgan had his prejudices. However, he had the preeminent ability to change his mind and admit he was wrong. Nearly every scientific worker performs an experiment or set of experiments with some hypothesis in mind. In most cases investigators have a distinct preference for which hypothesis they hope will not be disproved. Unlike many, Morgan was able to overcome his preferences when the evidence so demanded.

[82] Morgan to Driesch, July 15, 1901, pp. 1-2; Morgan-Driesch correspondence.

Once he had turned to experimental embryology, Morgan became one of the leaders in the field. His books *Regeneration* (1901) and *Experimental Zoology* (1907) established him as one of the leading experimental biologists in the United States, and, indeed, the world. His appointment as professor of experimental zoology at Columbia in 1904 was tangible recognition of his accomplishments. He was an acknowledged expert on the development of sea-urchin and frogs' eggs, as well as the most knowledgeable person in North America on the general problem of regeneration. Experimental embryology of the Roux-Driesch variety, or studies on regeneration such as those described above with the earthworm were by no means the only areas in which Morgan worked during this period. As we shall see in the next chapter, from the turn of the century on his interests lay increasingly in the areas of evolution, heredity, and sex determination. It was through these studies that he was led to his work on heredity.

Yet it is also true that the basic issues which Morgan later pursued with respect to the process of heredity were fundamentally related to the central issues of the *Entwicklungsmechanik* school. Morgan and other *Entwicklungsmechaniker* were convinced from the outset that development and regeneration were of factors internal to the organism itself. While Roux and Driesch differed on exactly how the internal mechanism operated (i.e., mosaic or regulative), they did agree that the process was internal to the developing embryo—at least from the fertilized egg stage onward (some embryologists, however, thought the entire process was determined by the structure of the egg cytoplasm itself, and thus was set prior to fertilization). Such a view commits its holders inextricably to the notion of heredity—of the transmission of some internal organizing factor or factors from parents to offspring. The next question was what *kind* of factors were actually inherited? It is important to emphasize here that Morgan's association with the *Entwicklungsmechanik* school prepared him not only to become an experimentalist; it also prepared him to think in new, and different ways about the process of heredity. Morgan's later focus on heredity was a logical (though not necessary) outgrowth of his earlier studies in experimental embryology.

CHAPTER IV

The Experimentalist Looks at Theory: Evolution and Heredity (1904-1910)

Two persistent problems stemming from the work of Darwin continued to plague biologists at the turn of the century. One centered around natural selection: was Darwin's theory sufficient to account for the origin of the many intricate adaptations known to exist among plants and animals? The other problem focused on the issue of heredity. Which characteristics were inherited and which not? Were there any laws governing the transmission of heritable traits? And, perhaps the most important of all to those concerned with the origin of species: how do new variations arise to provide the raw material for evolution?

From the mid-1890s on, evolutionary and heredity problems had been of great interest to Morgan, but they became increasingly so from 1904 to 1910. Through his work on regeneration he had become interested in the problem of the origin of adaptations, the ability to regenerate, of course, being one remarkable example. Furthermore, through his work in experimental embryology he had become fascinated with the problem of the determination of sex. The former brought him face to face with Darwinian evolution by natural selection and with de Vries's concept of speciation by large-scale "mutations." The latter, in turn, brought him into direct contact with the newly rediscovered Mendelian theory of heredity.

Morgan's views on both Darwinian and Mendelian theory offer considerable insight into his mode of approaching biological problems. Initially he opposed both theories rather vigorously, for a short time embracing de Vries's mutationism as a viable alternative for understanding both heredity and evolution. Eventually, however, Morgan changed his mind and became a strong, even zealous, supporter of both Mendelism and Darwinism. Morgan's initial opposition to these theories, especially in the early 1900s, reflects his abiding distrust for broad, large-scale hypotheses. On the other hand, his enthusiasm for the de Vriesian theory reflects, in part, his continuing commitment to the *Entwicklungsmechanik* tradition.

EVENTS, 1904-1910

Morgan left California after his summer's work and wedding trip in late August 1904. With Jacques Loeb and Loeb's close friend Hugo de Vries (1848-1935), he and Lilian traveled east to St. Louis, Missouri, for scientific meetings. De Vries and Morgan became firm friends. For Morgan the journey with Loeb and de Vries was stimulating, and he reports having gained a great deal from discussions with both men.[1] From St. Louis, the Morgans went on to Lexington, Kentucky, to visit Morgan's parents, while Loeb returned to California and de Vries went east. Before leaving New York for Amsterdam, de Vries stayed for a couple of days with the Morgans in their recently rented house near Columbia University.

During Morgan's first year at Columbia, he and Lilian rented the house of a classics professor at the university who was away on sabbatical. It was a large house, and with it came servants so that Morgan could write that all in all "life has been made easy for us."[2] Renting was only temporary, however, and it appears that he and Lilian soon decided to buy a house of their own. Lilian, an orphan from the age of three, had lived most of her life with grandparents in a large house in Germantown, Pennsylvania. She would have preferred to begin housekeeping in a small way so as to enjoy caring for her husband herself, but Morgan preferred the comfort of a large house, run with the help of servants. One evening in early October of 1905 Morgan returned late with the unexpected announcement, "We have bought number 409."[3] He was referring to one of the large graystone houses, 409 West 117th Street, on what amounted to Columbia's Faculty Row. The house, which is still standing, was a five-floor structure, which Morgan obtained for $20,000.[4] It was spacious and well built,

[1] Morgan to Charles Zeleny, September 24, 1904, p. 2; Zeleny Papers, University of Illinois Archives.

[2] Morgan to Driesch, December 13, 1904, pp. 2-3; Morgan-Driesch correspondence, APS.

[3] Isabel Morgan Mountain, "Notes on T. H. Morgan," p. 8.

[4] See Title and Deed, United States Trust Company; Morgan Papers, Caltech. The house was sold when the Morgans left New York for Pasadena in 1928 for $35,000.

with gas-lit fireplaces in many rooms, and Morgan furnished it with one of his earliest possessions, an old player piano. The middle floors contained bedrooms, and on the top floor was Morgan's study where he could work undisturbed into the night, after the family had gone to bed. The Morgans were to live in this house for the next twenty-three years. All of their children were to be born either in New York City or New Bedford, Massachusetts (across Buzzards Bay from Woods Hole) and raised in New York City during the winters and Woods Hole during the summers.

Considering that both Lilian and Tom had been living in the relatively small town of Bryn Mawr for a number of years, they seem to have adjusted well to New York City. Morgan wrote a few years later to Driesch: "It is undoubtedly distracting at times, but one can be alone as much as he likes in a big city. And then there is the opportunity to see a great deal if one cares now and then to come out of his shell."[5]

The Morgan's first child, and only son, Howard Key, was born the next year, on February 22, 1906. Morgan's enthusiasm was unbounded. A few days after his son's birth, he wrote to his friend E. G. Conklin:

> You will be glad to hear, I know, that the anxiety is over and that we have a boy. He is not a pretty baby I am glad to say —I looked at what the nurses called a pretty baby and didn't like it—but a sturdy eight-pounder with lots of character which he gets from his mother—apparently it dominates in the first hybrid, but perhaps ¼ of his offspring will be recessive like the grandfather.
>
> Mrs. Morgan came through it all finely and is very well and you may guess more than happy. It is lonesome for me now at home with *my family* all in the hospital, but of course I realize that my day is near.
>
> [Signed] Pater familias[6]

The Morgans were to have three more children: Edith Sampson, born on June 25, 1907; Lilian Vaughan, born January 5,

[5] Morgan to Driesch, November 23, 1910, p. 3; Morgan-Driesch correspondence.

[6] Morgan to Conklin, February 26, 1906; Morgan Papers, Caltech; quoted with permission.

1910; and Isabel Merrick, born August 20, 1911.[7] To Morgan, his children were a constant source of enjoyment and pleasure. There were, of course, some drawbacks: gone for a while were the days in which Morgan could journey freely to Europe for the whole or part of the summer. "And how can we come abroad with such a 'raft' of kiddies," he wrote to Driesch in 1912. "We are chained down to this side of the water for years to come."[8] But these disadvantages were far outweighed by the advantages that family life conferred. Morgan continued: "My Sabbatical is now

[7] Morgan's four children (the "F_1's" as he sometimes referred to them) each had their own distinct career. Howard Key Morgan received a B.S. in electrical engineering from the University of California, Berkeley. He subsequently became a specialist in aeronautical engineering, working for both TWA and the Bendix Corporation. He is married to Bernadine Buck and is the author of a book on aviation engineering.

Edith Sampson Morgan received her B.A. from Bryn Mawr, shortly thereafter marrying Douglas Whitaker (1904-1973) whom she had known since adolescence in California. She was trained as a physiotherapist at Stanford University and worked with cerebral palsied children for many years. Whitaker earned a Ph.D. in biology from Stanford (he coauthored a paper with T. H. Morgan in 1930), where he subsequently taught, rising ultimately to the position of provost. Whitaker's last position was as vice-president for academic affairs at Rockefeller University. The Whitakers have one son and one daughter.

Lilian Vaughan Morgan received a B.A. degree from Pomona College in California and a diploma from the New York School for Social Work, specializing as a medical social worker throughout her career. She married Henry W. Scherp (1908-1974), a research worker in oral microbiology and until his retirement associate director for the National Caries Program at the National Institute of Dental Research. He later served as a visiting professor at the University of Alabama Medical School in Birmingham. The Scherps had no children.

Isabel Merrick Morgan received a B.A. from Stanford, an M.A. from Cornell, and a Ph.D. from the University of Pennsylvania, specializing in microbiology. In a distinguished career at the Rockefeller Institute and later at Johns Hopkins, she worked on central nervous system immunity to poliovirus and established for the first time on a quantitative basis in monkeys the efficacy of a formalin-inactivated poliovirus vaccine (forerunner of the Salk vaccine). She later entered the field of biostatistics (M.S., Columbia), working as a consultant at Sloan-Kettering Institute for Cancer Research. She married Joseph D. Mountain (1902-1970), a colonel in the United States Air Force, an aviator, photographer, and electronic data processor. His son from a former marriage, their only child, was killed in a mid-air plane crash over New York City in 1960.

[8] Morgan to Driesch, January 1, 1912, p. 2; Morgan-Driesch correspondence.

two years overdue, but I cannot get away. But why should I, for there is nothing so absorbing or interesting as one's own children."[9] The fact that the female-male ratio among his children was 3:1 did not escape him, a confirmed Mendelian by 1912. He wrote to Driesch in that year: "We now have four children—a new baby born last summer. Their names are Howard (male), Edith (female), Lilian (female), Isabel (female). You will observe the dominance of femaleness in my family in a strict Mendelian proportion."[10]

Family life in the Morgan household was organized along rather traditional lines. The bulk of the responsibility for the care and maintenance of the children, as well as for all practical household problems, fell on Lilian. Although both Morgans talked over large problems, and shared many of the more important decisions, Lilian took care of nearly all the small problems. Morgan was not practically oriented and was anything but a man about the house. As one of Morgan's daughters wrote much later: "From the beginning of their marriage it was established that he didn't know what a hammer or a screw-driver was for."[11] Morgan was thoroughly dependent on his wife for all day-to-day organization and for maintenance of the household.

Morgan took keen pleasure in sharing his children's pastimes. He was always gentle and tender with young children and very concerned about their problems and illnesses. On Sundays, for example, Morgan frequently took the children to the Columbia University greenhouse across 117th Street to check the roses. Morgan was very interested in growing flowers both for pleasure and for experimental purposes. On the way to the greenhouse the children placed bets, encouraged in this activity by Morgan himself, as to whether the Irish Elegance or Fire Flame would be the first of the roses to bloom. Morgan was not a meticulous gardener, and often the plants he raised were not kept in the most tidy condition, but they seemed to thrive nonetheless. While he inspected his plants, Morgan often set the children to washing flower pots at the sandstone sink—offering the material incentive of a penny for each clean pot. Not to be fooled, some of the children would demand an additional penny for particularly dirty

[9] Ibid.

[10] Ibid. Of course, femaleness is *not* a dominant trait, nor is it attributable to a single Mendelian gene.

[11] Mountain, "Notes on T. H. Morgan," p. 8.

pots that had to be washed more than once. And, not insensitive to the growing demands of labor in the real world, Isabel would periodically remind her father that "wages are going up." Undoubtedly Mrs. Morgan would have enjoyed going on these jaunts, but she deferred to Morgan's wish to spend some time alone with his children. In his tactful way, Morgan would generally bring his wife a special flower from the greenhouse.[12]

Perhaps nothing better illustrates both Morgan's love of children and his childlike enthusiasm at being with them than what turned out to be his traditional Christmas antics. When the children were still young, Christmas involved a veritable ritual. Everyone got up and opened stockings and had breakfast. Morgan continually asked the children whether they thought Santa Claus had come or not and what they thought their presents might be. When the children's enthusiasm had reached a peak, Morgan would matter-of-factly rise from the table and say, "Well, I think I'll go over to the laboratory for a little while." To the children's dismay, he disappeared out the front door. A few minutes later, however, the tingling of a bell from the top floor was a signal to the children to start a mad dash up the stairs. Just as they reached the top, they could see emerging through the skylight from the roof next door Santa Claus carrying a huge pack loaded with gifts. After distributing the gifts, making jokes, and playing with each child, Santa would disappear out the skylight again. A few minutes later, Morgan would come back through the front door, at which time all the children would vie to show him what they had received, and to tell him how he had missed Santa Claus.[13]

The structure of the Morgan family was very much in the prevailing Victorian model. The father took a less active role then in day-to-day affairs, but nonetheless maintained final authority over family policy. Emotions were kept low-key and were seldom expressed openly. The mother took care of most details, and any professional goals of her own took second place to her husband's career. In addition, like many of the well-to-do middle class, the Morgans had servants both in New York and in Woods Hole, including a cook and maid. It was partly this highly structured organization of his family and personal life that made it possible for Morgan to have such an enormous scientific output. Morgan's

[12] The description of the greenhouse jaunts, remembered by all his daughters, was supplied by Isabel Mountain, "Notes on T. H. Morgan," pp. 11-12.
[13] Ibid., pp. 12-13.

creative genius and his genuine interest in studying biological problems could be effectively put into action because he was freed from much of the day-to-day activity necessary to keeping himself and his family alive and active.

In understanding the social role of science in modern times it is important to keep in mind the relationship between the individual and the day-to-day productive effort necessary to maintain life. In western science the work of "great men," the geniuses we extol, rests usually, like the trip of a pyramid, on the work—the hard and sometimes uninteresting work—of many others. This includes laboratory technicians, secretaries, and various professional assistants. But it also includes the families, particularly the wives, or husbands of those investigators.

In Morgan's case, it would be impossible to understand how he could have had such a successful and productive career along with a family life without taking into account the vital and necessary contribution of other people, especially his wife and servants. This is not to take credit away from Morgan but rather to recognize the immense contribution of those who surrounded and helped him throughout his long and active career.

There is a lesson here for the practice of contemporary science as well. The structure of the family in the 1970s is considerably different from the early 1900s. Contemporary society will never go back to the Victorian type male-dominated family relationship (even though male dominance has not vanished). Under such conditions, it is necessary for a readjustment of male-female roles to occur in modern families. Through research grants and large supplies of money research scientists may be able to push off operational details onto assistants in the laboratory. But a male scientist's family life is likely to demand more time than in the past in dealing with day-to-day problems, problems to which he must devote some attention and whose emotional burden he must be willing to share. This means that contemporary male research workers may be less able to achieve the degree of scholarly output, quantitatively at least, that was characteristic of many investigators in Morgan's day.

It was during the period 1904 to 1910 that the Morgans established the pattern of spending their summers at Woods Hole. In the first summers after their marriage (1905 and 1906), they boarded at a fisherman's house, within walking distance of the laboratory. However, it became apparent, especially in the summer of 1906 with a new baby, that some more permanent arrangement would be desirable. Thus, in the winter and spring of 1906-1907 the Morgans designed and had built at Woods Hole a large, spacious summer house on Buzzards Bay Avenue for a total of $4,000.

During this period, Morgan continued his work at Woods Hole on various aspects of experimental embryology. For example, he studied the effects of external agents such as gravity and various salt solutions on the development of the eggs of marine organisms. He also studied regeneration in planaria, bony fish, and the coelenterate tubularia, and prepared a long series of papers (ten in all) on the relations between normal and abnormal development in frog embryos. These papers dealt with the effects of chemicals (such as lithium chloride), temperature, injuries to various of the blastomeres, and lack of oxygen on embryonic development. Finally, Morgan carried out a number of experiments investigating the problems of polarity in planaria and the earthworm—extensions of his regeneration studies discussed in chapter III. In addition, Morgan also became interested in several new topics. One was the mechanism of sex determination, especially the chromosomal relationships. Another was the efficacy of natural selection as a means of bringing about the development of species and adaptations. A third was the problem of heredity, with special reference to the Mendelian model based on analyses of breeding results. In addition to his research, Morgan was also active in a variety of administrative and organizational capacities during the period 1904 to 1910. These included trusteeship of the MBL and membership on the editorial board of the *Journal of Experimental Zoology*, founded in 1904.

Probably the most important of Morgan's new activities during this period was his role as program chairman for the section on experimental zoology at the Seventh International Congress of Zoology, held in Boston and New York during August of 1907. Having accepted this position in 1906, Morgan wanted to organize a series of papers under the auspices of the congress to promote the new lines of experimental work embodied in the *Entwicklungsmechanik* school. To this end, Morgan invited his friend Driesch to give the opening address for the section on experimental zoology at the congress.[14] Driesch suggested several topics, three of which dealt with various experimental topics, and

[14] Morgan to Driesch, January 10, 1907; Morgan-Driesch correspondence.

the fourth with philosophical issues of vitalism. Morgan wrote back urging Driesch to offer an experimentally based paper, explaining that the philosophical topic "would be over the heads of us poor, creeping zoologists with our noses on the ground looking for bugs and worms."[15] It was with great enthusiasm that Morgan learned of Driesch's acceptance; at last, Morgan hoped to fulfill his wish to get Driesch to the United States. He urged Driesch and his wife to come two weeks before the congress and stay with the Morgans at Woods Hole in their new house. Driesch never made the trip, however, much to Morgan's disappointment. Mrs. Driesch became seriously ill in early August, and Driesch did not feel that he would be able to leave Germany.[16] Driesch did, however, send on a copy of the paper he had prepared for his address. Morgan asked one of his former teachers at Johns Hopkins, E. A. Andrews, to read the paper before the assembled zoologists. Morgan later reported to Driesch that the paper was well received and that the congress as a whole was a success.[17] However, to Morgan, much of the excitement was taken out of the events by Driesch's absence.

During the period 1904 to 1910 Morgan published two important books which indicated the growing breadth of research problems in which he was interested. The first was *Evolution and Adaptation* (actually published in 1903) and the other *Experimental Zoology* (1907). The former summarized much of Morgan's thinking on the evolutionary process in general and the Darwinian model of natural selection in particular. It also contained discussions of many of the then current alternatives to Darwinism, including the inheritance of acquired characteristics, evolutionary momentum, and the role of isolation. *Experimental Zoology* was a compendium of much of the recent literature, principally on experimental embryology, but also including regeneration, problems of fertilization, and parthenogenesis (natural and artificially induced). The book was an extension of the principles of the *Entwicklungsmechanik* to areas of biology broader than embryonic development. So influential was *Experimental Zoology* that it was translated into both Russian and German within the next two years.

However, it was Morgan's interest in evolution, rather than in experimental zoology per se, that led to a search for mutations in animals of the type Hugo de Vries had claimed to observe in

[15] Ibid., March 13, 1907, p. 2.
[16] Ibid., August 15, 1907.
[17] Ibid., September 15, 1907.

plants. And it was through the search for mutations that Morgan took up work with the fruit fly, *Drosophila melanogaster*. In turn it was in his cultures of *Drosophila* that he first observed the white-eyed male which initiated his conversion to Mendelism and the chromosome theory of heredity.

MORGAN AND THE PROBLEM OF NATURAL SELECTION

Morgan's exposure as a graduate student to W. K. Brooks and the morphological tradition had generated within him considerable interest in the problem of evolution. Although Morgan was annoyed with the speculative approach to evolutionary problems that characterized so many of the morphologists, he was nonetheless highly intrigued by the problem of adaptation, of which one of the most fascinating examples to him was regeneration: "One of the general questions that I have always kept by me in my study of regenerative phenomena is how such a useful acquirement as the power to replace lost parts has arisen and whether the Darwinian hypothesis is adequate to explain the result."[18] Morgan then answered his own question: "The conclusion I have reached is that the theory is entirely inadequate to account for the *origin* of the power to regenerate; and it seemed to me therefore desirable to re-examine the whole question of adaptation for might it not prove true here, also, that the theory of natural selection was inapplicable? This was my starting point."[19] How did Morgan come to this conclusion? Why should he oppose the theory of natural selection so strongly? What does Morgan's objection reflect about the general attitude of biologists around the turn of the century toward Darwin's model for evolution?

Although the terms "evolution" and "natural selection" are often used synonymously, they refer, of course, to quite different ideas. The concept of evolution maintains that presently existing groups of animals and plants have originated by modification from previously existing animals and plants. Over the centuries many

[18] Morgan, *Evolution and Adaptation* (New York: Macmillan Co., 1903), p. ix.

[19] Ibid., pp. ix-x. Morgan had begun looking at the means by which natural selection could explain the origin of the regenerative power as early as 1898; see "Some Problems of Regeneration," in *Biological Lectures Delivered at the Marine Biological Laboratory of Woods Hole in the Summer Session of 1897 and 1898* (Boston: Ginn & Co., 1899), pp. 193-207.

theories have been advanced as to how evolution might be brought about, of which Darwin's was only one. In the late nineteenth and early twentieth centuries many biologists were confirmed evolutionists yet were not Darwinians. Morgan was unquestionably an evolutionist from his earliest days, but he was *not* always a Darwinian. His objection to Darwinism was that he did not see how specific mechanisms of selection acting upon very small individual (but supposedly inherited) variations could produce adaptive structures over numerous generations.

To understand Morgan's objections to the Darwinian theory of natural selection, it is first necessary to understand his concept of species, for the view that a biologist holds about the nature of species will determine much of what he believes about the mechanism of evolution—the means by which species arise. Like many other biologists at the turn of the century, Morgan was a nominalist: he held that species were only arbitrary units created for convenience by taxonomists. The only real unit in nature, he contended, was the individual: "We should always keep in mind the fact that the individual is the only reality with which we have to deal, and that the arrangement of these into species, genera, families, *etc.* is only a scheme invented by man for purposes of classification. Thus, there is no such thing in nature as a species, except as a concept of a group of forms more or less alike."[20] Adaptation was a key aspect of Morgan's view of species. To him, a species was a group of forms more or less alike because they shared a number of common adaptations—not because they were alike in trivial details such as number of bristles or petals. When taxonomists spoke of species, they spoke of groups differentiated on the basis of nonadaptive characters, and thus to Morgan their concept of evolution had no relation to the development of real adaptations in nature.[21] Thus, he argued, it was futile to try and devise a theory to account for the origin of groups that have only subjective, rather than objective, reality: "If, then, the systematists' definition of species is what we mean when we speak of species, and this definition does not concern adaptive characters (or only incidentally), clearly it is futile to attempt to explain the origin of species by the theory of natural selection."[22]

[20] Morgan, *Evolution and Adaptation*, p. 33.

[21] T. H. Morgan, "Chance or Purpose in the Origin and Evolution of Adaptation," *Science* 31 (1910): 203-204.

[22] Ibid., p. 203.

Like many of his contemporaries, Morgan tended to see species in two contradictory ways. On the one hand, his experience with organisms in nature (particularly marine invertebrates) suggested that all forms exist in an infinitely graded series, with no sharp lines possible between discrete groups. On the other hand, he also tended to see groups (such as starfish, earthworms, or maple trees) as types, bounded by a limit on their range of variability. Neither view is acceptable to modern taxonomists, who see species as real, natural groups, but with no fixed type with a limit to the range of variability. Morgan's view that species are not real prevented him from understanding the basic population level on which natural selection inevitably works (evolution does not occur with individuals, only with groups). His "typological" view of species was a strong deterrent to understanding how new groups can be formed from old.[23] Given his view of species, it was difficult for Morgan to really understand any mechanism for evolutionary change. But he also held a number of specific objections to Darwinian theory that went beyond his general confusion about the nature of species.

MORGAN'S OPPOSITION TO THE THEORY OF NATURAL SELECTION

The version of Darwinian theory with which most biologists around the turn of the century were familiar had been set forth in the sixth and final edition of *The Origin of Species.* Since more offspring were always born than the environment could support, Darwin maintained, some would have to perish. The slight inherited differences among organisms gave some individuals a slightly greater survival value than others. However, the key to evolution by natural selection was not simply longevity, but rather what Darwin referred to as "differential fertility," the ability to leave a greater number of offspring in the next generation. Evolution could thus be marked by the gradual change in frequency of certain traits in a population of organisms over time.

The one persistent problem Darwin had left unsolved was the

[23] The typological concept of species has been discussed by Ernst Mayr in "Agassiz, Darwin and Evolution," *Harvard Library Bulletin* 13 (1959): 165-194; and "Footnotes on the Philosophy of Biology," *Philosophy of Science* 36 (1969): 197-202, especially pp. 198-199.

mechanism by which new characters could arise and be passed on in a population. Lacking a workable explanation of the origin of variations, Darwin and his followers had frequently fallen back on a revised version of Lamarckian theory.[24] Lamarck had explained the appearance of specific adaptive variations as the result of environmental influences on the germ tissue of the organism, or as the result of use and disuse of parts. In earlier editions of the *Origin*, most notably the first, Darwin had consciously avoided reliance upon Lamarckism. However, criticism from a variety of directions had forced him to adopt more of this older idea about how variations could arise.

Morgan's objections to Darwinian theory provide considerable insight into his view of scientific methodology. These objections, which were all enunciated by Morgan between about 1900 and 1915,[25] are summarized below:

1. *Selection does not act on continuous variations.* Morgan claimed that Darwin was confused about the types of variation on which natural selection acted. Darwin distinguished between two types of variation: large ones which appeared suddenly (what he called "sports"), and small ones (what he referred to as "individual differences"). The former Darwin considered to be relatively unimportant for evolution; it was the latter that formed the raw material on which selection acted.

Objections to Darwin's view that selection acted on small, individual, "continuous variation" had gained considerable prominence in England after the publication of William Bateson's *Materials for the Study of Variation* in 1894 and Hugo de Vries's *The Mutation Theory*, between 1901 and 1903. Both espoused theories of *discontinuous variation*. Discontinuous variations were much larger than individual differences (i.e., continuous variation) occurring at one jump and showing no intermediates between two types. Bateson claimed that discontinuous variations were the only ones that were inherited and therefore the only ones on which

[24] Named for the French invertebrate zoologist and evolutionist, Jean Baptiste Lamarck (1744-1829), who postulated environmental influences, principally use and disuse, as the major cause of variation. Because the environment stimulated certain nerves, the organism supposedly willed variation in the direction of a needed variation.

[25] A more detailed study of Morgan's objections to Darwinian theory can be found in G. E. Allen, "Thomas Hunt Morgan and the Problem of Natural Selection," *Journal of the History of Biology* 1 (1968): 113-139.

natural selection could act. De Vries went further and claimed that very large, discontinuous variations (what he called "mutations" and what we today might call "macromutations") could create a new species in one generation. De Vries's mutations were something akin to Darwin's sports. More will be said on de Vries's theory later in this chapter.

In the period before 1910, Morgan accepted both the idea of discontinuous variation as proposed by Bateson and the mutation theory proposed by de Vries. Morgan felt that there were two reasons why continuous variations could not be considered sources of evolutionary change: first, there was no proof that continuous variations could be inherited;[26] second, continuous variations could not occur in a large enough single step to prevent the new form from being lost through outbreeding in the next and subsequent generations. To Morgan, continuous variations represented quantitative rather than qualitative differences among organisms, i.e., a normal distribution about a mode. Selection of one extreme or the other could not ultimately produce a new species. According to Morgan, the most that selection could accomplish would be to keep the type at the upper limit attained in each generation by selection of small, individual variations. He thought of a species as having a fixed mode, a form like a rubber ball which could be altered temporarily by an outside agent (in this case selection) but which always returned to its original shape. Only a distinct, qualitative break with the type could produce a new species. In light of this view of species, it is not surprising to find that Morgan became a strong supporter of de Vries's mutation theory (see below).

2. *Ideas of discontinuous variation are also inadequate.* While he could not accept evolution by continuous variation, Morgan was also skeptical about the idea of evolution by discontinuous variation. He claimed that new variations would be quickly lost through "swamping." Swamping was based on a blending theory of inheritance. The germinal material from each parent was pictured as mixing together like two cans of paint, so that the offspring represented an intermediate between the parental forms. Thus, in each succeeding generation the appearance of the new variation would be diluted. Because in most species offspring can intercross with parents, swamping was seen as unavoidable. Morgan wrote:

[26] Morgan, *Evolution and Adaptation*, p. 267.

We see then that discontinuity in itself, unless it involved infertility with the parent species, of which there is no evidence, cannot be made the basis for the theory of evolution, any more than individual differences, for the swamping effect of intercrossing would in both cases soon obliterate the new form. If, however, a species begins to give rise to a large number of individuals of the same kind through a process of discontinuous variation, then it may happen that a new form may establish itself, either because it is adapted to live under conditions somewhat different from the parent form, so that the dangers of intercrossing are lessened, or because the new form may absorb the old one."[27]

There was one way around the dilemma that swamping posed to those who tried to understand heredity and evolution. Morgan assumed that if enough of the same discontinuous variations occurred at one time, the possibility of two organisms with the same variation mating would be greatly increased. Thus, the variation could be passed on in full strength to the next generation. Morgan admitted to having no evidence that large numbers of the same variation could occur at once in the same population, so his suggestion remained only hypothetical.

3. *Natural selection is only a negative factor.* To Morgan natural selection was only a negative factor in the origin of adaptive characters. It could select out the unfit, to be sure, but could not create the new variations from which new adaptations could be derived.[28] In other words, selection could not explain the *origin* of the fit, but only the failure of the unfit to propagate. Thus, for Morgan, the theory of natural selection was incomplete concerning the very point that Darwin felt it explained: the origin of species. The problem of how distinct and inherited variations arose and were passed on was to Morgan still the critical and unsolved problem in understanding evolution.

Morgan argued vehemently against those Darwinian supporters who claimed that selection could be a *creative* factor in evolution. To Morgan selection created nothing; it only weeded out the unfit. Thus to him the theory of natural selection was silent on the most crucial element of evolution: how new, adaptive variations arise.

[27] Ibid., pp. 286-287. Morgan made the same objection to the doctrine of sexual selection (see p. 195).

[28] Ibid., p. 462.

4. *Selection cannot account for the incipient stages of highly adaptive organs.* Morgan's fourth objection to natural selection was essentially the same as that voiced by St. George Mivart (1827-1900) in 1871: if selection acted on small individual differences which occur by chance, fine levels of adaptation (the vertebrate eye was always cited as the ultimate adaptation) could never have become established because their incipient stages would not provide enough advantage for favorable selection.[29] Darwin tried to refute this argument in the sixth edition of *The Origin of Species*[30] by arguing that even the slightest variation, as long as it was in any way favorable, would be preserved by selection. It was not necessary, Darwin argued, for a complete eye to arise at once in order for some adaptive value to be conferred on the organism. Morgan accepted Mivart's argument and countered Darwin's. Especially with regard to the problem that had originally brought him to the study of Darwinism, regeneration, Morgan claimed that the power to regenerate would be useless in intermediate stages. The ability to replace only part of a limb would be no more advantageous than replacing none of it. Thus, Morgan argued, how could regenerative ability evolve by selection acting on minute individual variations? To be of any adaptive significance, regeneration must be complete.[31]

5. *Darwin's theory is based on Lamarckian principles.* Morgan's fifth objection to natural selection was that Darwin himself had leaned too heavily on Lamarckian principles, relying in many places on the concept of use and disuse to explain hereditary novelties. As Morgan wrote of one example: "By falling back on the theory of inheritance of acquired characters, Darwin tacitly admits the incompetence of natural selection to account for the evolution of the flatfish."[32] This indicated to Morgan that the whole idea of selection acting on minute variation was insufficient even in Darwin's eyes. What he had not realized was that Darwin had not seen natural selection and Lamarckian use and disuse as an either-or choice, but had opted for the middle road and accepted both.

[29] St. George Mivart, *The Genesis of Species* (New York: Appleton, 1871), especially chapter 2 on the development of incipient structures.

[30] Charles Darwin, *The Origin of Species* (New York: Appleton, 1889), especially pp. 177-181.

[31] Morgan, *Evolution and Adaptation*, pp. 380-381.

[32] Ibid., p. 138. Although Morgan had not followed the various editions of *The Origin of Species*, he was aware that Darwin had moved closer to the Lamarckian views between the first edition (1859) and the sixth (1872).

Morgan was disconcerted by the school of neo-Lamarckism which had arisen in the latter part of the nineteenth century, feeling that there was no evidence to suggest that the effects of use or disuse could be passed on to the next generation. In 1903 he discussed the trend toward Lamarckism in the thinking of a number of American naturalists, including A. D. Cope, Alphaeus Hyatt, and J. A. Ryder:

> A number of evolutionists, more especially of the American school, have tried to show that the evolution of a number of groups can best be accounted for on the theory of acquired characters. . . . Despite the large number of cases that they have collected, which appear to them to be most easily explained on the assumption of the inheritance of acquired characters, the proof that such inheritance is possible is not forthcoming. Why not then spend a small part of the energy that has been used to expound the theory, in demonstrating that such a thing is really possible: One of the chief virtues of the Lamarckian theory is that it is capable of experimental verification or contradiction, and who can be expected to furnish such proof if not the neo-Lamarckians?[33]

Neo-Lamarckians spent a good portion of their time devising elaborate hypotheses and trying to explain away objections to their theory.[34] To Morgan all of this was beside the point, for neo-Lamarckians failed at the most critical part of their scientific work: they did not put their theory to the experimental test. To Morgan the notion of inheritance of acquired characteristics, while possible, remained in the category of "not proven."[35]

6. *Darwinian methodology was not rigorous.* Perhaps the most general criticism Morgan leveled at the Darwinian theory concerned scientific methodology. As we have seen, Morgan was a rigor-

[33] Morgan, *Evolution and Adaptation*, p. 260 (see also pp. 230-231).

[34] For a discussion of the neo-Lamarckian movement in America see the study by Edward J. Pfeifer, "The Genesis of American Neo-Lamarckism," *Isis* 56 (1965): 156-167, where the works of Hyatt, Cope, Dall, LeConte, and Clarence King are treated at some length. For European neo-Lamarckism, one of the best sources for the nineteenth century is Herbert Spencer's *The Principles of Biology* (New York: Appleton, 1866), vol. 1, pp. 184-200; 402-431. See also P. G. Fothergill, *Historical Aspects of Organic Evolution* (New York: Philosophical Library, 1953), and Y. Delage and M. Goldsmith, *The Theories of Evolution*, trans. A. Tridon (New York: B. W. Huebsch, 1912).

[35] Morgan, *Evolution and Adaptation*, p. 260.

ous thinker and an experimentalist, concerned first and foremost that hypotheses should be testable and should be in agreement with all (or nearly all) the facts. He felt that many of the strict Darwinians, neo-Darwinians, and particularly the neo-Lamarckians, had engaged in flights of fancy and speculation that had no basis in fact.[36]

To Morgan one of the greatest offenders along these lines was the German biologist August Weismann (1834-1914) who enjoyed widespread popularity and enormous influence around the turn of the century. Weismann had tried to unite five important areas of biology—evolution, heredity, cytology, physiology, and development—into one comprehensive conceptual scheme. To Morgan and his generation, however, Weismann's hierarchy of hypothetical hereditary particles and their hypothetical "struggle for existence" during development presented more problems than it could solve. Morgan objected to the speculative and fanciful nature of Weismann's theories, and he chastised Weismann for failing to distinguish between an assumption and a fact.[37] But it was on the nature of theory formation that Morgan leveled his most vehement attack:

> Thus, Weismann has piled up one hypothesis on another as though he could save the integrity of the theory of natural selection by adding new speculative matter to it. The most unfortunate feature is that the new speculation is skillfully removed from the field of verification and invisible germs (particles) whose sole functions are those which Weismann's imagination bestows upon them, are brought forward as though they could supply the deficiencies of Darwin's theory. This is, indeed, the old method of the philosophizers of nature. An imaginary system has been invented which attempts to explain all difficulties, and if it fails, then new inventions are to be thought of. Thus, we see where the theory of selection of fluctuating germs has led one of the most widely known disciples of the Darwinian theory.[38]

[36] See, for example, Morgan's article, "Regeneration and Liability to Injury," *Science* 14 (1901): 235-248; also "The Origin of Species through Selection Contrasted with Their Origin through the Appearance of Definite Variations," *Popular Science Monthly* 67 (1905): 54-65; and *Evolution and Adaptation*, pp. 126-163; 165 ff.; 171-172; 180.

[37] Morgan, *Evolution and Adaptation*, pp. 163; 165-166.

[38] Ibid., pp. 165-166.

Weismann's and Morgan's conceptions of the role of theory in biology were direct opposites. To Weismann, working in the tradition dominated by Darwin's *On the Origin of Species*, a theory served primarily as an umbrella under which the facts could be assembled.[39] To Morgan, on the other hand, a theory served primarily as a means for giving direction to further research, so that for a theory to be of any value in science it had to be framed in such a way as to be subject to verification.

7. *Darwinians confused the problems of chance and purpose in evolutionary development.* Philosophically, Morgan found that one of the most difficult problems in Darwinian evolutionary theory was the question of chance versus purpose. Always opposed to teleological explanations in science (i.e., those explanations that assume some predetermined purpose or goal in organic processes) he maintained firmly that variations in organisms did not arise because they were needed, but occurred solely by chance.[40] In 1910 Morgan pointed out that the term "chance" has two different meanings in relation to evolution and that confusion of these meanings had led many writers to adopt some sort of teleological explanation to account for adaptation.[41] The first meaning was that of the occurrence of one out of many possible events: for example, of the many possible variations that could occur, a particular one actually did take place, and whether it was variation A, B, or C was merely a matter of chance. The second meaning was that of the occurrence of a chance event at a particular time and place (unconnected with other events at that time or place), such as in the statement, "I chanced to be there." In evolutionary terms, the one chance event, a single variation, will be favorably selected only if it occurs in a particular environment at a particular time that happens to make it favorable. The occurrence of the variation and the set of conditions composing the environment are two independent sets of chance events, both of which must be considered in discussing the role of chance in evolution.

Morgan made this distinction in order to get to a more basic point: that the concept of chance did not imply anything mysteri-

[39] August Weismann, *Vorträge über Descendenztheorie*, 2 vols. (Jena: Gustav Fischer, 1902), vol. 2, p. 4.

[40] Morgan, *Evolution and Adaptation*, pp. 391-394.

[41] T. H. Morgan, "Chance or Purpose in the Origin and Evolution of Adaptations," *Science* 31 (1910): 201-210.

ous, outside of normal cause and effect.[42] In opposing those who held that chance events could not be studied scientifically, he pointed out that, in fact, it was the opposing idea—that of purposefulness—that was obscure and not subject to scientific study.[43] After all, Darwin himself had used the term "chance variation" as synonymous with "fluctuating variations."[44] Morgan saw in this confused use of the term "chance" the ugly specter of teleological explanations, which had no place in science:

> The origin of adaptive structure and the purpose it comes to fill are only chance combinations. Purposefulness is a very human conception for usefulness. It is usefulness looked at backwards. Hard as it is to imagine, inconceivably hard it may appear to many, that there is no direct relation between the origin of useful variations and the ends they come to serve, yet the modern zoologist takes his stand as a man of science on this ground. He may admit in secret to his father confessor, the metaphysician, that his poor intellect staggers under such a supposition, but he bravely carries forward his work of investigation along the only lines that he has found fruitful.[45]

Despite his many objections to Darwinian theory, Morgan believed that Darwin had served the admirable function, historically, of putting the question of evolution, particularly that of adaptation, on a sound scientific basis, collecting many observations and arguing his theory directly from them. Morgan could only object that many of the neo-Darwinians had not followed their leader's example in this regard, but had wandered too far from the facts.

MORGAN'S ACCEPTANCE OF DE VRIES'S MUTATION THEORY: AN ALTERNATIVE TO DARWINISM

Morgan became a strong proponent of Hugo de Vries's mutation theory when it was published between 1901 and 1903, since it appeared to resolve his objections to the theory of natural selec-

[42] Ibid.

[43] T. H. Morgan, "For Darwin," *Popular Science Monthly* 74 (1909): 367-380.

[44] Morgan, "Chance or Purpose," pp. 201-210.

[45] Morgan, "For Darwin," p. 380.

tion. De Vries's theory was put forward as a direct alternative to Darwinian theory and achieved an amazing popularity for over a decade.[46] It was intended to supersede both the theory of natural selection in accounting for the origin of species, and the Mendelian theory (of which de Vries had been one of the rediscoverers) in accounting for the inheritance of specific traits. That Morgan accepted de Vries's theory more readily than either Darwinian or Mendelian theory between 1900 and 1910 indicates much about his conception of the two theories, as well as his philosophy of science.

The mutation theory was a particulate concept of heredity according to which new species originated from quick jumps instead of from the minute gradations of Darwin's theory. De Vries's theory was developed out of his earlier concept of "intracellular pangenesis," originally published in 1899.[47] Pangenes were hereditary particles "in many respects analogous to the molecules of the chemist," though having a more complicated structure. Although de Vries was an experimentalist and would have liked to study the transmission of pangenes in the rigorous way chemists study molecules, he realized this was impossible. It was necessary to deduce hereditary and evolutionary laws by the more indirect route of breeding organisms and observing the kinds of offspring produced.[48]

De Vries's commitment to a particulate theory of heredity enabled him to interpret some very curious observations he had made in the early 1890s. During a field excursion near Hilversum, just outside of Amsterdam, de Vries noted what appeared to be several species of the evening primrose, *Oenothera*, growing side by side. One seemed to be a parental strain which had given rise to two offspring types, different in enough characters to be regarded by de Vries as separate species. De Vries transplanted all three types in his laboratory garden and grew them to maturity. He recognized that these plants, under controlled breeding and cultivation, offered an opportunity for laboratory and experimen-

[46] Hugo de Vries, *Die Mutationstheorie*, 2 vols. (Leipzig: Von Veit & Co., 1901-1903). All references to this work, unless otherwise noted, are from the English translation, *The Mutation Theory*, by J. C. Farmer and A. D. Darbishire (Chicago: Open Court, 1910).

[47] Hugo de Vries, *Intracelluläre Pangenesis* (Jena: Gustav Fischer, 1899); English translation, *Intracellular Pangenesis*, by C. S. Jaeger (Chicago: Open Court, 1910).

[48] De Vries, *The Mutation Theory*, vol. 2, pp. 567-668.

tal study of evolution in a way that was impossible in the field. What de Vries thought he had observed with *Oenothera* were examples of the obvious discontinuity that existed in nature and that was ultimately responsible for evolutionary change. According to de Vries, new species arose all at once, with no gradual transition and no obvious interconnecting forms between two types. De Vries propounded his theory of evolution by large jumps or "mutations" as an alternative to the Darwinian theory:

> The origin of species has so far been the object of comparative study only. It is generally believed that this highly important phenomenon does not lend itself to direct observation and, much less, to experimental investigation. This belief has its root in the prevalent form of the conception of species and in the opinion that the species of animals and plants have originated by imperceptible gradations. These changes are indeed believed to be so slow that the life of a man is not long enough to enable him to witness the origin of a new form.
>
> The object of the present book is to show that species arise by saltation [i.e., jumps] and that the individual saltations are occurrences which can be observed like any other physiological process. Forms which arise by a single saltation are distinguishable from one another as sharply and in as many ways as most of the so-called small species and as many of the closely related species of the best systematists, including Linnaeus himself.
>
> In this way we may hope to realize the possibility of elucidating, by experiment, the laws to which the origin of new species conform. The results of these studies can then be compared to those which have been obtained with systematic, biological, and particularly with paleontological data. A most remarkable agreement will be found to exist between these and my new results.[49]

De Vries made a point of distinguishing mutations from individual, fluctuating variations. The latter could be regarded merely as temporary adaptations to the environment, having nothing to do with the origin of species. Selection could not make permanent any changes that depended on this type of variation.[50] Muta-

[49] Ibid., vol. 1, p. viii.

[50] Ibid., pp. 5-6.

tions, on the other hand, occurred less frequently than individual variations and could not be regarded merely as extremes of some spread around the mean established for the species as a whole. Mutant forms had a newly established mean of their own, which represented a distinct break from the mean of the parent species. A consequence of the large difference between the mutant and original forms was that the offspring would not be able to interbreed with the parents.

De Vries went on to distinguish between several types of mutations.[51] However, it is important to point out that the term "mutation" as used by de Vries was quite different from the term as used today, or even in 1915 or 1920. De Vries's term might more appropriately be called today "macromutation," and was similar in some ways to Darwin's conception of "sports" or "monstrosities." Today the question of whether new species can arise "overnight," between parents and offspring, is still a controversial issue. All biologists recognize that breeding barriers can arise more or less in one jump, through such mechanisms as doubling or tripling the number of chromosomes between parent and offspring (a process known as polyploidy). However, there is considerable disagreement as to whether such forms, although not interfertile, are actually new species. In modern perspective, de Vries may have been more correct than many of the early twentieth century Darwinians were willing to admit. However, as will be explained later, the specific examples de Vries himself worked with (*Oenothera*) are not accepted by modern biologists as demonstrating the creation of new species by single jumps. Like Morgan and many others, de Vries used the term "species" in two different ways. He felt that the species groups designated by systematists were largely arbitrary, based for the most part on differences in trivial and nonadaptive characters. However, "real" species did, in fact, exist and were distinguished from one another on the basis of marked differences in character. These were the true or elementary species; it was these units that underwent evolution by mutations.

Selection played a role in de Vries's theory, but less so than in Darwin's. New mutations might be favorable or unfavorable in a given environment, and thus would be subject to the action of

[51] A detailed study of de Vries appears in G. E. Allen, "Hugo de Vries and the Reception of the 'Mutation Theory,'" *Journal of the History of Biology* 2 (1969): 55-87. See also Peter W. van der Pas, "The Correspondence of Hugo de Vries and Charles Darwin," *Janus* 57 (1970): 173-213.

selection. Not every mutation survived, according to de Vries—only those that conferred favorable advantages on their bearer. Thus, de Vries preserved the one element of Darwin's theory that most biologists at the time could agree on: that selection existed and served to weed out unfit variations.

Morgan became a strong advocate of the mutation theory because it seemed to answer many of his longstanding objections to Darwin's handling of several problems: the nature of variation, the role of selection, the utility of incipient stages of new characters, the evolutionary time scale, discontinuities in the geological record, the questionable role of isolation, and the rigor of selection. Specifically, there were two explicit problems in Darwin's work that de Vries's theory seemed to account for. First, de Vries's theory was one of discontinuous variation, relegating individual and minute variations to a negligible evolutionary role. The large mutations de Vries observed in *Oenothera* represented qualitatively new variations, which were wholly distinct from the parental form. Here was a new form, arisen fully blown, ready to take its chances in the world. Second, as a result of de Vries's emphasis on discontinuity, his theory avoided the problem of swamping, which had so plagued Darwin and his followers. Not only was a mutant form infertile with its parents, thereby preventing its being lost through backcrossing, but also de Vries hypothesized "periods of mutation" in which the same mutation would very likely occur in a number of organisms simultaneously. Thus, a mutation could be preserved by cross-fertilizing with another like itself. The mutation theory answers objections that one or another group of biologists had brought against the Darwinian theory. For example, de Vries's theory did away with Mivart's criticism that incipient stages of highly adaptive organs would not in themselves be adaptive. Because a new structure could appear in one step more or less fully developed, the problem of utility of incipient stages became less important.

Morgan felt that the mutation theory also answered the objections of paleontologists concerned with gaps and apparent sudden jumps in fossil types between strata of rocks. What appeared to be sudden, qualitative differences between species in adjacent strata were in fact the natural result of mutations. Morgan further noted that those who felt there had not been enough geological time for Darwinian selection to produce the many new species observed today, could easily accept de Vries's alternative. Since de Vriesian mutations produced much more rapid changes in structure than would be possible in the same period with the Darwinian model, the evolutionary time scale could be considerably telescoped. De Vries's theory could thus solve one of the most persistent problems that Darwin's critics, especially Lord Kelvin, had raised.[52]

Morgan noted that de Vries's theory appeared to receive considerable corroboration in the "pure-line" experiments of the Danish botanist Wilhelm Johannsen (1857-1927) published in 1903. These experiments spoke to the persistent evolutionary question of the types of variation, large or small, on which selection acted. Johannsen had shown that selection of fluctuating, individual variations only separated out the pure lines already existing in a heterogeneous population of organisms. He emphasized that there was thus a limit to the change selection could produce in a population of organisms. The longer selection was continued, the less progress that could be made in any one line. Johannsen also noticed that if selection were relaxed, that is, if the pure lines were allowed to hybridize again, the differences between them would soon disappear. Selection seemed to produce changes in a population that persisted only so long as the selection process was rigorously maintained. It would not cause the species to transcend that threshold level of variation between one species and another.[53]

Johannsen's conclusion opposed the Darwinian theory on two accounts. First, it held that selection of fluctuating variation was not able to produce a new species; second, it held that the effects of selection by itself were purely negative. The pure-line experiments provided important experimental evidence confirming the anti-Darwinian prejudices of many investigators. Thus, Johannsen's work contributed significantly to the acceptance of de Vries's mutation theory. A number of biologists, including Morgan, saw in the combined arguments of de Vries and Johannsen the best evidence to date against the Darwinian model of selection acting on small individual differences.[54]

[52] For a description of Kelvin's criticism of Darwinism, and Darwin's response, see Gavin de Beer, *Charles Darwin* (London: Thomas Nelson, 1963), chap. 8.

[53] Wilhelm Johannsen, *Über Erblichkeit in Populationen und in reinen Linen* (Jena: Gustav Fischer, 1903).

[54] See, for example, Morgan, *Evolution and Adaptation*, p. 298. Also, Thomas L. Casey, "The Mutation Theory," *Science* 22 (1905): 307-309;

The idea that selection could not produce new species was strengthened by the assertion that Darwin erred in equating artificial and natural selection. Several writers in the later nineteenth century claimed that the most rigorous artificial selection practiced in animal and plant breeding had never produced a single case of a wholly new species.[55] De Vries made use of this point in maintaining that the forms produced by artificial selection were not as permanent as those produced by natural selection. Acting on slight individual variations, artificial selection produced only temporary types which, if allowed to hybridize, would quickly revert to their original form à la Johannsen. To de Vries, the creation of the fit was the result only of new mutation. The species differences created by mutation, de Vries argued, were more in agreement with those differences found among species in nature than with those forms produced through artificial selection.

De Vries's theory was well accepted by Morgan and many of his contemporaries, especially the younger biologists with an experimental orientation. Vernon Kellogg wrote that "on the whole the theory has been warmly welcomed as the most promising way yet presented out of the difficulties into which biologists had fallen in their attempts to explain satisfactorily the phenomena of the origin of species through Darwinian selection."[56] An even more enthusiastic worker, F. C. Baker, wrote that "no work since the publication of Darwin's *Origin of Species* has produced such a profound sensation in the biological world as *Die Mutationstheorie* by Hugo de Vries."[57] One indication of the interest raised by de Vries's theory was the attempt in the early years of the twentieth century to find mutations in a variety of organisms other than *Oenothera.* For example, in 1905 and 1906 C. B. Davenport set about to find examples of animal species that differed by discon-

---

E. G. Conklin, "Problems of Evolution and Present Methods of Attacking Them," *American Naturalist* 46 (1912): 121-128; C. B. Davenport, "Species and Varieties, their Origin by Mutation by Hugo de Vries. A Review," *Science* 22 (1905): 369-372; and D. T. MacDougal, "Discontinuous Variation in the Origin of Species," *Science* 21 (1905): 540-543.

[55] See, for example, Ludwig Plate, *Selectionsprinzip und Probleme der Artbildung*, 3rd ed. (Leipzig: W. Engelmann, 1908), chap. 1, part 3, pp. 36-56; Morgan, *Evolution and Adaptation*, pp. 19-20.

[56] Vernon L. Kellogg, *Darwinism Today* (New York: Henry Holt & Co., 1907), p. 348.

[57] F. C. Baker, "Application of de Vries' Mutation Theory to the Mollusca," *American Naturalist* 40 (1906): 327-334.

tinuous variation that would fit de Vries's concept of mutation.[58] Davenport recorded a number of examples in animals that he felt exemplified the de Vriesian concept of mutations. Both Morgan and Jacques Loeb, independently, were interested in testing de Vriesian mutations in animals, particularly the fruit fly *Drosophila.* D. T. MacDougal and his colleagues at the New York Botanic Garden tried to repeat de Vries's production of mutations in *Oenothera* itself, as well as in several other varieties of plants.[59] The list of those who supported the mutation theory goes on and on.[60]

Ranged against Morgan and the experimentalists who supported de Vries were a number of distinguished biologists (mostly, although not exclusively, naturalists) both in the United States and in Europe. Naturalists found the mutation theory naive and an insufficiently documented concept to add anything serious to evolutionary thought. Among these opponents were Ludwig Plate and August Weismann in Germany, William Bateson in England, and C. Hart Merriam, C. O. Whitman, O. F. Cook, W. E. Castle, and H. L. Boley in the United States. Merriam, for example, made a survey of over 1,000 species and subspecies of North American mammals and birds to look for differences that might have originated by de Vriesian mutation. His conclusion was succinct and definite: "My own conviction is that the origin of species by mutation among both animals and plants is so uncommon that as a factor in evolution it may be regarded as trivial."[61] Merriam accused his fellow biologists of being fickle: "Are we, because of the discovery of a case in which a species appears to have arisen in a slightly different way [from the Darwinian theory]—for after all the difference is only one in degree—to lose faith in the stability of knowledge and rush panic-stricken into the sea of unbelief, unmindful of accumulative observations and conclusions of zoologists and botanists?"[62]

By and large those favoring de Vries were young experimentalists located in universities and research institutes. They found in

[58] C. B. Davenport, "The Mutation Theory in Animal Evolution," *Science* 24 (1906): 556-558.

[59] D. T. MacDougal, "Mutation in Plants," *American Naturalist* 37 (1903): 737-770.

[60] See Allen, "Hugo de Vries," p. 67, for further references to those who supported de Vries.

[61] C. Hart Merriam, "Is Mutation the Factor in the Evolution of the Higher Vertebrates?" *Science* 23 (1906): 241-257.

[62] Ibid.

de Vries one of the first examples of what they thought to be an experimental approach to evolution. Furthermore most of those who favored de Vries were opposed to the old-style morphological speculation, and particularly to the work of taxonomists. Those opposing de Vries's theory included older workers, especially the strong neo-Darwinians (and even the neo-Lamarckians). Many of these workers were trained in the German tradition. Plant and animal breeders also comprised a significant portion of the opposition to de Vries. Many of these were not university trained but had gained most of their knowledge through practical experience. In England, opposition to de Vries was strong among the biometricians, who as followers of Darwin tended to emphasize evolution by small variations. Furthermore, opponents of de Vries were generally unfavorable to any biological theories that attempted to reduce complex organic processes to the interaction of particles or to single mechanistic explanations (such as postulating a single, large variation). Thus support for or opposition to the mutation theory fell along the lines of, respectively, the experimentalist and naturalist camps. In many ways the predispositions of the two groups determined the way in which they reacted to the new theory.[63]

[63] The opponents of de Vries's mutation theory were to have their heyday beginning shortly after 1910. In a series of papers published between 1910 and 1912, Bradley M. Davis showed that the so-called mutations of *Oenothera* itself were actually the result of an unusual hereditary pattern which basically followed Mendelian laws. In 1914 O. Renner showed specifically that *Oenothera* was a permanent heterozygote between two complexes, the pure homozygotes failing to survive. For a summary, see A. H. Sturtevant, *A History of Genetics* (New York: Harper & Row, 1965), pp. 63-64. In work extending from 1923 to 1950, R. E. Cleland of Indiana University conducted a thorough investigation into the cytology of *Oenothera*, demonstrating that chromosomal behavior during gamete formation could explain de Vries's peculiar variations as well as the persistent recovery of parental types from hybrid crosses. It would be impossible to describe here in detail the complex chromosomal patterns in *Oenothera* that Cleland's careful work has uncovered. For further references see: R. E. Cleland, "Chromosome Arrangements during Meiosis in Certain *Oenotheras*," *American Naturalist* 57 (1923): 562-566; and "Some Aspects of the Cytogenetics of *Oenothera*," *Botanical Review* 2 (1936): 316-348. Much of this work has been summarized in Cleland's posthumous monograph, *Oenothera: Cytogenetics and Evolution* (New York: Academic Press, 1972). Thus the many variant forms of *Oenothera* were not new species at all, but rather, complex recombinations of much smaller individual variations. De Vries kept defending the generality of his theory until his death in 1935, but by 1915 the

Morgan's fascination with the mutation theory, as well as his great personal admiration for de Vries, made him one of the very strongest proponents until early in 1910. In fact, it was Morgan's attempt to find de Vriesian mutations in *Drosophila* that led him to begin breeding the fruit fly, *Drosophila*, in 1908. As the results of Davis's, Renner's, and Cleland's work became public, Morgan abandoned the enthusiastic support he had originally given to de Vries. However, he always maintained that de Vries's work was enormously stimulating and provided an important alternative at the time to the difficulties encountered by Darwinian theory.

MORGAN, SEX DETERMINATION, AND THE MENDELIAN THEORY

The problem of heredity as it related to evolution and the origin of variations was obviously a major topic in the early 1900s, drawing the attention of many workers, including Morgan. In the period after 1900, Morgan found his interest in the problem of heredity itself growing. This derived from two sources. One was his interest in evolution, particularly mutation theory. The other was his interest in a problem that had long concerned embryologists: the nature of sex determination.

The theory of heredity that Morgan ultimately did so much to establish after 1910 was a combination of two previously separate lines of thought: the chromosomal theory, which maintained that the cell structures known as chromosomes were directly involved in hereditary transmission, and the Mendelian theory, which postulated hereditary factors that segregate and assort themselves in the production of eggs and sperm. The chromosomal theory was based on extensive cytological analysis of plant and animal cells starting in the 1850s. The Mendelian theory was based on analysis of breeding results and had grown out of a long tradition of plant and animal hybridization. Morgan had encountered both these theories before 1910, primarily because of his interest in sex determination.

On December 28, 1906 the American Society of Naturalists held a symposium at Columbia University on the topic "The Biological Significance and Control of Sex." Five papers were presented at

---

mutation theory, especially in its original form, had passed out of the serious biological literature.

this meeting by some of the leading botanists and zoologists of the day. One of the most important issues to come out of this symposium was the nature of the factor or factors responsible for sex determination. Of the papers presented, two represented particularly divergent views: one by the cytologist Edmund B. Wilson, the other by his colleague and close friend at Columbia, T. H. Morgan. Wilson advocated the Mendelian-chromosomal hypothesis as the best guide for research on the problem of sex determination. Morgan, on the other hand, was skeptical of both the chromosomal theory and the interpretation of sex by Mendelian factors.

The problem of sex determination raised a number of issues that illustrate Morgan's early views on these theories dramatically. Was sex determination the result of internal factors established at the moment of fertilization or the result of external factors during embryonic growth (such as the amount of nourishment, temperature, etc.)? If sex is established at fertilization, by what mechanism does it occur? Was the egg or the sperm the more important germ cell in determining sex? By what mechanism is the 1:1 sex ratio, observed in so many organisms, established? How can parthenogenesis (development from an unfertilized egg), sexual mosaics, and other unusual sexual phenomena be accounted for? These and many other questions had to be answered by any theory of sex determination. As in the case of evolution, Morgan's views on the Mendelian and chromosome theories represent something more than his own peculiar attitudes. There was considerable skepticism toward both theories during the early part of the century.

Morgan was led to the problem of sex determination initially through his work in embryology. Many theories current in the late 1890s claimed that factors external to the egg and the sperm determined the sex of an embryo after development had begun. It was of considerable concern to embryologists to discover whether sex was in fact the result of environmental influences acting on the embryo or whether it was determined by hereditary factors at fertilization.

While the question of sex determination was considered by many to be a part of the more general problem of heredity, it nevertheless presented some difficulties of its own. There were the anomalies, parthenogenesis and gynandromorphism. Bees, ants, wasps, rotifers, and a number of other forms are known to reproduce parthenogenetically under some circumstances and sexually under others. Gynandromorphs are a type of mosaic organism composed of a combination of male and female tissues. In the early 1900s it was not known whether the male and female tissues differed in hereditary composition, since the hereditary basis of sex determination was not yet established. How was this occasional appearance of gynandromorphism to be explained? In addition, there was a third problem—the well-known and documented effect of external conditions such as temperature or amount of food in modifying observable sex ratios.

Morgan's first published paper dealing with heredity was a review of the various theories of sex determination current in 1903.[64] These theories fell into two general groups. One maintained that external conditions were the primary factors in determining sex. In the years before 1905, even E. B. Wilson held this view. Wilson and others maintained that the fertilized egg exists in a sort of balanced state; whether it develops into a male or a female depends ultimately on factors (amount of food or temperature) in the environment.

An opposing school maintained that sex was determined only by internal factors. The ideas of this school received much support form the work of the French biologist Lucien Cuénot who, in 1899, provided distinct evidence that external conditions in no way affect sex ratios in mice.[65] The hereditary concept of sex determination placed strong emphasis on the determination of sex at fertilization, or shortly thereafter. One of the chief difficulties with the hereditarian view, as Morgan was quick to point out, was the lack of supporting evidence. The same was true, of course, of the environmental view. Of all the attempts to prove one point or the other, Morgan was most impressed with Cuénot's work as rigorous and convincing. The proliferation of numerous theories concerning sex determination, and the existence of two opposing schools, marked the unsettled state of opinion.

One attempt to formulate a theory of sex determination based on hereditarian principles was made by W. E. Castle in 1903. Castle attempted to treat the determination of sex as a Mendelian phenomenon. To explain sex determination, Castle suggested that there existed two classes of eggs and two classes of sperm—that

[64] T. H. Morgan, "Recent Theories in Regard to the Determination of Sex," *Popular Science Monthly* 64 (1903): 97-116.

[65] Lucien Cuénot, "Sur la determination du sexe chez les animaux," *Bulletin Scientifique de la France et de la Belgique* 32 (1899): 462-535.

is, there were both male- and female-producing sperm, as well as male- and female-producing eggs. Castle realized that such a system in terms of Mendelian inheritance would yield three classes of sexual individuals in a 1:2:1 ratio. Either the middle group would be composed of "sexless" individuals, or else, if dominance were invoked, the normal 1:1 sex ratio would be violated. Castle was thus forced to invent a new hypothesis, that of "selective fertilization." Selective fertilization implied that only certain sperm and certain eggs could unite (or at least produce a viable zygote). A sperm containing a female-determining factor could only fertilize an egg that also contained a female-determining factor. In this way, the expected sex ratio would be preserved. It is important to point out that Castle's theory of sex determination was based only on Mendelian principles. It did not involve any assumptions about the relationship between Mendelian "factors" and chromosomes.

Morgan severely criticized Castle's theory on a number of grounds.[66] In the first place, Morgan found the hypothesis of selective fertilization highly questionable. There was, he concluded, little or no evidence in support of such an idea and on that basis he had to reject Castle's theory. In the second place, Morgan enunciated clearly his skepticism of any attempts to explain sex determination by simple Mendelian factors. In 1903 Morgan did not believe that the sperm and the egg were different in their sex-determining ability: both male and female elements were present in all types of eggs. He felt that factors such as the size of the egg or other conditions (wisely left unspecified) determined which sex would develop. In 1903 it wasn't clear whether Morgan subscribed to either the hereditarian or environmental school. By 1905, however, Morgan was leaning more firmly toward the hereditarian viewpoint. In that year he suggested that either sperm or egg nucleus alone (he did not mention chromosomes at all) would determine a male offspring; a sperm plus an egg nucleus would produce a female.[67] But Morgan, typically, never made anything of this suggestion except as an alternative to the view that there are specific male and female determinants in the egg and sperm. Frequently in his writings Morgan would put forward a hypothesis for which he himself had no concrete experimental evidence. He

[66] Morgan, "Recent Theories in Regard to the Determination of Sex."

[67] T. H. Morgan, "Ziegler's Theory of Sex Determination and an Alternative Point of View," *Science* 22 (1905): 839-841.

seldom considered these hypotheses as having much value, except insofar as they might lead to further experimental or investigative work.

It is important to recognize that the Mendelian theory and the chromosome theory were really separate ideas during much of the first decade of their coexistence (1900-1910). Each derived from a quite separate tradition, and each attracted a separate group of workers and supporters. As the decade went on, however, it became increasingly apparent that the two theories had much in common and could be treated in similar terms. In the period prior to 1905, Morgan's major objections to hereditary interpretations of sex determination were limited to the Mendelian theory. After 1905, his objections centered more and more on the combined theory, which saw the chromosomes as possible bearers of the Mendelian factors. Morgan saw the two theories as closely interrelated and in some cases as variations of the same theme: namely, the attempts to explain heredity by postulating discrete particles (whether invisible "factors" or visible chromosomes) in the germ cells.

An important contribution to the problem of sex determination came in 1905 with the independent work of two people closely associated with Morgan. One was E. B. Wilson, Morgan's colleague at Columbia and the head of the zoology department there; the other was Nettie M. Stevens, a former graduate student of Morgan's, now a research assistant at Bryn Mawr and a cytologist of considerable skill. They had studied the so-called "accessory chromosome" and its relation to sex determination. The accessory chromosome was first described in 1891 by Hermann Henking (1858-1942) in the insect *Pyrrhocoris*, as a "peculiar chromatin element" observed in the second cell division leading to sperm production.[68] The same element was described shortly thereafter in a number of other organisms.[69] The accessory chromosome was an odd-shaped chromosome observed largely in the sperm-pro-

[68] Hermann Henking, "Über Spermatogenese und deren Beziehung zur Entwicklung bei Pyrrhocoris apterus L.," *Zeitschrift für wissenschaftliche Zoologie* 51 (1891): 685-736.

[69] See, for example, T. H. Montgomery, "A Study of the Germ Cells of Metazoa," *Transactions of the American Philosophical Society* 20 (1901): 154-236; C. E. McClung, "The Accessory Chromosome—Sex Determinant?" *Biological Bulletin* 3 (1902): 43-84; and W. S. Sutton, "On the Morphology of the Chromosome Group in *Brachystola magna*," *Biological Bulletin* 4 (1902): 24-39.

ducing cells of male insects. Unlike all the other chromosomes, it did not seem to have a similarly shaped partner chromosome. Today, of course, the accessory chromosome for most species is called the Y-chromosome, and is normally paired with a differently shaped chromosome, the X. The first suggestion that this accessory element was concerned with sex determination was given by McClung in 1901. He emphasized the parallel between two classes of sperm (equal in number), differentiated by the accessory chromosomes, and two equal classes of sexual adults. Supportive evidence for his suggestion was difficult to obtain, however, and conclusive results had to wait until the work of Mulsow in 1912.[70] At about the same time, in 1902, the astounding similarity between the cytologically observed separation of members of each chromosome pair (including, but not limited to, the accessory chromosomes) during sperm or egg production, and Mendel's postulated segregation of factors, was pointed out by both Boveri and Sutton.[71] There was thus evidence suggesting a *possible* link between chromosomes and Mendelian factors on the one hand, and between the accessory chromosome and sex determination on the other.

In 1905 Wilson and Stevens presented strong evidence that the accessory chromosome was the sex-determining element. Their separate papers offered cytological evidence demonstrating that, in regard to the accessory chromosome, two patterns could be found in the animal kingdom: those species in which the male has only one X-chromosome, designated the XO type, and those in which the male has an X and Y chromosome, the XY type. Despite the clear parallel between these cytological observations and the determination of sex, Wilson himself, however, was not ready to accept the view that the chromosomes were specifically male or female determiners. Near the end of his 1905 paper, Wilson wrote: "The foregoing facts irresistibly lead to the conclusion that causal

[70] Mulsow observed two classes of sperm, those carrying an X and those lacking an X. Furthermore, he observed that one or the other type of sperm entered the eggs in equal numbers.

[71] Th. Boveri, "Über mehrpolige Mitrosen als Mittel zur Analyse des Zellkerns," *Verhandlungen der Physikalisch-Medizinischen Gesellschaft zu Würzburg* 35 (1902): 67-90; translated by Salome Gluecksohn-Waelsch in B. H. Willier and Jane Oppenheimer, eds., *Foundations of Experimental Embryology* (Englewood Cliffs, N.J.: Prentice-Hall, 1964), pp. 76-97; W. S. Sutton, "On the Morphology of the Chromosome Group"; and "The Chromosomes in Heredity," *Biological Bulletin* 4 (1903): 231-251.

connection of some kind exists between chromosomes and the determination of sex. . . . Analysis will show, however, that great, if not insuperable, difficulties are encountered by any form of the assumption that these chromosomes are specifically male or female sex determinants. It is more probably . . . that the difference between eggs and spermatozoa is primarily due to differences of degree or intensity, rather than kind, in the activity of the chromosome groups in the two sexes."[72] Wilson's view was cautious and underwent several modifications in the next few years. However by 1906 he was ready to say that the inheritance of sex could most profitably be treated from both the Mendelian and chromosomal points of view. Like Castle, Wilson was still unclear about how the Mendelian concepts of dominance and recessiveness could be applied to sex inheritance without upsetting the normally observed 1:1 sex ratio. Nonetheless, Wilson took a bold step: "It seems to me that the most available stepping-stone towards the investigation of this problem [sex determination] is afforded by recently acquired evidence that sex production stands in some definite casual relation with the chromosomes and can be treated from the standpoint of Mendelian phenomena, as interpreted by the Sutton-Boveri chromosome theory. . . . A very definite material basis, therefore, exists for a treatment of the sex characters as if they were Mendelian alternates, sex determination being a matter of Mendelian dominance, more specifically of chromosome dominance."[73]

Morgan was, of course, familiar with Wilson's work. In fact, he seemed not to be able to get away from it. In a letter to Driesch from Woods Hole in July of 1905, Morgan wrote that Wilson is "wild over chromosomes." Then followed a list of Morgan's objections to the chromosome theory.[74] In October of 1905, Morgan wrote to Driesch that he was "in the thick of it" regarding chromosomes. Again the following spring Morgan referred in a letter to Driesch about "this modern way of referring everything to the chromosomes"; in fact, Morgan claimed that he was always in hot water at Columbia because he lived "in an atmosphere satu-

[72] E. B. Wilson, "The Chromosomes in Relation to the Determination of Sex in Insects," *Science* 22 (1905): 500-502.

[73] E. B. Wilson, "Sex Determination in Relation to Fertilization and Parthenogenesis," *Science* 25 (1907): 376-379.

[74] Morgan to Driesch, July 25, 1905 (erroneously dated 1902); Morgan-Driesch correspondence.

rated with chromosomic acid."[75] To Morgan, Wilson's recent observations seemed quite sound, but he still could not accept Wilson's conclusions; the chromosome and Mendelian theories seemed unconvincing in relation to sex determination.

What were Morgan's specific objections to interpreting heredity in general, and the inheritance of sex in particular, in Mendelian and chromosomal terms? First, Mendelism itself seemed to imply the existence of only a dominant or recessive condition. Morgan felt that the facts of nature, in both the field and the laboratory, showed that for every character there was a large variety of intermediates between what appeared to be a purely dominant or purely recessive expression. Referring to the work of Bateson, Saunders, and Hurst in England, Morgan observed that they had pointed to numerous examples where dominance was not complete in the first generation. From this Morgan went on to observe: "In other words, there may be almost a continuous series in this [the first, or F_1] generation. Such results are difficult to account for on the basis of 'pure' gametes, although a tendency toward segregation may be recognized."[76] The continuous variation that Morgan felt characterized any population of offspring, even from supposedly "pure" parents, and which especially characterized the offspring of heterozygous parents, was an indication that inheritance was due to factors other than rigid dominant or recessive particles. Morgan even went so far as to accuse the Mendelians of willfully ignoring the range of variation among their offspring in order to get the expected ratios predicted by their theory.[77]

Morgan's objection rests on two misunderstandings that were current among many biologists at the time. One is the nature of dominance and recessiveness itself. In the early days after the rediscovery of Mendel, there was a tendency to look at all traits as being either pure dominant or pure recessive. In fact, as more and more organisms and their characteristics were studied, it became apparent that other conditions, such as incomplete dominance (where the hybrid offspring are an intermediate mix between the two parent strains for a given character), or various other types of gene interactions, were more the rule than the exception. A

[75] Ibid., April 17, 1906, p. 4.

[76] T. H. Morgan, "Review of 'Inheritance in Poultry' by C. B. Davenport," *Science* 25 (1907): 464.

[77] T. H. Morgan, "Some Books on Evolution," *Nation* 95 (1912): 543-544.

second misunderstanding arose out of the confusion still prevalent in the first decade of the twentieth century over the nature of continuous versus discontinuous variations. As Johannsen pointed out, there is considerable variation among the offspring of even pure homozygous parents. Some of these variations may be due to unknown modifier genetic elements, but others may be due to environmental influences such as amount and kind of nourishment. Until biologists had a clear understanding of the differences between genetically caused and environmentally caused variability, it was difficult to reconcile the wide range of observed character differences in a population of organisms with the simple assumption of purely dominant or purely recessive hereditary factors.

A second criticism Morgan leveled specifically against the Mendelian theory was that it was based on a variety of assumptions for which there was very little experimental evidence. One of these was the assumption of selective fertilization, which was again invoked in 1905 in Lucien Cuénot's studies on coat color in mice.[78] Wishing to obtain pure yellow mice, Cuénot had crossed two heterozygotes,[79] whose coat color was designated as gray. He expected one-fourth of his offspring to be pure yellow. To his surprise, however, he obtained no yellows at all, but two impure yellows to one gray. This was especially confusing since in other combinations than with gray yellow seemed to follow the expected Mendelian rule.[80] Like Castle, Cuénot had also assumed the idea of selective fertilization: that sperm containing the factors for yellow never combine with eggs containing the yellow factor. Morgan again objected to invoking a purely hypothetical idea as a

[78] Lucien Cuénot, "Les Races Pures et leur Combinaisons chez les Souris," *Archives de Zoologie Experimentale*, 4th series, 3, Notes et Revue (1905): cxxiii-cxxxii.

[79] The terms "heterozygote" and "homozygote" were introduced by Bateson in 1902 in *Mendel's Principles of Heredity: A Defense* (Cambridge: Cambridge University Press). The term homozygote refers to a combination of a matching pair of alleles (either both dominant or both recessive) for the same character. (For example, an offspring that receives two dominant factors for tallness would be considered a homozygote dominant.) Heterozygote refers to an organism in which the pair of alleles for a character are contrasting (as, for example, an organism that receives a dominant gene for tallness from one parent and a recessive gene for shortness from the other).

[80] The yellow condition in mice was later discovered to be a "lethal" in which the homozygous dominant (both alleles for the yellow condition) are not viable, and the embryos do not develop.

necessary component of a more general theory.[81] To Morgan it seemed highly improbable that such trivial differences as the color of hair could in any absolute way prevent the conjugation of gametes carrying those colors. This was especially true, Morgan felt, when it was apparent that the gametes must carry thousands of other "unit characters" that were identical.[82]

The Mendelian theory also rested on other assumptions of which Morgan was skeptical. One was that the two factors for any given character always separated from each other during the cell divisions leading to gamete production and came to reside in different gametes. Why, Morgan asked, should the alternative characters (the alleles) be constantly "turning their backs on each other and fleeing to opposite germ cells"?[83] To Morgan there seemed to be no viable mechanism by which this regular and repeated separation of contrasting characters would be brought about. In fact, Morgan felt that results such as Cuénot's yellow mice could be better explained by assuming that certain alleles always remained together rather than segregating in a random fashion. He developed a highly complex scheme for explaining Cuénot's results by assuming that the yellow and gray "unit characters" stayed together in 50 percent of the offspring, giving rise to a combination that, with only a few further assumptions, could be made to fit Cuénot's observed ratio.

Another assumption was the concept of the "purity of the gametes," an essential ingredient of Mendelian theory. This idea had been developed to explain how parents themselves showing a dominant trait could, as heterozygotes carrying a recessive allele, give rise to offspring that resembled either the original dominant or original recessive grandparent. "Purity" meant first of all that the two alleles invariably separated (segregated) from each other in the formation of the germ cells from hybrid parents. Thus the dominant and recessive factor *never* came to reside in the same egg or sperm. It also meant that while the two contrasting factors existed side by side in the heterozygote, they did not contaminate

[81] For Morgan's view on Castle see G. E. Allen, "Thomas Hunt Morgan and the Problem of Sex Determination, 1903-1910," *Proceedings of the American Philosophical Society* 110 (1966): 49. On Cuénot, see T. H. Morgan, "The Assumed Purity of the Germ Cells in Mendelian Results," *Science* 22 (1905): 877-879.

[82] Morgan, "The Assumed Purity," p. 877.

[83] T. H. Morgan, "What are Factors in Mendelian Inheritance?" *American Breeders Association Report* 5 (1909): 365-368.

each other. Morgan thought that, since there was no mechanism to assure segregation of contrasting alleles, the concept of purity of gametes was necessarily speculation. Morgan felt he either had to accept the doctrine of segregation and purity of the gametes on faith, or reject it. Never one to accept anything on faith if he could avoid it, Morgan chose the latter course and questioned the very basis of Mendel's first "law," i.e., the absoluteness of segregation.[84]

Morgan further argued that there were too many exceptions to both the Mendelian and chromosome theories, especially in relation to sex determination, to make either theory an attractive explanation for heredity. Neither theory could explain the normal 1:1 sex ratio; since some species had males determined by XY (or XO), and others had males determined by XX, it seemed unlikely that the same result should be attributed to opposite causes. Finally, the unusual phenomena of gynandromorphism and parthenogenesis seemed to defy explanation by either theory.[85] In

[84] Morgan was not the only one to question the principles of segregation and gametic purity. W. E. Castle, from 1900 a staunch Mendelian, nevertheless raised questions about the prospects of continually recovering contrasting alleles after generations of passing back and forth in the hybrid state. Both W. E. Castle and C. B. Davenport argued that Mendelian factors might well alter one another permanently by existing side by side in the heterozygous condition. See L. C. Dunn, "William Ernest Castle," *Biographical Memoirs, National Academy of Sciences* 38 (1965): 31-80; and Davenport's *Inheritance in Poultry* (Washington, D.C.: Carnegie Institution of Washington, 1906), p. 80.

[85] Working from an earlier idea of Boveri, Morgan suggested a possible way in which at least gynandromorphism might be explained on the chromosome theory (T. H. Morgan, "An Alternative Interpretation of the Origin of Gynandromorphous Insects," *Science* 21 (1905): 632-634). Boveri had hypothesized that gynandromorphs could result from the failure of the sperm nucleus to unite with the egg nucleus, but subsequently uniting with one of the products of the egg nucleus' first division. Thus all the cells deriving from the unfertilized nucleus would be expected to give rise to male tissue (having only one X) while those deriving from the cell that originally combined with the sperm would produce female tissue (having two X chromosomes). Morgan proposed an alternative hypothesis, namely that the original egg might be fertilized by two or more sperm, only one of which fused with the egg nucleus, the other remaining unpaired, but developing without combining with any parts of the egg nucleus. The products of division of the paired nucleus would give rise to female parts of the embryo, while products of the unpaired nucleus would account for the male parts. Morgan pointed out that examples of multiple sperm entry into eggs were well known for a variety of organisms, including bees and

short, such fundamental exceptions argued strongly against the generality of the Mendelian and chromosome theories as explanations of heredity.

In 1905 Morgan found himself on the brink of being convinced of the chromosome theory of sex determination by Wilson's cytological work. To Morgan, the survey Wilson had carried out on numerous species seemed highly convincing. However, there appeared to be a logical flaw in the generalizations Wilson was prepared to make from his results, which Morgan described in a letter to Driesch in October of that year:

> Wilson's recent discovery that in certain bugs the spermatozoon that has the extra chromosome makes the *female every time*, and the one without it makes the male (exactly the reverse of McClung's supposition) makes it look at first sight as though the chromosomes were *the thing*. But work it through as he has done and you will find that it lands you in an absurdity for the same chromosome will be a male determining one in the following generation. Wilson is going to indulge in generalities about it. I will confess that when he first showed me his results I was somewhat staggered, especially as I had just sent a little paper to the press in which I had attacked the well-known assumption that the nucleus must be the bearer of the hereditary qualities of the male. On the contrary I argue that the protoplasm [i.e., cytoplasm] may account for the results. . . . Now however when it is evident that the chromosome theory will not even explain the case of sex determination I feel assured of my position once more. In fact as I have told Wilson the simplest way to account for the

---

wasps. Morgan then went on to suggest that whether his view or Boveri's was correct could actually be tested. If two strains of honey bee, for example, an Italian and German, were crossed and gynandromorphs appeared, the composition of tissues in the gynandromorphs would differ according to one or the other hypothesis. According to Boveri's hypothesis the male characters would be derived from the egg nucleus and the female from the combined egg and sperm; on Morgan's the male characters would derive from the unpaired sperm nucleus, and the female from the paired. Therefore, if a queen of the Italian strain were fertilized by a drone of the German strain and a gynandromorph resulted, the male parts should be Italian in character according to Boveri's view, and German according to Morgan's (p. 633). Morgan was not committed to his own hypothesis, but saw the value of it as an alternative that could be tested experimentally.

> two kinds of sperm is that there are two kinds of protoplasm and this determines the cell that shall get the accessory.[86]

What did Morgan mean by saying that the same chromosome that determines a female in one generation would be a male-determining one in the next? The basis of Morgan's objection lies in considering the class of organisms that have males with the XO chromosome complement, and females with the XX type. These males will produce two classes of sperm: 50 percent will bear an X-chromosome, and 50 percent will bear no X. The eggs will all bear X's. Now, a sperm bearing an X always produces a female (by combining with an egg, producing the complement XX). The X in the sperm can thus be said to be female producing. However, if that same X-chromosome comes to reside in an egg which, in the next generation is fertilized by an O-sperm, the result is a male. To Morgan the fallacy lay in assuming that the X-chromosome itself has anything directly to do with determining sex, since in one generation it determined a female while in the next it could produce a male. It was obvious to Morgan that sex was determined in some way by more complex interacting factors, perhaps some combination of chromosomes, rather than by the action of a single element like the X-chromosome.[87]

[86] Morgan to Driesch, October 23, 1905, pp. 1-2; Morgan-Driesch correspondence. What Morgan means by "McClung's supposition" is the following: Through a miscount of the number of chromosomes in the female, McClung thought a sperm bearing an accessory (X) chromosome was male determining; in fact the accessory was, according to Wilson's theory (and verified by all subsequent work), female determining.

[87] By present-day understanding Morgan was at least partly correct, though for reasons he could not fully understand at the time. The problem of sex determination is very complex, and the process does not appear to follow a consistent pattern throughout the animal and plant kingdoms. Moreover, the determination of sex must be distinguished from the inheritance of specific sexual traits. For example, in many species, such as *Drosophila*, sex appears to be determined by the ratio of X-chromosomes to autosomes, with the Y-chromosome playing little if any role. However, the genes governing the development of specific sexual traits, such as male or female sex organs, are located throughout the genome including the X- and possibly even the Y-chromosomes and not necessarily at specific loci on the X. In other species, such as human beings, the Y-chromosome appears to have a distinctly male-determining function, although no genes for specific male traits have been localized on the Y-chromosome to date. A more complete discussion of the modern conception of sex determination can be found in any genetics text, as for example, W. F. Bodmer and L. L.

Still another objection Morgan brought against the Mendelian and chromosomal theories centered around the problem of coupling. He pointed out that if adult characters are determined by hereditary particles on chromosomes, then a large number of traits inherited together, i.e., "coupled," should result. Bateson had just begun to observe some examples of coupling in 1906,[88] but they seemed to be few and far between. To Morgan they were not as frequent as would be expected, given the small number of chromosomes compared to the large number of inherited traits in most organisms. In reality, Morgan pointed out, the number of cases of coupling was so small that it was doubtful that Mendelian unit characters could be considered as carried by chromosomes.

Morgan objected also to the ideas of "chromosomal individuality" and "chromosomal integrity." The idea of chromosomal individuality had been propounded in the 1880s by several workers, most notably Boveri. In a penetrating cytological study published in 1888, Boveri had purported to demonstrate that the members of each pair of homologous chromosomes are qualitatively different in their heredity determinants from members of all other pairs. Thus, each pair of chromosomes was unique; it had "individuality." This individuality, according to Boveri, persisted from one generation to the next. On the other hand, there seemed to be compelling evidence against chromosomes persisting as material bodies from one generation to the next. For example, when there was no cell division, the chromosomes became invisible. It had been assumed by many workers that the chromosomal components dissolved away from the threadlike linear arrangement and were dispersed within the cell's nucleus during the nondividing period. Boveri attempted to show that the chromosomes retained their integrity from one generation to another, despite passing through

---

Cavalli-Sforza, *Genetics, Evolution and Man* (San Francisco: W. H. Freeman, 1976), pp. 152 ff.; Curt Stern, *Principles of Human Genetics* (San Francisco: W. H. Freeman, 1976), pp. 503 ff.

[88] See William Bateson, "The Progress of Genetics Since the Rediscovery of Mendel's Papers," in *Progressus Rei Botanicae* (1906): 368-418. Bateson wrote "sometimes, however, there is evidence of a linking or a *coupling* between distinct characters. When such a coupling is complete, the two characters, of course, can be treated as a single allelomorph . . . but besides this simpler phenomenon of complete coupling, we now know that the usual ratios are liable to disturbance by a *partial coupling* between distinct characters." Bateson had stumbled upon the phenomenon of linkage and crossing-over, which Morgan investigated so thoroughly after 1912.

many phases of cell division.[89] In regard to individuality, Morgan wrote to Driesch: "I am glad you are going to examine Boveri's experiment. I have always distrusted it, but until it is cleared up, the chromosomal people will find it convincing."[90] Morgan intended, apparently, to test Boveri's results himself, but never actually did. As for the hypothesis of chromosomal integrity, Morgan felt that there was little evidence supporting it. He argued that chromosomes might well fuse completely during the process of intertwining (synapsis) in sperm or egg production. He felt that the union of two homologous chromosomes during synapsis was as complete as when two drops of water fuse into one.[91] Thus, to Morgan, the chromosomes did not appear to contain hereditary information for individual characters of the embryo. Rather, the

[89] Boveri, "Über mehrpolige Mitosen als Mittel zur Analyse des Zellkerns." Boveri's support for the concept of chromosomal integrity was based largely on observational studies of the shape, size, and general structure of chromosome pairs from one generation to the next. Boveri showed that despite the apparent disappearance of the chromosomes during the nondividing phase, when they reappeared at the beginning of the next cell division, the exact same structural characteristics were observed for each pair. Proof for the "individuality" concept was more indirect. Boveri had caused sea-urchin eggs to be fertilized by two sperm so that the resulting zygote nucleus contained an uneven number of chromosomes. In the subsequent embryonic cleavages, some cells ended up with incomplete chromosome sets, in some cases two or more copies of one identical chromosome, while lacking any copies of some others. Those embryos that lacked certain chromosomes developed abnormally; furthermore, the abnormalities were usually of a very similar kind for a given abnormality in chromosome complement.

[90] Morgan to Driesch, April 17, 1906, pp. 3-4; Morgan-Driesch correspondence.

[91] T. H. Morgan, "Chromosomes and Heredity," *American Naturalist* 44 (1910): 449-496, especially p. 472. In this paper Morgan diagrammed the ways in which the chromosomes might split apart again after they have intertwined during synapsis. Depending on the plane of split, a whole variety of different kinds of combinations of their elements might be produced. Morgan felt that this was just as possible as Boveri's claim that the two chromosomes separate themselves completely like two strands of rope unwinding. In reviewing Boveri's work, Morgan maintained that too many chromosomal and Mendelian adherents were reading too much into Boveri's limited results. As he wrote (p. 461): "It should indeed be pointed out that Boveri's evidence seems to prove too much for that form of the particulate theory that ascribes unit characters to chromosomes, or it indicates, I think, that individual chromosomes do not in any sense contain either preformed germs or determinants, or unit characters, or even stand for the production of particular organs in any sense."

embryo's hereditary traits were produced by the interactions of materials produced by the "entire constellation of chromosomes.[92]

Morgan's final set of objections to the Mendelian chromosome theory was based largely on philosophical and methodological grounds. Chief among these was Morgan's general opposition to any particulate theory of inheritance as being "too morphological." Concentrating on structure alone, the morphological approach assumed that by attributing a function to some discrete structure, the function was thus "explained":

> The nature of present Mendelian interpretation and description inextricably commits to the "doctrine of particles" in the germ and elsewhere. It demands a "morphological basis" in the germ for the minutest phase (factor) of a definitive character. It is essentially a morphological conception with but a trace of functional feature. Although heredity is quite truly a functional process of major complexity, it may be recalled that the primary and fundamental Mendelian conception of this process utilizes not a single finding of the science of biochemistry. . . . With an eye seeing only *particles* and a speech only symbolizing them, there is no such thing as the study of *process* possible.[93]

To Morgan the particle theory not only overlooked functional processes, but in fact even tended to discourage further work. Particulate theories, he claimed, posed a kind of finalistic solution that is both rigid and static. Further investigation is discouraged because the "explanation" is neatly defined. He wrote:

> It may be said in general that the particulate theory is the more picturesque or artistic conception of the developmental process. As a theory it has in the past dealt largely in symbolism and is inclined to make hard and fast distinctions. It seems to better satisfy a class or type of mind that asks for a finalistic solution, even though the solution be purely formal. But the very intellectual security that follows in the train of such the-

[92] Ibid., pp. 460-461.

[93] The quotation itself comes from an article by Oscar Riddle, "Our Knowledge of Melanin Color Formation and its Bearing on the Mendelian Description of Heredity," *Biological Bulletin* 16 (1909): 316-351. Morgan quoted Riddle's statement in his own article, "Recent Experiments on the Inheritance of Coat Colors in Mice," *American Naturalist* 43 (1909): 449-510; quotation, pp. 509-510.

ory seems to me to be less stimulating for further research than does the restlessness of spirit that is associated with the alternative conception.[94]

Morgan's "romantic" or imaginative character could not accept any solution that seemed too all-encompassing and final without adequate experimental demonstration.

On other philosophical grounds, Morgan argued that the Mendelian theory was nothing much more than a logical construct, a conceptualization dealing in formalistic symbols which had no basis in reality. Morgan wanted a theory of heredity and development "without recourse to hypothetical 'particles' or to immutable and immortal factors."[95] So hypothetical were the hereditary particles that Mendelians seemed to shuffle them about in any way that made theoretical sense, without any attempt to make their ideas conform with reality. In 1909, at a meeting of the American Breeders Association in St. Louis, by and large a staunchly Mendelian group, Morgan delivered one of his most searing attacks on the formalism of Mendelian theory:

> In the modern interpretation of Mendelism, facts are being transformed into factors at a rapid rate. If one factor will not explain the facts, then two are invoked; if two prove insufficient, three will sometimes work out. The superior jugglery sometimes necessary to account for the results are often so excellently "explained" because the explanation was invented to explain them and then, presto! explain the facts by the very factors that we invented to account for them. . . . I realize how valuable it has been to us to be able to marshal our results under a few simple assumptions, yet I cannot but fear that we are rapidly developing a sort of Mendelian ritual by which to explain the extraordinary facts of alternative inheritance. So long as we do not lose sight of the purely arbitrary and formal nature of our formulae, little harm be done; it is only fair to state that those who are doing the actual work of progress along Mendelian lines are aware of the hypothetical nature of the factor assumption.[96]

[94] Morgan, "Chromosomes and Heredity," pp. 451-452.

[95] Morgan, quoting from Riddle in "Recent Experiments on the Inheritance of Coat Colors in Mice," p. 509.

[96] Morgan, "What are Factors in Mendelian Inheritance?" p. 365.

Although the line of cytological work represented by Boveri, Wilson, Stevens, and others had become well developed by 1908, the close correlation between cytological and Mendelian results was not so obvious as it might appear in retrospect. To be sure, the evidence was available by which it was possible to claim that the Mendelian assumptions were something more than pure formalism. But it was equally true that there was no conclusive evidence that Mendelian factors had any material basis in the chromosomes themselves. While Morgan's opposition to the Mendelian theory as a formalism might appear overly conservative, in 1909 it involved an act of faith to make the transition between one theory and the other. Moreover, Morgan was not alone in his claims that the Mendelian theory was at best symbolic. Such claims were also made by R. A. Emerson, E. M. East, and E. G. Conklin between 1908 and 1912.[97]

At the heart of Morgan's objections to the particulate theory was a view characteristic of embryologists. If the Mendelian factors were admitted to be real particles in the germ cells, then to Morgan the ominous specter of preformation was raised once again in biology. Popular in the seventeenth and eighteenth centuries, the preformation theory was the doctrine that the individual adult organism was already "preformed" either in the unfertilized egg or in the sperm. Embryonic development was simply a matter of growth of the tiny, but otherwise perfectly formed adult to a larger size. Since the early nineteenth century, embryologists had almost universally come to accept an opposing view, epigenesis, which stated that the embryo developed after fertilization out of formless undifferentiated material in the egg. Under the epigenetic view, development was more than a quantitative change, it was also qualitative. To Morgan and others, the invocation of material particles as the bearers of heredity information from one generation to another seemed too reminiscent of the old idea of a preformed "character" in the germ plasm. Mendel's assumption of "factors" or *Anlagen* suggested that the particle contained the character that was assigned

[97] R. A. Emerson and E. M. East, "The Inheritance of Quantitative Characters in Maize," *Bulletin of the Agricultural Experiment Station of Nebraska* 2 (1913): 5, n.; also East, "The Mendelian Notation as a Description of Physiological Facts," *American Naturalist* 46 (1912): 633-655, especially p. 635; E. G. Conklin, "The Mechanism of Heredity," *Science* 27 (1908): 92.

to it, and that development was merely an unfolding of that character:

> In Mendelian inheritance we also have to face the alternatives of preformation and epigenesis. The currently accepted interpretation of Mendelian inheritance is strictly one of preformation. Alternative characters are treated as entities in the germ cells that may be shuffled but seldom get mixed. With each new deal the characters are separated, one germ cell getting one character, and another the contrasted character. If we take the opposite point of view, that of epigenetic development, the outcome, whether alternative or contrasted characters were involved, is not due to separation, but to alternative dominance or recession. . . . Which of these general points of view, preformation or epigenesis, we may think more profitable as a working hypothesis is, I believe, the question of the hour.[98]

From the point of view of an embryologist, the most important question was not so much how hereditary information was *transmitted* from one generation to the next, but rather how that information was *translated* into adult characters. Between 1900 and 1910, Morgan continued to look at the process of sex determination in terms of his own theory of "reaction." This theory held that the egg and the sperm exist in a kind of balanced state and can progress in various ways according to the combination of internal and external factors. The initial state of the fertilized egg would not be fixed; rather, it would be dynamic and always subject to change. It was the total interaction between the entire body of chromosomes and the cell cytoplasm, chemical and physiological in nature, that determined the ultimate character of the adult. Viewing development in this way, Morgan could not accept any theory that simply postulated the adult character as somehow residing in a particle within the cell.

Morgan also objected to the particulate and preformationist nature of the Mendelian and chromosome theories because both reminded him too closely of the speculative ideas of the previous generation of zoologists, such as Haeckel and Weismann. Morgan

[98] T. H. Morgan, "Sex Determining Factors in Animals," *Science* 25 (1907): 382-384; quotation p. 384. See also Morgan, "Recent Experiments on the Inheritance of Coat Colors in Mice."

had repeatedly attacked Weismann's speculation in regard to both heredity and evolution and contrasted Weismann's exaggerated claims for the chromosomes and their hierarchy of particles[99] with the work of more modest and respectable cytologists such as Boveri and Wilson.[100] It was the tendency toward rampant speculation, with an eye for neither material reality nor experimental testing, that Morgan most disliked. He felt that if biology were to become a sound, experimental science, it would have to abandon the all too prevalent tendency to invent structures and particles with no regard to the question of their actual existence.[101]

MORGAN'S CHANGE OF VIEW ON THE MENDELIAN AND CHROMOSOME THEORIES

Considering the fact that Morgan received the Nobel Prize in 1933 for studies which showed the compatibility of the Mendelian and chromosome theories, it is ironic that during the first decade of the twentieth century he was such a strong opponent of both views. It is pertinent, therefore, to investigate what caused his change of mind, when it came about, and the new evidence or

[99] Weismann's concept postulated a hierarchy of hereditary particles, from more general down to highly specific. It was the release of the highly specific particles, the biophors, after a certain number of embryonic cell divisions, that caused the differentiation of each cell type. Weismann had no proof that such a hierarchy existed, yet his theory of heredity and development was based on this assumption.

[100] Morgan, "Chromosomes and Heredity," p. 449-496.

[101] A number of other workers during the same period raised doubts about the general applicability of Mendel's laws. The zoologist F. T. Lewis, for example, rejected Mendelism as a universal law in 1907 because it in no way seemed to account for blending inheritance. (F. T. Lewis, "Mendelism," *American Naturalist* 41 [1907]: 329-332.) Raymond Pearl, who studied hereditary problems among invertebrates and in biometrical analyses, felt that Mendel's laws in many cases had no application to lower organisms. (Raymond Pearl, "Mendel's Principles of Heredity, a Review," *The Independent* 67 [1909]: 762.) The British zoologist G. Archdall Reid devoted a chapter in his book *The Laws of Heredity* to showing how Mendel's laws are untenable as a general theory of heredity. (G. A. Reid, *The Laws of Heredity* [New York: Macmillan Co., 1910], pp. 142-168.) Reid was assisted in his work by a number of the English anti-Mendelians, including E. Ray Lankester, A. D. Darbishire, and Sir William Thistelton-Dyer (see Reid's preface). Thus, while Bateson in England, Cuénot in France, and Castle in the United States (among others) waged a strong, vocal, and persistent battle for the acceptance of Mendel's work on general laws of heredity, other segments of the biological community were skeptical of Mendel's basic premises because they seemed to cover so few actual cases.

occurrences that were responsible for Morgan's acceptance of both the chromosome and Mendelian theories as generally applicable to heredity.

Two lines of work contributed to Morgan's change of view. One was additional evidence, published between 1900 and 1910, suggesting that chromosomes really did have a significant role in sex determination. The other was Morgan's own studies with the small fruit fly *Drosophila melanogaster* and recognition of a new phenomenon, what Morgan called "sex-limited" inheritance.

The work of Richard Hertwig in 1906 and 1907 is an example of the first line of evidence. It had been a contention of Wilson's theory of sex determination that the sperm was the sex-determining element in those species where the male was XY or XO and the female XX. Sex determination was due to a single difference between the classes of sperm: the presence or absence of an X-chromosome. Hertwig fertilized female frogs from different localities (different geographic races) with sperm from a single male. The result was the normal sex ratio among the offspring. When he did the reciprocal experiment, fertilizing one female with sperm from several male frogs from different populations, the results were quite variable (i.e., ratios of males to females varied, and some individuals were of indeterminate sex). Morgan was highly impressed with these results. It appeared that the sperm really might be the sex-determining element after all, and that put the spotlight on the chromosomal element in the sperm.[102]

By far the most important line of evidence, however, came from Morgan's own work with *Drosophila*; this work led him to accept first the Mendelian, and shortly thereafter, the chromosome interpretations of heredity. As early as 1906, Morgan had been interested in breeding mice and pigeons, especially in conjunction with his analyses of Cuénot's Mendelian experiments.[103] Yet these or-

[102] T. H. Morgan, "The Determination of Sex in Frogs," *American Naturalist* 42 (1908): 67-70. It is in this paper that Morgan first mentioned Wilson's work of 1905 in a completely favorable light. Although Morgan at this time seemed to accept the chromosome theory of sex determination as having some validity, he still remained skeptical of the idea that Chromosomes contains specific units or particles that determine adult characters. In other words, he accepted the usefulness of the chromosome theory without accepting the reality of Mendelian factors.

[103] Morgan to Driesch, August 15, 1906, p. 3; Morgan-Driesch correspondence. Results of these experiments are reported in "Some Experiments on Heredity in Mice," *Science* 27 (1908): 493; and "Recent Experiments on Inheritance of Coat Colors in Mice," especially pp. 503-504.

ganisms had serious drawbacks for laboratory breeding experiments. The animals took a long time to produce a new generation and required considerable space and constant maintenance. In 1908 or 1909 Morgan began breeding the small fruit fly *Drosophila*, an organism that had all the advantages the mice and pigeons lacked: a short generation time (twelve days), little need for space (they could be bred by the thousands in milk bottles), and once the culture was started, little need for maintenance.

As a laboratory organism, *Drosophila* was becoming something of a vogue among many workers between 1900 and 1910.[104] It is a common insect, found virtually everywhere (one species even lives on the gills of certain marine crabs), and has been known to man since antiquity.[105] *Drosophila* was being used in W. E. Castle's laboratory starting around 1900, and with considerable success by 1905 or 1906. Castle actually carried out a number of breeding experiments with the fly and even claimed that one of its traits, low fertility, was inherited as a Mendelian recessive![106] At Indiana University W. J. Moenkhaus was also breeding *Drosophila* as part of a series of experiments on the heredity of sex.[107] At the Carnegie Institution Laboratory for the Study of Experimental Evolution at Cold Spring Harbor, New York, F. E. Lutz had begun experiments on *Drosophila* to test Castle's results. And, somewhere between 1906 and 1907, Nettie M. Stevens, at Bryn Mawr, had raised *Drosophila* for cytological purposes. Thus, many

[104] Details on the use of *Drosophila* between 1900 and 1910 and Morgan's early experiments with it are given in G. E. Allen, "The Introduction of *Drosophila* into the Study of Heredity, 1900-1910," *Isis* 66 (1975): 322-333.

[105] It has gone under a number of different scientific and common names. In the early 1900s biologists often called it *Drosophila ampelophora* or *Drosophila ampelophila*. After 1910, and especially through Morgan's publicized work with the organism, the standard name came to be *Drosophila melanogaster* for the commonly occurring wild-type form. There are, in addition, a number of common names for the fly: the vinegar fly (the term Morgan seemed to prefer), the fruit fly, the pomace fly, and the banana fly. Throughout the remainder of this book we will refer to the organism scientifically as *Drosophila melanogaster* and commonly as the fruit fly.

[106] W. E. Castle, "Inbreeding, Cross-Breeding and Sterility in *Drosophila*," *Science* 23 (1906): 153. Since he claimed he had been doing his experiment for five years, it appears Castle began using *Drosophila* for hereditary studies around 1901.

[107] W. J. Moenkhaus, "The Effect of Inbreeding and Selection on the Fertility, Vigor and Sex Ratio of *Drosophila ampelophila*," *Journal of Morphology* 22 (1911): 123-154. Moenkhaus appears to have gotten the idea of using *Drosophila* from Castle.

investigators had begun to recognize the value of this small insect for a variety of laboratory purposes.

Morgan was an avid reader of the literature, and knew of all these studies by 1906 or 1907. He was particularly impressed with Castle's work on inbreeding in *Drosophila*, and discussed it in his course on experimental zoology at Columbia.[108] A direct stimulus to Morgan's own use of the fly, however, appears to have come from one of his graduate students, Fernandes Payne, who entered the zoology department in 1907. Payne had taken his B.A. and M.A. degrees at Indiana University in 1905 and 1907, respectively, and had known of Moenkhaus's work with *Drosophila*. In the fall of 1907 he enrolled in Morgan's experimental zoology course, where each student had to carry out an independent laboratory experiment throughout the year. Payne reported that he discussed various topics with Morgan and finally hit on one that Morgan liked. While a student at Indiana, Payne had carried out some independent investigations on the evolution of blindness in cave fauna, a phenomenon often cited as an example of Lamarckian evolution, i.e., the inheritance of the effects of disuse of an organ. As we have seen from Morgan's own work on evolutionary theory, he was very much intrigued by such problems. Knowing of Payne's previous work on blindness in a field situation, Morgan suggested carrying out a rigorous laboratory study. Why not breed some organism in the dark for a number of generations and see if there was any reduction in size or functioning of the eyes? Payne's suggestion that they use *Drosophila* undoubtedly made the project practical, since it would be possible, in a nine-month period, to obtain over ten generations.[109] According to Payne, Morgan told him to collect his own starting cultures of *Drosophila* by leaving ripe fruit on the windowsill.[110]

In the period 1906 to 1907, Morgan was very much excited

[108] T. H. Morgan, "Genesis of the White-Eyed Mutant," *Journal of Heredity* 33 (1942): 91-92; here, Morgan states, "I was, of course, familiar with the important paper of Castle, Carpenter, Clarke, Mast and Barrows [in *Proceedings of the American Academy of Arts and Sciences* 41 (1906): 729-786] and used it in my lecture on experimental zoology."

[109] Payne to A. H. Sturtevant, October 16, 1947; Sturtevant Papers, California Institute of Technology Archives. Payne's results were ultimately reported in his paper, "Forty-nine Generations in the Dark," *Biological Bulletin* 18 (1910): 188. The results showed no diminution in eye shape or function.

[110] Payne to Sturtevant, October 16, 1947; Sturtevant Papers.

by the prospect of finding or inducing *Oenothera*-like mutations in animals.[111] Beginning in 1906 he raised a number of insect species in the laboratory and attempted to induce mutations by injecting salts, sugars, acids, alkalis, and other substances into pupae in the regions of the reproductive cells.[112] It was apparently around 1908 that he began using *Drosophila* for this purpose. By this time he had hit upon another method for possibly inducing mutations: exposing the larvae to radium.[113] The culture Morgan used for these radium experiments was probably a pure, i.e., inbred, culture of *Drosophila* obtained from F. E. Lutz, who in 1909 had moved to New York City to the American Museum of Natural History. Morgan could not recall whether Lutz had given him the starting culture, but Edith Wallace, Morgan's longtime assistant, claimed that Lutz brought the culture up to the Columbia laboratory one day.[114] It is likely that Morgan used a pure culture for his mutation studies, since wild-type flies contain a considerable amount of hidden variability, which would have made detection of any newly occurring mutation extremely uncertain. Whatever the source of his original culture of flies, it is clear that by 1908 or 1909 Morgan was actively breeding *Drosophila* in the laboratory in conjunction with his evolutionary studies.

MORGAN AND THE WHITE-EYED MALE

For about a year Morgan was unable to induce de Vriesian mutations in *Drosophila*. Then a rare event totally changed the

[111] A. H. Sturtevant, "Thomas Hunt Morgan," *Biographical Memoirs, National Academy of Sciences* 33 (1959): 290.

[112] Morgan to Driesch, August 15, 1906; Morgan-Driesch correspondence.

[113] Morgan to A. F. Blakeslee, May 2, 1935; Morgan Papers, Caltech. Morgan claims that he got the idea of using radium from a suggestion made by de Vries in 1904, in a speech at the dedication of the Cold Spring Harbor Laboratory for the Study of Experimental Evolution (see also Blakeslee to Morgan, May 22, 1935; Morgan Papers, Caltech). Quite independently Jacques Loeb had begun similar experiments irradiating *Drosophila* with radium and even claimed to have gotten some mutations. Morgan criticized Loeb's work in his letter to Blakeslee, cited above, but that was many years later. It was a temporary bone of contention between Morgan and Loeb as to who had priority in these radium experiments (see Loeb to Morgan, May 17, 1911; Loeb Papers, Library of Congress).

[114] A. H. Sturtevant to J. Walter Wilson, September 22, 1965; Sturtevant Papers.

direction of his work. Sometime early in 1910, Morgan observed a curious variation in one of his *Drosophila* stock bottles. He found a single male fly with white, rather than the normal wild-type, red eyes.[115] Since this variation represented a distinct, discrete change which appeared to have occurred by chance in the culture, it was to Morgan an enormously exciting find. The eye-color variation was obviously a mutation, but not a species-level macromutation of the de Vriesian type. Morgan bred this male with a red-eyed female and found that the first generation (F_1) produced all red-eyed flies. The results with the F_1 suggested to Morgan that the white-eyed condition might be a Mendelian recessive. Matings between F_1 siblings produced an F_2 in which the white-eyed condition reappeared in a ratio of three red-eyed to one white-eyed fly. This would appear to confirm the hypothesis that white eye was a Mendelian recessive, since this is exactly the ratio the Mendelian scheme would predict for a cross between two hybrid forms. However, Morgan noted something that was not in strict accordance with the Mendelian scheme: all of the white-eyed flies were male! The red-eyed flies were found to occur in a ratio of two females to one male. Still another cross of white-eyed males to F_1 red-eyed females gave both red-eyed and white-eyed progeny in equal ratios of male to female. The white-eyed condition was obviously a stable genetic variation, which Morgan called a "mutation," but of a completely different sort from what de Vries had encountered. Morgan's mutation was a small, discrete, Mendelian recessive which did not in any way make the offspring a different species from the parents.

The appearance of the white-eyed mutant suddenly shifted Morgan's interest from evolution to heredity. What appeared to attract Morgan's attention was not simply that the white-eyed condition acted according to Mendelian laws, but further that it appeared in some fundamental way to be associated with sex inheritance. Why, after all, should all the white-eyed flies in the F_2 have been males?

[115] The term "wild type" is used by geneticists to refer to the predominant phenotype (appearance) in a population for any given trait. Most fruit flies found in the wild have red eye color as well as long wings which fold back horizontally over the body. These two hereditary forms are referred to as the wild-type condition in regard to eye color and wing shape and position. Almost every heritable characteristic has a wild-type form, as well as a large number of variant, mutant forms.

To explain these results, Morgan not only accepted the basic tenets of the Mendelian scheme, but combined these notions with the suggestions of Wilson and Stevens that the determination of sex had a chromosomal basis. In a now famous paper reporting his initial results, Morgan wrote:

> Assume that all of the spermatozoa of the white-eyed male carried the single "factor" for white eyes (W), that half of the spermatozoa carry a sex factor (X), the other half lack it, i.e., the male is heterozygous for sex. Thus the symbol for the male is "WWX," and for his two kinds of spermatozoa WX and W. Assume that all of the eggs of the red-eyed female carry the red-eyed "factor," R; and that all of the eggs (after reduction) meiosis carry one X each. The symbol of the red-eyed female will be therefore RRXX and that for her eggs will be RX-RX.[116]

Note that Morgan did not use the normal Mendelian notation (where, more commonly, red eye might be designated as R, and white eye as r). The significance of this will be apparent shortly.

For his results to work out, Morgan assumed that his original white-eyed male was homozygous for W, but heterozygous for the sex factor, X. Thus, normal, wild-type strains of *Drosophila* should be made up of males that are homozygous for R (RR) and heterozygous for X (RRXO or RRX). A red-eyed male of this genetic composition, when crossed to one of the rare white-eyed females (some white-eyed females did occur in later generations) would be expected to produce all red-eyed offspring. But, when such a "backcross" was made, Morgan found "a most surprising fact: The anticipation was that wild males and females alike carry the factor for red eyes, but the experiments showed that all wild males are heterozygous [i.e., hybrid] for red eyes, and that all the wild females are homozygous. Thus when the white-eyed female is crossed with a red-eyed male, all of the female offspring are red-eyed, and all of the male offspring white-eyed.[117] To explain these results, Morgan modified his original hypothesis, and concluded that all red-eyed males in a wild population are heterozygous for W. However, this asumption contradicted predictions based on the Mendelian hypothesis, for, if all red-eyed males were heterozygous for

[116] T. H. Morgan, "Sex Limited Inheritance in *Drosophila*," *Science* 32 (1910): 120-122.

[117] Ibid.

white, then white-eyed females should appear relatively frequently in the wild population. Morgan then made the further assumption that the X factor was coupled with R in the spermatozoa of the F_1 hybrid males. Making the assumption of coupling would keep all the males and females in the F_1 red, because both would contain at least one X factor, coupled with R (thus males would be RX-W, and females RX-RX). The mutation, or sport, must have originally occurred in one of the eggs of the red-eyed females, in which RX changed to WX.[118] As Morgan stated explicitly in the conclusion of his paper: "The fact is that this R and X are combined, and have never existed apart."[119]

Morgan now proceeded to alter his previous symbolic representation. At the end of the paper he symbolized the white-eyed male as OX-O and the hybrid F_1 female as RX-OX. By this scheme, wild-type females would be written RX-RX, wild-type males as RX-O, and white-eyed females as OO-XX.[120] The reason for this change was based on Morgan's acceptance of Bateson's popular notion that the appearance of some trait, such as color, was due to the *presence* of a gene, and white, i.e., no color, to the *absence* of a gene.[121] The substitution of O for W indicated to Morgan the probability that the white-eye condition resulted from the absence (loss) of the red factor (R). By this scheme, Morgan could write the R and X as always associated with each other (or coupled, as Morgan wrote it).

It was for more than theoretical reasons, however, that Morgan changed his symbolic representations. He was particularly troubled in 1910 by the apparently conflicting evidence of Wilson and Stevens on the one hand, and Bateson and his coworkers on the other. Wilson and Stevens had studied mostly insects (excluding moths and butterflies), and concluded that males were heterozygous (XY) and females homozygous (XX) for the accessory chromosome. Punnett and Raynor in England had worked mostly with moths and chickens and had found just the opposite: i.e.,

[118] See E. A. Carlson, *The Gene: A Critical History* (Philadelphia: W. B. Saunders, 1966), p. 45.

[119] Morgan, "Sex Limited Inheritance in *Drosophila*," p. 122.

[120] Ibid.

[121] See William Bateson and R. C. Punnett, "A Suggestion as to the Nature of the 'Walnut' Combs in Fowls," *Proceedings of the Cambridge Philosophical Society* 13 (1905): 165-168; an analysis of the presence-and-absence theory has been presented by R. G. Swinburne, "The Presence-and-Absence Theory," *Annals of Science* 18 (1962): 131-145.

males were homozygous (XX) and females heterozygous (XY). This discrepancy had formed one of the mainstays of Morgan's objections to the chromosome theory of sex determination throughout the preceding decade. However, with the inheritance patterns observed in the white-eyed *Drosophila*, Morgan saw that the discrepancy was not as profound as it appeared. Punnett and Raynor had found that sports appeared exclusively in females! This would be expected if, in fact, the females in moths and chickens were the heterozygous sex (XY or XO). Thus, far from contradicting each other, the results of Wilson and Stevens's work on the one hand, and of Punnett and Raynor's on the other, were complementary. Assuming only the presence or absence of a factor for eye color, Morgan thus explained rationally what had otherwise been a major stumbling block for understanding the inheritance of sex in terms of chromosomes.

Morgan spoke of the coupling of the eye-color factor with the X-chromosome as a case of "sex-limited" inheritance.[122] However, throughout his 1910 paper he rigorously avoided stating that the factor for red eye was structurally a part of the chromosome. Morgan remained an agnostic on the chromosome theory. One case of sex-limited inheritance did not seem to him enough to warrant discarding his previous skepticism. Thus, although he hinted at the close physical relationship between Mendelian factors and material structures such as chromosomes, he refrained from making that association explicit. However, in the ensuing months of 1910 and early 1911, Morgan's skepticism vanished almost completely. He found more mutants in 1910, and two of these, yellow body color and miniature wings, were also sex limited.[123] There appeared to be no question in Morgan's mind that the appearance of so many

[122] Morgan later referred to traits determined by genes on the X-chromosome as "sex-linked" inheritance, the term that is still used today. In modern genetic terminology, sex-limited inheritance refers to conditions that are determined by genes on one or another autosome, but whose expression is influenced by whether the individual is a male or female (presumably by the presence of male and female hormones).

[123] T. H. Morgan, "The Method of Inheritance of Two Sex-Limited Characters in the Same Animal (Abstract)," *Proceedings of the Society for Experimental Biology and Medicine* 8 (1910): 17-19; "An Attempt to Analyze the Constitution of the Chromosomes on the Basis of Sex-Limited Inheritance in *Drosophila*," *Journal of Experimental Zoology* 11 (1911): 365-412; and "The Origin of Nine-Wing Mutations in *Drosophila*," *Science* 33 (1911): 496-499.

factors, all coupled with each other and all appearing as sex-limited traits, was substantial evidence in favor of the chromosome theory of Mendelism and sex determination.

Morgan was very excited about his results and the broadening horizons for further research that *Drosophila* represented. He discovered the original white-eyed male in early 1910, about the time his third child, Lilian, was born (January 5). The story is told that when Morgan went to the hospital to see his wife just after Lilian's birth, she greeted him with the question, "Well, how is the white-eyed fly?" Morgan reported the most recent results with great enthusiasm, but suddenly stopped and asked, "And how is the baby?"[124] His first paper on *Drosophila* was published in July 1910 in *Science*. In June of 1910 he wrote an ecstatic letter to his friend and collaborator H. B. Goodale, with whom he had worked on breeding experiments in mice and rats: "My white-eyed fly gave a splendid case of sex limited inheritance: the F_1 gives white-eyed only in the males."[125] And in a letter in November of the same year to Driesch, Morgan wrote that although he had been overwhelmed by work, mostly breeding experiments with fruit flies, "it's wonderful material."[126] Already, Morgan claimed, he had more than a half-dozen mutants, some of which were sex limited and others not. Morgan confessed to Driesch that the sex-limited cases interested him most; "they may throw some further light on the process of heredity."[127] Morgan saw quickly and clearly that *Drosophila* provided the basis for answering, in an experimental way, questions about the physical nature of the hereditary process.

[124] H. K. Morgan, "Notes on Thomas Hunt Morgan's Life," p. 4.

[125] Morgan to Goodale, June 15, 1910, p. 3; H. B. Goodale Papers, American Philosophical Society.

[126] Morgan to Driesch, November 23, 1910, p. 2; Morgan-Driesch correspondence.

[127] Ibid.

CHAPTER V

Development of The Mendelian-Chromosome Theory (1910-1915)

In his book, *The Structure of Scientific Revolutions*, Thomas Kuhn describes the ways in which new ideas are introduced into the scientific community.[1] Written in sociological, psychological, historical, and philosophical terms, Kuhn's work is a brilliant analysis of the factors that promote and retard intellectual change among natural scientists. He proposes that new ideas emerge in science during periods characterized either by the lack of any comprehensive theory to explain a certain set of phenomena, or by the inability of a long-established theory to continue explaining new observations. New ideas, representing radical and wholly different views, thus begin to gain a foothold as the result of the breakdown of older explanations. However, for sociological and economic reasons, radical ideas gain acceptance relatively slowly, and only come to predominate after considerable time has elapsed.

Kuhn describes new ideas in science as "paradigms," and periods in which no general theory or paradigm holds sway as "preparadigm periods." Paradigms are "universally recognized scientific achievements that for a time provide model problems and solutions to a community of practitioners."[2] Paradigms are more than a simple statement of a generalization or "law." They involve a complex of ideas and fundamental assumptions that provide an entire point of view from which a group of workers may approach a set of problems. In preparadigm periods there may be several competing theories, each with its adherents, but with none winning overall acceptance. In other cases, there may be no theories at all of a scientific sort. In the study of heredity, for example, the period prior to the rediscovery of Mendel's laws in 1900 can be

[1] T. S. Kuhn, *The Structure of Scientific Revolutions*, 2d ed. The International Encyclopedia of Unified Science, vol. 2, 2 (Chicago: University of Chicago Press, 1970).

[2] Ibid., p. viii.

considered a preparadigm period in which a number of competing theories existed side by side (e.g., Darwin's theory of pangenesis, Weismann's theory of ids and biophors, Nageli's "micellae," Galton's "stirps," de Vries's "pangenes"). The introduction of Mendel's concepts was the first stage in the development of the new paradigm—but Mendel's theory was not, as we have seen, in itself complete enough to act as a paradigm on its own.

The decade between 1900 and 1910 was another stage in the preparadigm period. Here, one theory, Mendel's, emerged as more adequate to many investigators than most of the others; yet, as Morgan pointed out, it still had major drawbacks, which reduced its support. Only after 1910, with the union of the Mendelian and chromosome theories, did the emergence of a more nearly adequate paradigm in the study of heredity become apparent.

One of the characteristics of a new paradigm in any area of science is that it opens up for workers a whole host of new problems which can be worked out (or, "articulated" as Kuhn puts it) in a new theoretical framework. Investigatory work carried out within the framework of a paradigm Kuhn calls "normal science."[3] The new paradigm is a revolution because it sweeps aside old world views, old ways of looking at problems, and provides a fundamentally different framework, with different rules, different assumptions, and a different philosophical basis.

Morgan's work with *Drosophila*, beginning with the white-eyed fly in 1910, is a clear example of Kuhn's concept of scientific revolution in biology. By themselves neither the Mendelian theory nor the chromosome theory provided a thorough and adequate understanding of the problems of heredity and variation. By combining the two, Morgan and his coworkers introduced into biology a paradigm of powerful and far-reaching dimensions, the chromosome theory of inheritance or, more simply, the theory of the gene. At first the articulations of the theory were more or less limited to Morgan's lab. But gradually others took up the torch as well, first in the United States and then throughout the world. Acceptance and development of the theory varied considerably from one country to another, indicating the sometime crucial role that cultural, philosophical, institutional, and governmental views have on the spread of scientific ideas.

[3] Ibid., p. 10.

BEGINNINGS OF THE CHROMOSOME CONCEPT: LINKAGE AND CROSSING-OVER

In 1910, the discovery of two additional sex-limited mutants, yellow body (as opposed to normal brown body) and miniature wings (as opposed to normal-sized wings), strongly suggested the association of these factors with the X-chromosome. Morgan followed the coupling of these factors through several generations and made an astounding observation: when crosses were made between heterozygotes, coupling did not always appear to be complete! A recombination of factors could be observed to take place in a manner that should have been impossible if the factors were indeed physically attached to their respective X-chromosomes. Morgan noted that a recombination of factors occurred at rates that were different for each pair, but relatively constant for any one pair.[4]

The apparent anomaly that Morgan was observing had been encountered in the previous decade by Bateson and his coworkers in England. In 1905, Bateson, Elizabeth Saunders, and R. C. Punnett, working with sweet peas, had found that flower color and shape of pollen grains were coupled traits. They, too, had observed that coupling was not always complete, but that occasionally the coupled traits were separated and recombined. Bateson was an embryologist (recall he had studied for several summers with W. K. Brooks) and even more opposed than Morgan to morphological theories of inheritance and development. Although he was familiar with the cytological evidence for the movements of chromosomes in cell division, and even their possible role in sex determination, Bateson steadfastly refused to ascribe any hereditary role to the chromosomes. Bateson and Punnett had, rather, developed a complex theory of coupling and repulsion, as well as one of "reduplication," to explain their results.[5] These hypotheses were

[4] T. H. Morgan, "Random Segregation versus Coupling in Mendelian Inheritance," *Science* 34 (1911): 384.

[5] Because Bateson and his coworkers were committed at that time to the presence-or-absence theory as an explanation for dominance and recessiveness, their theory of coupling and repulsion was phrased only in reference to the dominant allele for any trait. It is thus important to keep in mind during the following discussion that the explanation of either coupling or repulsion always referred only to dominant factors. Bateson's symbolism was such that the dominant factor was represented by a capital letter (such as R) which was present as a Mendelian gene in the germ

never very successful, however, and maintained prominence only in the preparadigm period, i.e., before 1910. After the expansion of Morgan's work, Bateson and Punnett's elaborate and contorted ideas faded from view.

Ironically, Bateson and Punnett closed their 1911 paper, in

---

plasm; its recessive allele was written as a lower-case letter (such as r), which represented the absence or loss of the dominant Mendelian factor. To Bateson, the term "coupling" referred to the situation when, in the inheritance of two associated traits (such as flower color and pollen shape) the two dominant conditions appeared together (that is, when the two dominant conditions were apparently contributed to the offspring by the same parent). Thus, for example, a double heterozygote whose genotype was Aa Bb might produce only two types of gametes: AB and ab, instead of the customary four (i.e., Ab, AB, aB, ab) if random assortment occurred. The phenotypic ratio between offspring of a cross of two such heterozygotes would be as follows: three showing both dominant traits to one showing both recessive (neither dominant trait). This ratio would be expected for all cases of complete coupling. As long as they started out coupled, the dominant A and dominant B alleles would never be dissociated from each other. However, as Bateson and his coworkers observed, coupling of certain characteristics could be very strong while that of others could be very weak. Very strong (complete) coupling would be found wherever A and B always stayed together. Weak coupling (or as it was sometimes called, partial coupling) would be involved when the two dominant genes sometimes formed recombinations with their recessive alleles. Partial coupling would introduce two more phenotypic classes in addition to the two found with complete coupling: A and b and the reverse, a and B. The frequency of these two additional classes would reflect the strength of the coupling. For example, a weak coupling might produce the following ratio: 3 AB: 1 aB: 1 Ab: 3 ab; similarly, strong coupling might produce something like the following ratio: 63 AB: 1 aB: 1 Ab: 63 ab. In the cases of coupling he observed in a variety of plants, Bateson thought that he observed the appearance of certain regular ratios over and over again (such as 3:1, 1:7:7:1, or 15:1).

Bateson and his coworkers also observed a situation in which it appeared the two dominant conditions *always* separated from each other—i.e., would never be coupled—what they called spurious allelomorphism. (W. Bateson, E. R. Saunders, and R. C. Punnett, "Experimental Studies in the Physiology of Heredity," *Reports to the Evolution Committee of the Royal Society* 2 [1905]: 1-131.) For example, a double heterozygote with the genotype RrSs would produce only two types of gametes: Rs and rS. Because of what Bateson called "repulsion" there would be no RS gametes formed; by definition repulsion meant that the two dominant genes, R and S, would never go together into the same sperm or egg cell. The phenotypic ratio among offspring of a cross between two such double heterozygotes showing complete repulsion would thus be: two showing both dominant traits, one showing R and s, and one showing r and S (remember that while R and S

which they rejected the coupling and repulsion idea in favor of an even more hypothetical notion (reduplication) by claiming: "No case of coupling has been found in animals. . . . At present it seems not impossible that the two forms of life [plants and animals] are really distinguishable from each other in this respect." But at that

---

cannot go together in forming a germ cell, i.e., in any single egg or sperm, the sperm carrying an R is not restricted from fertilizing an egg carrying an S). At first Bateson, Saunders, and Punnett thought that repulsion was always complete, but in 1911 Bateson and Punnett discovered an exception to this, again in sweet peas. They found one example of a double recessive (written genotypically as rr ss) out of a large number of offspring from a cross between two double heterozygotes (that is, where the parental cross was Rr Ss x Rf Ss). (W. Bateson and R. C. Punnett, "On Gametic Series Involving Reduplication of Certain Terms," *Journal of Genetics* 1 [1911], 293-302.) They had only one specimen, but it was clear that this plant, with red flowers and round pollen, represented a double homozygote (i.e., in Bateson and Punnett's terms, having lost all dominant alleles for these two traits). They thus decided that repulsion, like coupling, must also be weak in some cases and strong in others. Bateson concluded that the frequency of partial repulsion, like coupling, ought to follow regular ratios. In studying a variety of cases of repulsion, Bateson and Punnett found certain ratios appearing relatively regularly. But Bateson and Punnett observed something else that was extremely intriguing. Considering those sweet pea characteristics they had studied in 1905 (associating flower color with pollen grain shape), they noted that coupling followed the ratio of: 7 BL: 1 Bl : 1 bL : 7 bl (where B is blue flower color and b is red flower color; L is long pollen grain, and l is round pollen grain). In 1910 they observed that the repulsion between B and L is just the converse: that is, it showed a ratio of 1 BL : 7Bl 7 bL : 1 bl. As Bateson and Punnett remarked, "It would be interesting if in such cases as these the coupling and repulsion systems for a given pair of factors were shown to be of the same intensity" (ibid., p. 297). Repulsion and coupling appeared to be, in fact, the converse of one another. As E. A. Carlson has pointed out, the five-year delay between the discovery of partial coupling and that of partial repulsion was most likely caused by the relative rarity of occurrence of the double recessive compared to the relative abundance of the double or single dominant. (E. A. Carlson, *The Gene: A Critical History* [Philadelphia; W. B. Saunders, 1966], p. 51.) Sturtevant later (1914) pointed out that Bateson and Punnett did not analyze a large enough quantity of data, that is, did not have large enough numbers of offspring in their sweet pea studies to overcome a natural amount of variation (avoid statistical problems) to make their observance of a constant ratio of any sort really reliable. (See A. H. Sturtevant, "The Reduplication Hypothesis as Applied to *Drosophila*," *American Naturalist* 48 [1914]: 535-549.) Bateson was particularly anxious to develop a theory that would account for the phenomena of coupling and repulsion, as well as the regularity he thought he observed in the frequency

very time, 3,000 miles away, new examples were turning up at a rapid rate. Coupling was none other than the "linkage" of characters that Morgan was observing with increasing frequency in his stocks of *Drosophila.*

To explain his observations on linkage and recombination, Morgan developed a novel suggestion. In 1909 or 1910 he had read the paper of the Belgian cytologist F. A. Janssens, "La théorie de la chiasmatypie."[6] Janssens studied meiosis in amphibians and

---

with which partial coupling or partial repulsion occurred. Since the meiotic process involved always a simple dichotomy (i.e., two homologous chromosomes were always separated from each other and always come to reside ultimately in different gametes), he felt it was impossible to account for any exceptions to either complete repulsion or complete coupling. As he wrote: "No simple system of dichotomies could bring about these numbers." (Bateson and Punnett, "On Gametic Series," p. 298.) Bateson assumed that during early embryonic development of the germ tissue (those cells that later would give rise to sperm, eggs, or pollen cells), certain cells with certain genotypic constitutions were produced in greater quantities than others. This was a theory of differential mitotic divisions in the presumptive germ tissue. Bateson assumed that this process occurred in a regular geometric way so that for any given pair of characteristics (such as flower color and pollen shape) that displayed coupling and repulsion, there could be only two sets of distributions of cell types in the germ tissue. Which pattern developed in any particular individual organism was a matter of chance, but the two would occur in roughly equal proportions. As a result of differential cell proliferation, certain genotypes would exist in the germ tissue of the adult plants in much greater quantity than in other cells. The specific numbers of each cell type would be predetermined by which ever geometric pattern of cell division took place (ibid., p. 300). Bateson and Punnett then went on to reject the original ideas of coupling and repulsion. They claimed that there was no actual association of factors in cells at all. It was wrong, in fact, to talk about coupled factors as though they were in some physical sense linked. The results of their earlier experiments could be explained, Bateson and Punnett pointed out, by the proliferation of certain cells in excess in the germ plasm. Such a process would alter the expected ratios in proportion to the rates at which certain cells proliferated relative to others. Bateson and Punnett had no cytological or embryological evidence to support the reduplication hypothesis. There was no known case of differential cell proliferation based on a sorting out of genetic material. Although the reduplication hypothesis explained Bateson and Punnett's ratios in sweet peas, it fared less well when applied to breeding data from other organisms.

[6] F. A. Janssens, "La théorie de la chiasmatypie," *La Cellule* 25 (1909): 389-411.

made careful chromosomal preparations of the germ cells. In observing these preparations, he described what appeared to be the physical twisting or intertwining that occurred between paired homologous chromosomes during the early stages of meiosis. This intertwining process was referred to as chiasmatypie or chiasmatype. Janssens interpreted these cross figures and intertwinings to mean that the members of a homologous pair of chromosomes broke and rejoined during chiasmatype, leading to an exchange of equal and corresponding regions between the two members of the pair. This process of intertwining with breakage came to be called "crossing-over." Morgan seized upon Janssens' idea as an alternative to the complex mechanism Bateson had offered in his theory of reduplication:

> In place of attraction, repulsions, and orders of precedence and the elaborate systems of coupling, I venture to suggest a comparatively simple explanation based on the results of the inheritance of color, body color, wing mutations, and the sex factor for femaleness in *Drosophila.* If the materials that represent these factors are contained in the chromosomes, and if these factors that "couple" be near together in a linear series, then when the parental pairs (in heterozygotes) conjugate, like regions will stand opposed. There is good evidence to support the view that during the strepsinema stage [an early stage of meiosis] homologous chromosomes twist around each other, but when the chromosomes separate (split), the split is in a single plane, as maintained by Janssens. In consequence the original materials will, for short distances, be more likely to fall on the same side of the split, while remoter regions will be as likely to fall on the same side as the last as on the opposite side. In consequence we find coupling in certain characters, and little or no evidence at all of coupling in other characters; the difference depending on the linear distance apart of the chromosomal materials that represent the factors. . . . The results are a simple mechanical result of the location of the materials on the chromosome, and of the method of union of homologous chromosomes, and the proportions that result are not so much the expression of a numerical system but of the relative location of the factors on the chromosome. *Instead of random segregation in Mendel's sense, we find "associations of factors"*

that are located near together in the chromosome. Cytology furnishes the mechanism that the evidence demands.[7]

The cytological mechanism did not in any way contradict Mendelian principles but simply offered a mechanism for accounting for what otherwise appeared to be anomalies of the Mendelian system.[8]

Morgan developed a concrete model to show how the effects of crossing-over could produce the partial coupling results that Bateson and others had observed. The diagram shown in figure 6 was taken from a paper Morgan published in 1915, but represents the basic mechanism for crossing-over which he developed as early as 1911.[9] In this diagram Morgan represented the chromosomes

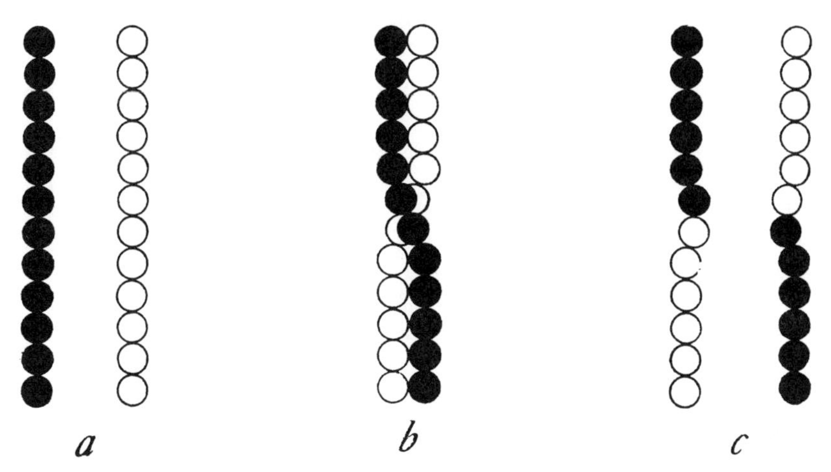

Figure 6. Morgan's simple diagram to show the mechanism of crossing-over between homologous chromosomes, published in 1915. Morgan envisaged the chromosome as a linear array of Mendelian factors (the circles), much like beads on a string. Black and white circles represent chromosomal factors of maternal and paternal origin. (From T. H. Morgan, "Localization of the Hereditary Material in the Germ Cells," *Proceedings of the National Academy of Sciences* 1 [1915]: 420-429; figure 7.)

[7] T. H. Morgan, "Random Segregation versus Coupling in Mendelian Inheritance," p. 384.

[8] T. H. Morgan and Eleth Cattell, "Data for Study of Sex Linked Inheritance in *Drosophila*," *Journal of Experimental Zoology* 13 (1912): 79-101, especially p. 79.

[9] T. H. Morgan, "Localization of the Hereditary Material in the Germ Cells," *Proceedings of the National Academy of Sciences* 1 (1915): 420-

in a schematic and very oversimplified way as a linear arrangement of Mendelian factors (the circles) arranged like beads on a string. The black strand and the white strand represent the two members of the homologous pair of chromosomes; one of these strands came from the individual's male parent and the other from its female parent. The two members of the homologous pairs of chromosomes undergo chiasmatype at the beginning of the meiotic divisions in the germ cells that give rise to sperm or egg. In some of these intertwinings, breaks will occur, with the consequent exchange of complementary parts between the two chromosomes (as shown in diagrams B and C). Thus, factors that were once linked together on the two chromosomes are now separated; since the two homologs separate during the later meiotic stages and pass into different gametes, the coupling is broken. It should be noted that neither Janssens nor Morgan thought that crossing-over and breakage between members of a homologous pair occurred during every meiotic division. Furthermore, it seemed quite plausible in 1911 that crossing-over could occur at any point along the length of the two intertwined homologs. Thus the frequency with which linkage between any two Mendelian factors was disturbed would be due both to the distance between the factors and the chance occurrence of a break between them.

Morgan's explanation of coupling and recombination was considerably less complex than Bateson's reduplication hypothesis. This gave it a considerable advantage in Morgan's own mind. More important, however, was the fact that it was based on what appeared to be an observed cytological process. Whereas Bateson had to invent a peculiar form of cell division to "explain" his numerical ratios, Morgan could provide a mechanism based on an observed occurrence. By 1914, A. H. Sturtevant, Morgan's student and colleague, could state clearly the distinction between Morgan's chromosome and Bateson's reduplication hypotheses as means of explaining the same phenomena:

> Thus we are forced to assume an enormously complex series of cell divisions, many of them differential, proceeding with mathematical regularity and precision, but in a manner for which direct observation furnishes no basis. It seems to me

429. A similar diagram is published in *The Mechanism of Mendelian Heredity* by Morgan, Sturtevant, Muller, and Bridges (New York: Henry Holt & Co., 1915), p. 60.

> that it is not desirable to assume such a complex series of events unless we have extremely strong reasons for doing so. I can see no sound reason for adopting the reduplication hypothesis. It apparently rests on two discredited hypotheses: somatic segregation, and the occurrence of members of the 3:1, 7:1, 15:1, etc. series of gametic ratios in more cases than would be expected from a chance distribution.
>
> The chief advantage of the chromosome hypothesis of linkage which has been proposed by Morgan . . . seems to be its simplicity. . . . It explains everything that any of the forms of the reduplication hypothesis does and in addition offers a simple mechanical explanation.[10]

It would be interesting to know what factors led Morgan to read Janssens's paper of 1909 in the first place. Morgan himself gives no indication of how he happened upon Janssens's unusual and provocative work. Some circumstantial evidence suggests an answer, however. As pointed out earlier, Morgan was closely associated with E. B. Wilson and his work on the chromosomal theory of sex determination. And, as Morgan's letters to Driesch showed, Morgan was subjected to strong doses of chromosome theory in the period between 1905 and 1910. Morgan probably talked over his findings with Wilson, who must have given a chromosomal interpretation of the sex-linkage data. Furthermore, Wilson was probably more versed than any other worker at the time in the literature of cytology. Even if Morgan had not known of Janssens's paper, it is likely that Wilson would have. But regardless of how Morgan found the Janssens reference, it is obvious that he very soon saw its application to his own observations on linkage.

Between 1911 and 1914 Morgan and his students Sturtevant and Muller identified about two dozen mutants, all of which fell into three "linkage groups," that is, groups of factors that appeared, in most instances, to be inherited together. It was tempting to think of linkage groups as being associated with particular chromosomes in *Drosophila*. However, the work of Wilson and Stevens in 1905 had shown that *Drosophila* had *four* chromosome pairs, and there were only three linkage groups. By 1914, however, pieces of the puzzle began to fall together. At that time Mul-

[10] Sturtevant, "The Reduplication Hypothesis as Applied to *Drosophila*," p. 548.

ler discovered the mutant "bent wings," which appeared to form the beginning of a fourth linkage group. Soon, several other mutants of the fourth group appeared. The breeding data fell perfectly into line with the cytological observations. There were as many linkage groups, determined by breeding data, as there were chromosome pairs determined by cytological observations. As a result of these two lines of work (the chiasmatype theory and the working out of linkage groups) by 1914 it became even more clear to Morgan that the chromosome interpretation of the Mendelian laws was a theory of enormous power. The two theories were not incompatible, but in fact seemed to be two sides of the same coin.

From this time on Morgan saw clearly that a combined Mendelian and cytological theory could open up expanded horizons. To explore these adequately, Morgan realized that a much greater effort was required. Beginning in the fall of 1910 he began to take into his laboratory an exceptional group of students, some graduate and some undergraduate. Without them and their persistence over the years, the *Drosophila* work could never have proceeded as far or as fast as it did.

It has been said that the *Drosophila* group represented a latter-day version of the Morgan's Raiders, fighting on a battlefield of ideas with the same spirit and cohesiveness as the old Morgan's Raiders in the Civil War. Certainly, there have been few research groups in modern biology that have functioned as effectively together as did Morgan's group in the "fly room" between 1910 and 1915. It was by virtue of working as a team that Morgan was able to develop his initial ideas of sex linkage and chiasmatype into the far-reaching chromosome theory of heredity.

It is interesting to look at the way Morgan formed his group. In the fall of 1909 he taught the introductory biology course at Columbia for the one and only time in his career. Sitting in that class were two students, one a sophomore, Alfred Henry Sturtevant (1891-1970), and the other a freshman, Calvin Blackman Bridges (1889-1938). From the outset both students were intrigued by and attracted to Morgan's style.[11] It was not Morgan's ability as a lecturer in the usual sense that fascinated these young men, for Morgan was not a dramatic speaker, nor were his lectures as well planned or logically developed as, for example, those of

[11] A. H. Sturtevant, *A History of Genetics* (New York: Harper & Row, 1965), p. 46.

E. B. Wilson. But to the student who possessed natural curiosity, Morgan conveyed a sense of discovery about new biological problems; he attracted those who would follow new leads into research. It was all in Morgan's personality, for otherwise the introductory biology course in 1909-1910 was rather traditionally organized (it consisted of a phylogenetic survey of animals). Ironically, Morgan did not discuss heredity in any organized way in the course; it was only later that Sturtevant and Bridges realized that Morgan himself was actively working in this field.

In the fall of 1910 both Bridges and Sturtevant asked Morgan if they could work in his laboratory. Sturtevant, the more advanced undergraduate (then a junior) was accepted as a research assistant, while Bridges was given a job washing bottles. Two years later a third student, H. J. Muller (1890-1967), was formally incorporated into the group, although he had already been closely associated with much of the work during his junior and senior years (1908-1910). Bridges proved so expert in observing mutants in the cultures he was cleaning up that he was soon given an official position as a part-time assistant. All three of these men received their B.A. degrees from Columbia (Muller, 1910; Sturtevant and Bridges, 1912), and they all stayed on to earn their Ph.D. degrees under Morgan's direction (Sturtevant, 1914; Muller and Bridges, 1916). In addition, many other investigators passed through the *Drosophila* lab between 1910 and 1920. Morgan's graduate students included Charles Zeleny, Fernandes Payne, Charles Metz, H. H. Plough, Franz Schrader, Donald Lancefield, Alexander Weinstein, and L. C. Dunn. Among those who worked for a year or more in the laboratory as research associates or postdoctoral fellows, were Otto Mohr from Norway and Theodosius Dobzhansky from the Soviet Union. Although all these men made important contributions to the *Drosophila* work, it was Bridges, Muller, and Sturtevant who participated with Morgan in the most dramatically productive period between 1910 and 1915.

Before continuing with a discussion of the work of the *Drosophila* group, it will be helpful to describe briefly the physical arrangement of Morgan's laboratory. The structure of this laboratory contributed significantly to the spirit and feeling that became part of the group's trademark. Affectionately called the fly room after 1910, Morgan's laboratory was located on the top floor of Columbia's Schermerhorn Hall (Biology). It was a small room, 16 x 23 feet, into which were crammed about eight desks. Dobzhansky, who came to work as a postdoctoral fellow with the Morgan group in 1927, described his amazement (and slight aversion) to finding the desk drawers in the fly room often filled with cockroaches, subsisting on spilled *Drosophila* food. Off to the left side of the laboratory was a small office occupied by Morgan and his longtime general assistant, Edith Wallace (1881-1964).[12] It was in this room that Morgan kept his large roll-top desk, and where Wallace worked painstakingly over the many fine and beautiful illustrations she prepared for the group's publications.

Physically crowded as the lab was, Morgan was careful in selecting those who had regular space in the fly room. This was partly a practical necessity. But it was also part of Morgan's approach to research. He was very serious about the progress of his work and had little or no time for those who did not share the same seriousness of purpose. Hard work and complete devotion to biology were virtual prerequisites to acceptance into the group. But also, Morgan was wise enough to recognize his own limitations. He wanted to include among those working with him individuals whose abilities complemented his own. He chose wisely when he selected Bridges, Sturtevant, and Muller. Morgan's creative imagination ranged over so many new ideas that he was not always interested in, or able to work out, the fine details.

A particularly cogent episode illustrates the working relationship that characterized the *Drosophila* group. In 1911 Sturtevant was sitting in Morgan's office, while the latter discussed the significance of Janssens's chiasmatype theory for the chromosome theory. Morgan suggested that the greater the distance apart of two genes on a chromosome, the greater the possibility of crossover between them.[13] Sturtevant recalled that he suddenly realized that if this

[12] A native of Boston, and of Nashua, New Hampshire, Edith Wallace was graduated from Mt. Holyoke College in 1903 and received an M.A. degree from Clark University. She taught biology at Western College for Women (Oxford, Ohio) and at the University of Maine (Portland) before joining Morgan's staff at Columbia in 1908. She retired in 1944. Most of the drawings of *Drosophila* published by the Morgan group were made by Wallace.

[13] In 1911 Morgan had written that the amount of crossing-over between any two genes depended "on the linear distance apart of the chromosomal materials that represented the factors" (Morgan, "Random Segregation versus Coupling in Mendelian Inheritance," p. 384). Sturtevant gives the date as 1911 (*A History of Genetics*, p. 47), but E. A. Carlson gives it as November 1912 (*The Gene*, p. 52). Since Sturtevant claims it was during his senior year, that would place the episode in 1911.

relationship were true, it would provide a basis for constructing maps of the relative distance apart of genes on their respective chromosomes. Sturtevant went home and "to the neglect of my undergraduate homework,"[14] made the first genetic map—for three genes on the X-chromosome. In essence, then, Morgan had suggested a novel idea; Sturtevant saw its implication for the pursuit of the *Drosophila* work and worked out the details that very night. This story should not be taken to indicate that Morgan was only an "idea man," tossing out new thoughts for others to work out. Morgan did his share of the tedious work, too, counting endless numbers of progeny from *Drosophila* matings. But he was less likely to work out all the exact details of an idea at once the way Sturtevant, for example, was. In turn, Bridges was given great credit by Sturtevant for his long, painstaking work keeping the *Drosophila* cultures genetically pure, a basic requirement for all the work.

In still another way, Morgan's original group complemented their teacher's abilities. Morgan had never been outstandingly good at mathematics; he had an exact mind, though not primarily a quantitative one. Sturtevant, Muller, and Bridges all had more ease with quantitative arguments than Morgan. They were all able to handle the complex reasoning that went into designing rigorous matings. Morgan would suggest what questions needed to be answered, but it was Sturtevant or Muller, most often, who would design the actual experiments (i.e., decide what males to cross with what females). Morgan was generally on the right track and often ahead of the others in sensing what needed to be known, but he often left it to the others to determine exactly how it could most effectively be found out. Furthermore, Morgan was described as frequently being the center of a whirlwind of activity: writing, talking, counting flies—always doing something and quite frequently several things more or less simultaneously (although he was able to concentrate when necessary). Sturtevant, Muller, and Bridges talked a great deal among themselves, but they concentrated largely on a single focus at a time. Thus, to a very considerable extent it was the variety and diversity of people involved that made the nuclear *Drosophila* group so effective.

Each of the three undergraduates Morgan admitted to the fly room had a unique personality, which contributed in numerous ways to the progress of the work. Sturtevant was raised in Alabama

[14] Sturtevant, *A History of Genetics*, p. 47.

and had developed an interest in heredity because of his father's hobby of raising pedigreed racehorses. After reading Punnett's book on Mendelism,[15] he knew that the study of heredity was what he most wanted to pursue in college. He fell naturally into a working relationship with Morgan after their first meeting in 1910. Sturtevant was slow and patient, quantitative and highly incisive. He seldom spoke; when he did his words were often obscured by the ever-present pipe in his mouth. Like Morgan, he was easy going on the outside, though active and persistent at his work. He shared with Morgan wide-ranging biological interests, including the topics of evolution, cytology, embryology, taxonomy, and many aspects of heredity. Unlike Morgan, he was less interested in publicizing and popularizing the *Drosophila* (or any other) work. His writing style was very factual, given to few expressions of emotion, and sometimes a little dry. Although he worked on many aspects of *Drosophila*, his major contributions over the years centered on mapping genes and developing a picture of the architecture of the germ plasm through analysis of breeding results. Sturtevant continually contributed to the design of new crosses to determine more accurately the relative positions of particular genes.

Calvin Bridges was an extraordinary investigator.[16] His professional work, winning personality, and unconventional personal life were a source of both inspiration and consternation to those who worked with him. Born and raised in upstate New York, largely by grandparents (his parents died early), he moved from place to place to such an extent that he did not graduate from high school until he was twenty. Yet his record was outstanding enough that he was awarded scholarships at both Cornell and Columbia in 1909. After receiving his Ph.D. with Morgan, Bridges remained at Columbia as a research associate of the Carnegie Institution of Washington, a position he continued to hold even after he moved with the group to California in 1928.

[15] R. C. Punnett, *Mendelism* (New York: Macmillan Co., 1905). The book was suggested to him in high school by his older brother.

[16] For a brief biographical account, see A. H. Sturtevant, "Calvin Blackman Bridges," *Dictionary of Scientific Biography* (New York: Charles Scribner's Sons, 1970), vol. 2, pp. 455-457; a more complete account is T. H. Morgan, "Calvin Blackman Bridges, 1889-1938," *Biographical Memoirs, National Academy of Sciences* 22 (1941): 31-49. An anonymous manuscript, "The Boyhood and Family Background of Calvin Blackman Bridges," is located in the Archives of the American Philosophical Society.

Bridges has been variously described as a nonconformist, a supporter of the political left, and an advocate of free love. Unlike Morgan and Sturtevant, he took active interest in the political movements of his day; for example, he supported the Bolshevik Revolution in 1917, and the insurgents in the Spanish Civil War in 1936. His ties with the Soviet Union were close enough that he was invited there for the year 1931-1932 by the Soviet Academy of Sciences.[17] Married with four children, Bridges always retained a flair for life and a childlike simplicity that made matters of conventional social mores meaningless to him. He was always making new friends, was frequently the center of social activities (in New York and Woods Hole), and appeared to have boundless energy (the photograph shown in figure 7 is characteristic). In the *Drosophila* group his main attention was given to cytological work: the study of chromosomes in cells from various offspring of *Drosophila* matings. It was largely Bridges's work that helped establish the close correlation between breeding data and chromosomal structure. With a mind less far ranging and less interested in all the broad problems of biology than Morgan's, Bridges had enormous patience with details and a manipulative and technical skill that surpassed that of all other members of the group. His mind was highly precise, and his personality open to such a degree that he would spend countless hours helping new students in the lab learn the rigorous cytological techniques involved in squashing and staining chromosome preparations.

Hermann Joseph Muller was born in New York City, the son of a first-generation metal worker.[18] Performing well in high school (he was valedictorian of his class), Muller entered Columbia on a scholarship, quickly becoming interested in biology. As an undergraduate, Muller took E. B. Wilson's advanced biology course (1908-1909), which used as a text R. H. Lock's *Variation, Heredity*

[17] See Sturtevant, "Calvin Blackman Bridges," p. 457; also, Bridges to Muller, March 1937; Muller Papers, Lilly Library, Indiana University, Bloomington. In this letter Bridges spells out some of his political feelings on the Spanish situation.

[18] For biographical detail, see E. A. Carlson, "Hermann Joseph Muller," *Yearbook, American Philosophical Society* (1967): 137-142; "H. J. Muller, a Memorial Tribute," *The Review* (Indiana University Alumni Publication) 2 (1968): 1-48; and "H. J. Muller," *Genetics* 70 (1972): 1-30; Tove Mohr has written a short, insightful account, "Hermann J. Muller, 1890-1967," *Journal of Heredity* 63 (1972): 132-134.

Figure 7. Calvin Bridges beside his rowboat at Woods Hole, Massachusetts, about 1920. Bridges was fond of boating and referred to his vessel as "my old canoe." Note the emblem on the hull, a fanciful modification of the chromosome grouping in *Drosophila*. (Courtesy of Tove Mohr.)

and Evolution,[19] a far-sighted book that treated Darwinian natural selection in terms of Mendelian heredity. It was this book, combined with Wilson's inquisitive mind, that made Muller intrigued with biology in general and heredity and evolution in particular. Muller helped to organize a biology club at Columbia in his senior year (1910), to which only undergraduates were invited. It was

[19] New York: E. P. Dutton, 1907.

here that he met Bridges and Sturtevant, who told him about the wonders they were just discovering and about Morgan and the fly room. Because he could not get a space in Morgan's lab immediately, Muller entered graduate school in physiology, first at Columbia (M.A.), then at Cornell Medical School. Physiology offered fellowships and teaching assistantships, which Muller needed to finance his further studies. In 1912 he entered the fly room as a full-time student, completing a Ph.D. in 1916. Like Morgan, Muller had broad-ranging interests in all aspects of biology, and his published writings consist of numerous articles on subjects from the hazards of radiation, to eugenics, science fiction, and evolution.[20] He was an intense person, taking himself and everything he did very seriously; he lacked the easy-going sense of humor that characterized Morgan, Sturtevant, and Bridges. Like Bridges, he was politically far to the left, being a strong supporter of the Bolsheviks, and spending three years (1933-1936) in the Soviet Union doing genetics research. He was the most politically active of the fly-room group. His work helping to distribute a socialist newspaper was partly responsible for his leaving the University of Texas in 1933.

Muller worked closely with the group but was never as close to its center as either Sturtevant or Bridges. Unlike them, Muller struck out on his own immediately after getting his Ph.D., his only subsequent period as a regular member of the fly-room group was from 1918 to 1920.[21] His research was also more independent, for soon after leaving Columbia in 1916 he became interested in the effect of radiation on gene mutation, a topic he pursued for many years; his demonstration that the mutation rate is directly proportional to X-ray dosage won the 1947 Nobel Prize in medicine and physiology. Muller's main contribution to the *Drosophila* work in the years between 1910 and 1916 was his analysis of

[20] For a discussion of Muller's interest in political and social aspects of biology, see his book, *Out of the Night: A Biologist's View of the Future* (New York: Vanguard Press, 1935); and G. E. Allen, "Science and Society in the Eugenical Thought of H. J. Muller," *BioScience* 20 (1970): 346-352.

[21] Muller's first position was at Rice Institute, where Julian Huxley was heading the new biology department in 1916. Muller did return to Columbia as an instructor (1918-1920), at which time he worked again with the *Drosophila* group, but still more loosely associated than the others. He continued his career at the University of Texas (1920-1933), the Soviet Union (1933-1936), Edinburgh (1937-1940), Amherst (1940-1945), and finally, Indiana University (1945-1967).

crossing-over, interference (a process by which crossing-over is inhibited), and the development of markers for following chromosomal transmission from parent to offspring. Muller's mind was highly inventive, and he had a good head for mathematics and quantitative work. It was often said that if someone in the group had a critical mapping problem, which required some special marker, Muller could quickly produce a *Drosophila* stock with just the right combination of chromosomes and appropriate mutations.

Between 1910 and 1915 Morgan, with this group of three young and enthusiastic workers, had gathered enough data on inheritance in *Drosophila* to make the first comprehensive statement of the chromosome theory of heredity. Because of the closely-knit nature of the group, all shared authorship in *The Mechanism of Mendelian Heredity*, published in 1915.[22] In the next section it will be necessary to look more closely at the specific contributions the *Drosophila* group made between 1910 and 1915 to see what instigated such a scientific revolution.

MAJOR LINES OF DEVELOPMENT OF THE *DROSOPHILA* WORK, 1910-1915

Between 1910 and 1915 Morgan and his group carried out work with *Drosophila* in three general directions. The first was the construction of chromosome maps. Mapping was an attempt to show the position of the various Mendelian genes, relative to each other, along the chromosome. Second, were various lines of work expanding, modifying, and refining the basic Mendelian rules. Concepts such as multiple alleles, lethal genes, and modifying factors, for example, made it necessary to revise some of the earlier Mendelian postulates such as random assortment or pure dominance and recessiveness. Third was the line of work initiated by the discovery of nondisjunction,[23] first in the X-chromosome and later in various autosomes as well. Nondisjunction led not only to the

[22] T. H. Morgan, A. H. Sturtevant, H. J. Muller, and C. B. Bridges, *The Mechanism of Mendelian Heredity* (New York: Henry Holt & Co., 1915), reprinted by the Johnson Reprint Corporation, New York in 1972, with an introduction by G. E. Allen.

[23] Nondisjunction is a process by which two homologous chromosomes fail to separate in meiosis, thus remaining together in one-half the gametes produced from the original parent germ cell. The importance of nondisjunction will be discussed more fully later in this section.

first direct evidence for the chromosome theory of inheritance, but also in later years to new views on the determination of sex.

Underlying all these approaches was the development of the basic outlines of the Mendelian-chromosome theory. Each line of work contributed something in support of the idea that Mendelian genes were, in fact, arranged in a linear fashion on specific chromosomes. Each line of work also helped to expand the original Mendelian concept in a more comprehensive and sophisticated way. The chromosome theory of heredity as elaborated by the Morgan group was far richer in experimental and conceptual detail than the simplified scheme Mendel had reported in 1865.

Mapping

The basic mapping technique Sturtevant and Morgan developed in 1911 was extremely simple. Two genes were selected within the same linkage group. The technique was to determine, from frequency of recombination of phenotypic traits in offspring of a cross, how relatively far apart the respective genes were on the chromosome.

Consider the case of two sex-linked genes. A special mating is required to carry out the experiment. The female parent must be heterozygous for the two traits, but both dominant genes would have to be on one X-chromosome and both recessives on the other. The male parent's one X-chromosome must be recessive for both genes. (If the mapping is done for the X-chromosome, crossing-over can only occur in the female, since the male's Y-chromosome is not homologous to the X and does not enter into synapsis and crossing-over with it.) If there were no crossing-over in the formation of egg cells by the female parent, then the offspring of such a cross, males and females alike, should show only the dominant traits for both genes. However, should there be crossing-over, then two combination types would be observed in addition to the double dominant: two recombination types (complementary to each other), each containing one dominant and one recessive gene (e.g., Ab, aB). This process is shown schematically in figure 8. The frequency of occurrence of recombination of observable characters in the offspring is thus a direct measure of the frequency of crossing-over in the germ cells of the parents. Furthermore, if the frequency of crossing-over is proportional to the distance between genes, then the frequency of observed recombination types is an

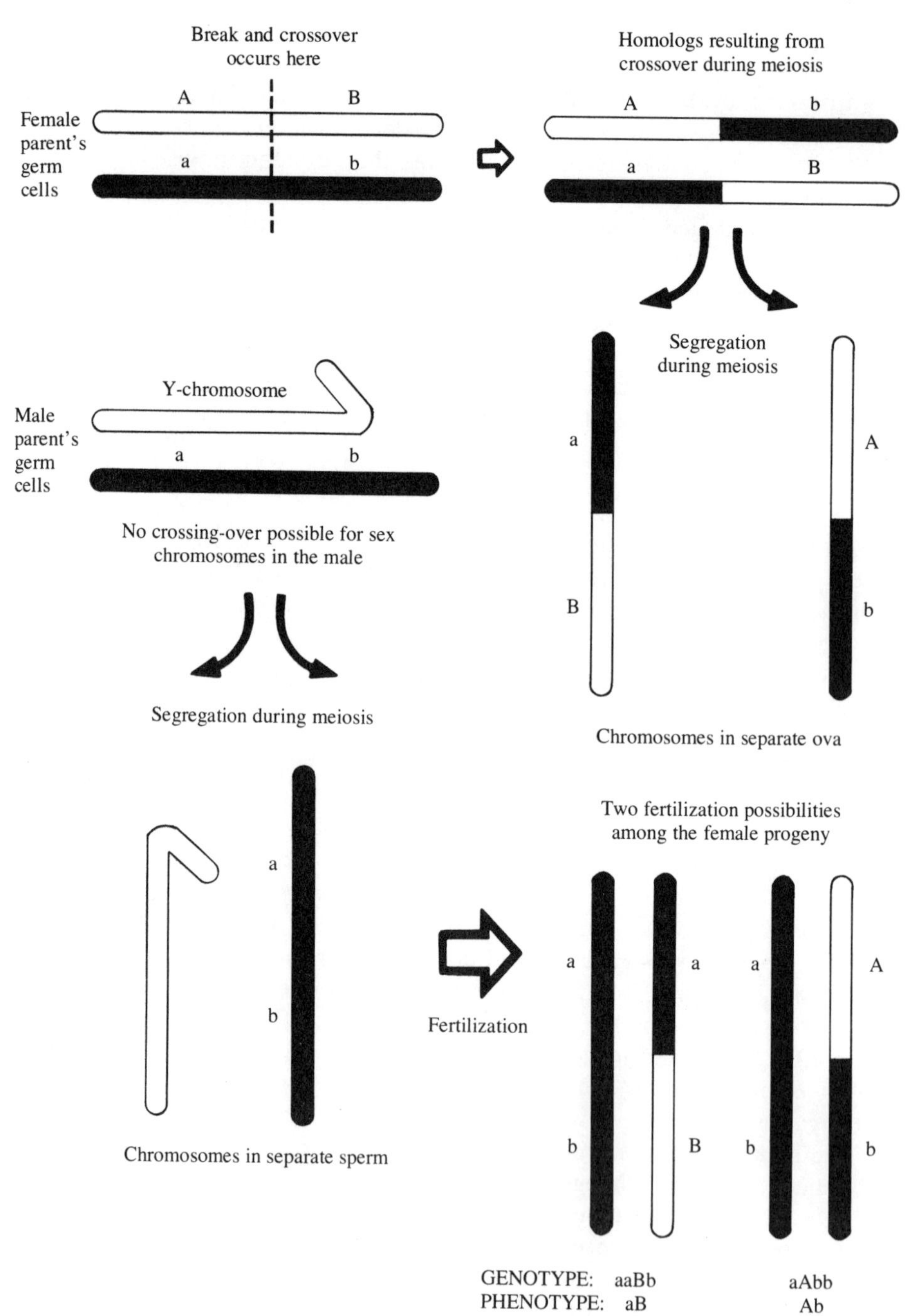

These are the two recombinant classes among female progeny resulting from the crossover. Frequency of recombination, as percentage of total progeny, provides a measure of distance apart of the two loci on the chromosome (map distance). Fertilization possibilities for male progeny are not shown.

Figure 8. Schematic diagram illustrating the technique of chromosome mapping based on the frequency of recombination between two genetic loci on the same chromosome. The frequency of recombinant types (usually a few percentage or less of the total progeny) is used to calculate the distance between the two loci. Shown here is a hypothetical experiment for two genes (A,a and B,b) linked on the X-chromosome. Starting in the upper left, crossing-over is shown in the early stages of meiosis in the germ cells (primary oocytes) of the female parent. The diagram at the top, right, shows the two recombinant types of chromosomes that can result from crossing-over. The further apart the two loci are on the X-chromosome, the greater the chance that crossing-over, and hence recombination, can occur between them. The two recombinant chromosomes are separated from each other during later stages of meiosis, and end up in different ova (egg cells). Each ovum can be fertilized by sperm from a male parent. Shown in the lower right are the two fertilization possibilities among the female progeny (for simplicity's sake the male progeny are not shown; they would display the same recombinant types). Counting the recombinant phenotypes as a percentage of total progeny gives the crossover frequency, which in turn is converted into relative map distance between the two loci on the chromosome.

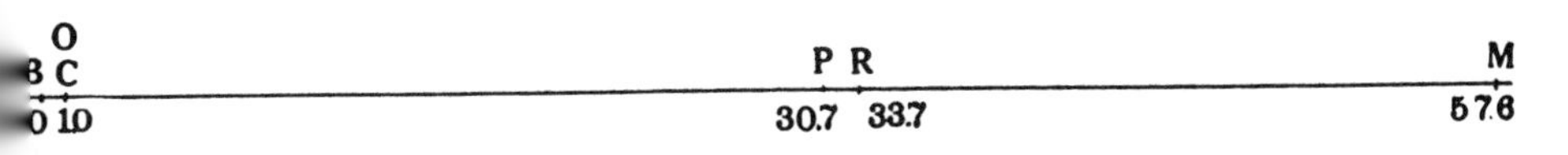

Diagram 1

Figure 9. The first published chromosome map for three sex-linked genes in *Drosophila*. The map distances between any two pairs of genes on this chromosome (the X) are calculated from frequencies of recombination as described in figure 8. The symbols shown here stand for: yellow body (B), white eye (C), wild-type red eye (O), vermilion eye (P), miniature wing (R), and rudimentary wing (M). (Later, different letters were used to designate each of these mutant phenotypes.) For the purpose of constructing a map, one end of the chromosome is designated as the starting, or zero (0.0) point, and all other loci are given as map units calculated from there. (From A. H. Sturtevant, "The Linear Arrangement of Six Sex-Linked Factors in *Drosophila*, as Shown by Their Mode of Association," *Journal of Experimental Zoology* 14 [1913]: figure 1, p. 49.)

indirect measure of the distance between the two genes on the chromosome. For the first genetic map, Sturtevant used five known sex-linked genes: y (yellow body), w (white eye), v (vermilion eye), m (miniature wing), and r (rudimentary wing). He used recombination data for two genes at a time and produced the map shown in figure 9. Despite the refinement in mapping procedures developed since that time, Sturtevant's original map, made when he was a college senior, is astonishingly accurate, being very close to newer maps constructed some twenty-five years later.[24]

Sturtevant and Morgan realized the power and significance of the mapping technique for exploring the architecture of the germ plasm (i.e., the structure of the chromosomes) in relation to Mendelian factors. At first the mapping technique was applied largely to sex-linked factors; thus the first map was of the X-chromosome. As crossing-over and recombination were studied in the other three chromosome groups, the mapping technique was extended to the autosomes as well.

However, Sturtevant recognized from the very beginning some of the uncertainties and assumptions associated with the mapping process. For one thing, the map positions represented only *relative* distances between the genes. There was no way to correlate any *specific* gene position with a physical part of the chromosome. It could not be assumed, for example, that the gene that occurred most frequently (at greatest distance) was actually located at the very tip of the chromosome itself. The chromosome map was thus an abstraction, tied at best only in a relative sense to the physical structure it was supposedly describing. The second objection that Sturtevant noted was that the mapping procedure rested on the assumption that chromosome breaks were equally likely to occur at any point along the length of the chromosome. Thus, if the chromosome had weak spots the distances shown on the chromosome map would not even be accurate relative to each other, let alone have anything to do with the actual chromosome position. As Sturtevant wrote in 1913:

> Of course, there is no knowing whether or not these distances as drawn represent the actual relative spatial distances apart of the factors [on the chromosome]. Thus the distance CP

[24] A. H. Sturtevant, "The Linear Arrangement of Six Sex-Linked Factors in *Drosophila*, as Shown by Their Mode of Association," *Journal of Experimental Zoology* 14 (1913): 43-59.

> may in reality be shorter than the distance BC, but what we do know is that a break is far more likely to come between C and P than between B and C. Hence, either CP is a long space, or else it for some reason is a weak one. The point I wish to make here is that we have no means of knowing if the chromosomes are of uniform strength, and if there are strong or weak places, then that will prevent our diagram from representing actual relative distances—but, I think will not detract from its value as a diagram.[25]

Since crossovers are observed only rarely, a large number of offspring are needed to obtain a reliable estimate of crossover frequency. Thus, it was often difficult to obtain accurate map distances because the small number of recombinations were affected by statistical sampling error. This problem became increasingly apparent as more workers in laboratories throughout the country and the world began constructing chromosome maps of *Drosophila* and other species. Studying the same two genes, for example, workers often obtained different map distances. In some cases the discrepancies appeared to be more significant than simple variation, in a statistical sense, would permit. Hence the foundation of the mapping technique was questioned since it appeared to give different results on different occasions.

Although they recognized these inherent limitations, Sturtevant and Morgan maintained a basic faith in the mapping technique. Initially, they never claimed that the chromosome map was anything more than a *model* for how genes might be arranged along the length of the chromosome. Morgan and Sturtevant were cautious in claiming that chromosome maps represented actual reality. In 1913, for example, Morgan described linkage in tentative language as "some sort of relation which we interpret in terms of a linear theory. We further interpret the theories in terms of chromosomes."[26] Morgan was excited about the implications of the mapping technique. However, perhaps recalling his own skepticism about chromosomes and particulate inheritance of a few years back, he and Sturtevant exercised considerable restraint in what they claimed for their scheme.

Some of the objections Sturtevant originally encountered in

[25] Ibid., p. 49. The symbols are the same as for figure 9.

[26] T. H. Morgan, "Simplicity versus Adequacy in Mendelian Formulae," *American Naturalist* 47 (1913): 372-374.

drawing up his first chromosome map were disposed of satisfactorily by 1914 or 1915, especially through the work of Muller. He suggested that between two genes that are very far apart on a chromosome, more than one crossover could occur simultaneously. Such double crossovers would make it appear as if no crossing-over had occurred at all (see figure 10).[27] The occurrence

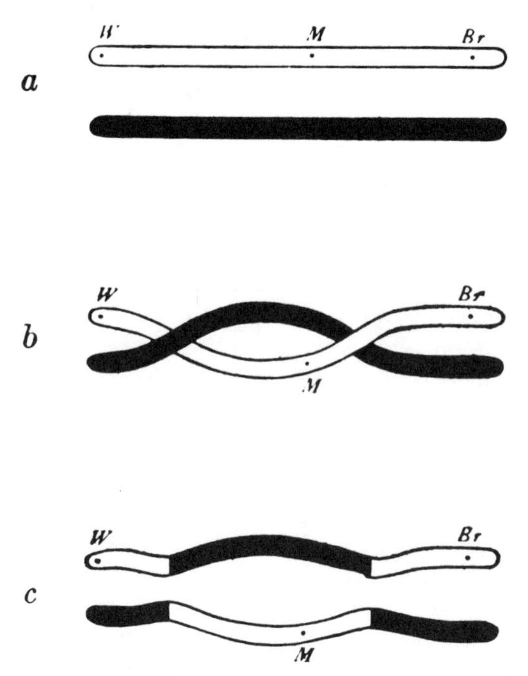

Figure 10. Diagram illustrating double crossing-over, as detected by examining three linked genes simultaneously (what is called a "three-point cross"). Morgan and his group found that the existence of double crossovers produced variations from the expected recombination frequencies in two-point crosses. (From Morgan, Sturtevant, Muller, and Bridges, *The Mechanism of Mendelian Heredity* [New York; Henry Holt & Co., 1915], figure 25, p. 62.)

of double crossovers could be detected by observing the inheritance of three genes at a time. As shown in figure 10, if a double crossover occurred, the two genes on the end would appear as if they had remained linked, while the gene in the middle would be recombined with traits from the homologous chromosome. Muller

[27] This figure is from Morgan et al., *The Mechanism of Mendelian Heredity*, p. 62.

also proposed about this time the idea of interference: that the occurrence of one crossover could sometimes actually interfere with the occurrence of a second crossover in the same area of the same chromosome. Muller worked out the details of both the double crossover and interference interpretations, showing that data from certain specific crosses he designed were consistent with these views. Muller published these ideas as part of his doctoral thesis on the mechanism of crossing-over.[28] The recognition of the occurrence of both double crossovers and interference made possible the more accurate analysis of recombination data, and thus more precise determination of map distances.

The mapping techniques allowed Morgan and his group to elaborate what appeared to be a consistent picture of the architecture of the germ plasm. That picture, beginning as the simple concept of genes arranged in a linear fashion on chromosomes, was necessary before more functionally oriented studies regarding concepts of the gene could arise. Without any knowledge of the nature of the genes, their location within the cell, or their mode of transmission, no theory could be built about how genes functioned or interacted.

Modification and Extension of the Basic Mendelian Rules

One of the consequences of the extensive breeding operations Morgan and his group were able to carry out with *Drosophila* was their realization that some of the tenets of pure Mendelism were too rigid and simple to apply to real-life breeding situations. Some of the basic assumptions of Mendelism that came under increasing scrutiny from the *Drosophila* group were the ideas that all hereditary characters are represented by only two contrasting states (factors) in the germ cells, such as tall or short in peas, red eye or white eye in fruit flies; that the factors for all characters segregate randomly; that every pair of factors functions independently of other pairs and that individual factors do not influence each other.[29] Because he had been unwilling to accept some of these

[28] H. J. Muller, "The Mechanism of Crossing-Over" *American Naturalist* 50 (1916): 193-221, 284-305, 350-366, 421-434.

[29] This last assumption meant two things to early Mendelians: (1) Every adult character is determined by one pair of alleles and only one pair; two or more pairs of alleles do not interact in forming the adult character. (2)

very assumptions prior to 1910, Morgan was prepared to deal seriously with each exception he encountered to the expected Mendelian results. Armed with the evidence from *Drosophila*, Morgan became prepared to accept the broad outlines of the Mendelian theory while admitting the necessity of extending and modifying some of Mendel's oversimplified conclusions.

What were some of these troublesome exceptions, and how did Morgan and his group bring them together within the Mendelian framework? Much work between 1900 and 1905 had shown that ideas such as complete dominance and the total independence of allelic pairs could not account for certain breeding results. As early as 1900, Carl Correns pointed out that many cases of inheritance showed "incomplete dominance"; that is, the offspring appeared intermediate between the two parents (for example, the crossing of a red and a white morning glory always produces pink offspring).[30] Bateson and Punnett had also found a similar situation for the inheritance of certain flower colors in sweet peas. By the time Morgan and his group began their work, a considerable body of evidence existed to suggest that genes must interact and influence each other's expression. Complete dominance and complete recessiveness were probably more the exception than the rule, yet breeding results might still be consistent with the Mendelian factor hypothesis. The various interacting pairs of alleles behaved according to Mendelian predictions, and thus could be shown to have an independent existence—to segregate like separate pairs of alleles. Still, they *also* interacted in a way that Mendel had not foreseen.

When Morgan and his group began extensive breeding of *Drosophila*, they encountered another problem that had not been ex-

---

Two different pairs of alleles, or dominant and recessive members of any one pair of alleles, do not alter, or "contaminate" each other by residing together in the same cell for one or many generations. Thus, the homozygous recessive individuals recovered in the second or later generations of a hybrid cross are as pure in phenotype as the original homozygous recessive ancestors from which they inherited their recessive genes.

[30] Carl Correns, "G. Mendel's Regel über das Verhalten der Nachkommenschaft der Rassenbastarde," *Bericht der Deutschen Botanischen Gesellschaft* 18 (1900): 158-168; translated as "G. Mendel's Law Concerning the Behavior of Progeny of Varietal Hybrids," by L. K. Piternick, "The Birth of Genetics," *Genetics* 35, supplement (1950): 33-41. That Mendel may have recognized incomplete dominance in his work on Mirabilis is indicated in his letters to Nageli (see Sturtevant, *A History of Genetics*, p. 31).

plicitly recognized in the preceding decade. The mutants Morgan first encountered were variations in one characteristic—eye color. Thus, in addition to red (normal, wild type) and white (mutant) eye color there was also vermilion, pink, eosin, and a variety of others. Sturtevant's early mapping studies showed that several of these appeared to occur at exactly the same location along the genetic map. In 1913, Sturtevant proposed the concept of "multiple alleles."[31] If red could mutate to white, then why not to eosin and other mutant variations? It appeared to Sturtevant that there could be a number of variants of a gene located at a single position (or locus, as it was called) on the chromosome. There could be many alleles for the same character, each representing a different form of mutation of an original wild-type gene. With the concept of multiple alleles, Sturtevant and Morgan rejected Bateson's presence-and-absence hypothesis (even though Morgan had invoked it to explain some of his initial findings in *Drosophila* between 1910 and 1913). It appeared that mutation, or variation, was considerably more complex than simply the presence of a factor (dominant) or its absence (recessive). Such a theory might account for the difference between wild-type and white eyes; it could hardly account for the whole range of eye-color mutations, which appeared to be alterations in a single wild-type gene. Mutations were real changes in the nature of genes, not simply the loss of function. Evidence for this view came from Morgan's encounter with the phenomenon known as "reversion." One mutant stock Morgan was studying in 1913 was called "wingless" and had originally resulted from a mutation of the normal wing(s). One day Morgan observed in this culture a winged fly; having convinced himself that the fly was a real mutant and not a stray one that had somehow crept into the bottle, he described the occurrence as a reversion of the mutant to the original, wild-type form.[32] These reversions or "back mutations" as they were later called, could hardly result if the original mutation had involved the loss of the gene for wings. Moreover, there was a logical difficulty to the absence view of mutations that Bateson never seemed to have noticed. If all mutations and variations from the wild-type condition result from the loss of a gene, then the original stock of

[31] A. H. Sturtevant, "The Himalayan Rabbit Case, With Some Consideration for Multiple Allelomorphs," *American Naturalist* 47 (1913): 234-248.

[32] T. H. Morgan, "Factors and Unit Characters in Mendelian Heredity," *American Naturalist* 47 (1913): 5-16, especially p. 11.

any species must have possessed all the genes ever possible in the species. Evolution was thus a kind of unfolding of continual loss. The presence-and-absence theory failed completely to account for where the original genes came from and how new genes might be introduced into a population. Thus, the consequences of the presence-and-absence hypothesis were in many ways absurd.[33] Bateson retained his belief in the fundamental correctness of the presence-and-absence hypothesis until 1917 or 1918 even though it had been largely rejected by most biologists by that time.[34]

Quantitative inheritance was another phenomenon that brought into question some of the fundamental tenets of Mendel's original laws. Quantitative inheritance had been observed since 1908 or 1909 by a variety of workers, including H. Nilsson-Ehle in Sweden and E. M. East in the United States.[35] Nilsson-Ehle and East had observed a number of traits that occurred in an almost continuous series of gradations, each differing from the other in a quantitative (rather than qualitative) way. One example was the inheritance of color character in wheat, which varied all the way from colorless to dark red with every conceivable intermediate color. Furthermore, it was found that by selection over a number of generations, a dark red strain could be converted into a colorless strain and vice versa. The alteration due to selection was permanent as long as inbreeding was maintained (i.e., as long as the selected strain was not bred back to one of the original ancestor types). To account for these results Nilsson-Ehle and East independently proposed that coloration was determined not by a single pair of alleles, but by a number of pairs which interacted in an additive way. This concept of *multiple factors* (not to be confused with multiple alleles), or as it later came to be called, *polygenic*

[33] Ibid., p. 5.

[34] For a general discussion of the rise and fall of the presence-and-absence hypothesis, see R. G. Swinburne, "The Presence-and-Absence Theory," *Annals of Science* 18 (1962): 131-134; for Bateson's reaction to criticism of the presence-and-absence hypothesis, see L. C. Dunn, *A Short History of Genetics* (New York: McGraw-Hill, 1965), p. 71.

[35] See H. Nilsson-Ehle, "Kreuzungsuntersuchungen an Hafer und Weizen," *Lunds Universitets Årsskrift* n.f. 5, 2 (1909): 1-122; E. M. East, "A Mendelian Interpretation of Variation that is Apparently Continuous," *American Naturalist* 44 (1910): 65-82; and E. M. East and H. K. Hayes, "Inheritance in Maize," *Connecticut Agricultural Station Bulletin* 167 (1911): 1-141.

inheritance, gained support from the fact that the various genes could be separated from each other and analyzed according to Mendelian principles. That is, the many pairs of alleles involved in contributing to any trait could be sorted out by selecting mutations of one or another locus. Each pair of alleles segregated according to Mendelian principles and was subject to linkage with other genes on a specific chromosome. According to Nilsson-Ehle and East, selection could act either to increase or decrease the number of dominant factors for a given trait present in any original starting population.

A case of what appeared to be quantitative inheritance showed up relatively early in Morgan's stocks. Morgan reported that in May of 1910 a peculiar wing mutation, known as "beaded," originated in a stock bottle of flies he had exposed to radium.[36] In September of 1912, J. S. Dexter, who was working with the Morgan group, began a series of selection experiments on relatively pure beaded-wing cultures. As Nilsson-Ehl and East had shown with plants, Dexter showed for beaded wings that almost any variety of intermediate conditions between purely normal and heavily beaded could be obtained by the appropriate selection methods (see the range of intermediate variants in figure 11). Dexter interpreted these results in terms of the multiple-factor hypothesis, especially as it had been recently proposed for Mendelian characters by E. C. MacDowell.[37] Dexter analyzed the genic components of the beaded condition and showed that it was due to what appeared to be at least several genes where linkage relations and positions could be accurately determined.[38] What Dexter pointed out was that there appeared to be a major gene for beaded wing which was acted upon in quantitative—i.e., additive—ways by a series of other genes. The other genes, called by Sturtevant *modifiers,* influenced expression of the major gene. Other cases of inheritance that appeared to be explained by the quantitative or modifier-gene hypothesis, including eosin eye and truncate wing, were also investigated by various members of the

[36] T. H. Morgan, "Origin of Nine Wing Mutations in *Drosophila,*" *Science* 33 (1911): 496-499.

[37] E. C. MacDowell, "Multiple Factors in Mendelian Inheritance," *Journal of Experimental Zoology* 16 (1914): 177-194.

[38] J. S. Dexter, "The Analysis of a Case of Continuous Variation in *Drosophila* by a Study of its Linkage Relations," *American Naturalist* 48 (1914): 712-758.

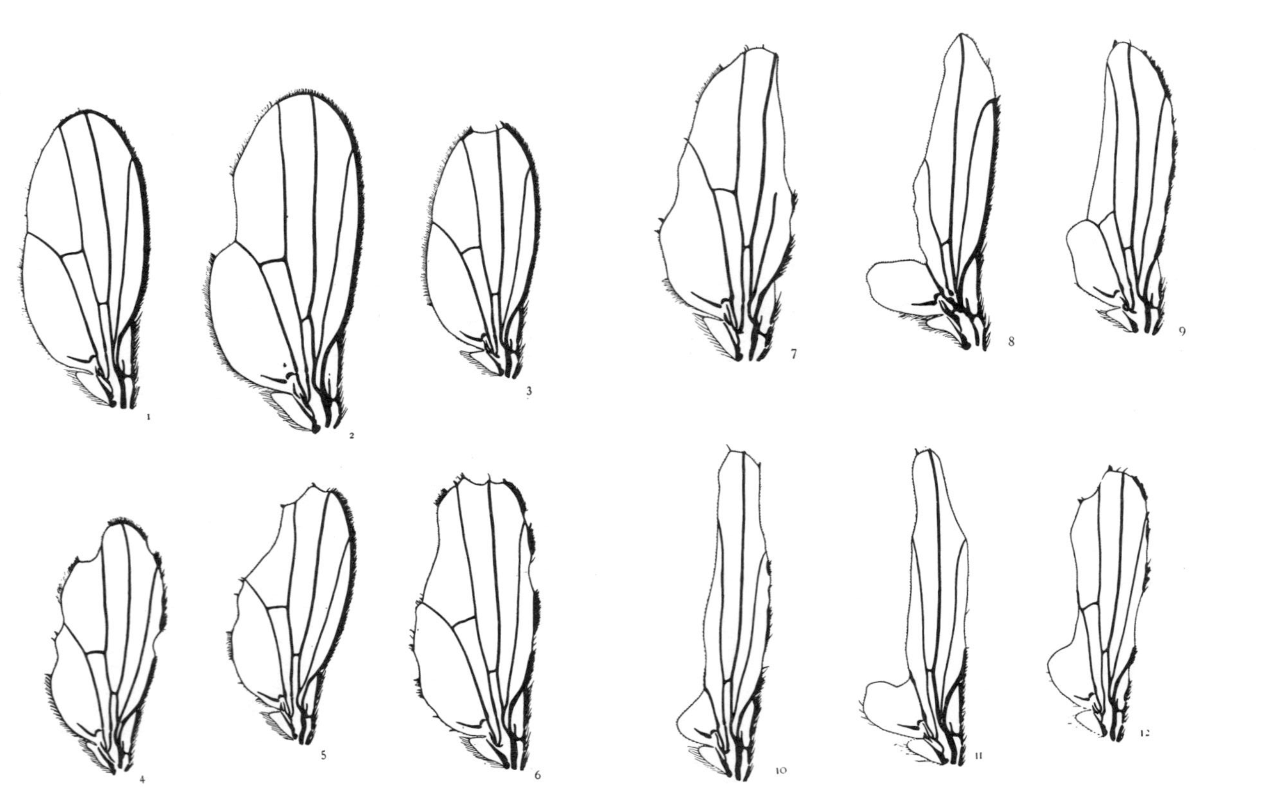

Figure 11. Variants in the beaded-wing phenotype in *Drosophila.* The spectrum of variants runs from almost normal (1) to a strongly narrow wing shape (12). In 1914 J. S. Dexter showed that the different phenotypes could be selected for, thus supporting the hypothesis that beaded wing was under control of a number of genetic factors. (From J. S. Dexter, "The Analysis of a Case of Continuous Variation in *Drosophila* by a study of its Linkage Relations," *American Naturalist* 48 [1914]: diagrams, pp. 714-715.)

Morgan group.[39] It was even found that modifying genes were not necessarily linked on the same chromosome, since the major gene could be mapped, and selection appeared to produce its effect by separating one or more modifiers from the major genes.

The modifier-gene concept represented one of the many apparent exceptions to the original Mendelian view of purity of gametes and the one-to-one correspondence of character and factor. Yet as work with *Drosophila* showed, the modifier hypothesis was, in fact, consistent with Mendelian rules of segregation. It did, however, point to a much more complex series of relationships between and among factors than had originally been proposed. The multiple-factor hypothesis and its various ramifications gradually helped to build the conception of genes as functional units, yielding products that interacted with each other to result in the overall traits of the adult.

Nondisjunction and the Chromosome Theory

Normal sex linkage produces what Morgan and his group called crisscross inheritance: that is, a male transmits his sex-linked traits to his grandsons through his daughters, never through his sons; thus, between generations, the trait seems to alternate or cross from one sex to another. However, in 1913 while working on his Ph.D. dissertation, Bridges found there were some exceptions to this pattern. In a cross between a female homozygous for a sex-linked mutation (e.g., white eye) and a wild-type male, about 2.5 percent of the sons resembled the father (i.e., were red eyed) and about 2.5 percent of the daughters resembled the mother (i.e., were white eyed).[40] To account for these results, Bridges hypothesized that in the original female parent, the two homologous X-chromosomes must have failed to separate from one another in a small number of meiotic divisions (during egg production). As a result of this nondisjunction, as Bridges called it, some egg cells got two X-chromosomes (when they should have gotten only one), while other egg cells got no X at all. The latter eggs, if fertilized by an X-bearing sperm (containing the normal eye-

[39] See, for example, F. Payne, "An Experiment to Test the Nature of the Variations on which Selection Acts," *Indiana University Studies* 5 (1918): 1-45.

[40] C. B. Bridges, "Non-disjunction of the sex chromosomes of *Drosophila*," *Journal of Experimental Zoology* 15 (1913): 587-606.

color gene), would produce a zygóte that was XO and hence a male (normally, of course, it would have been a female had the egg cell had its expected X-chromosome).[41] Bridges's hypothesis of nondisjunction would account for the observed results as follows: If the nondisjunction egg (with two X-chromosomes, both containing the recessive white allele) were fertilized by an X-bearing sperm (containing the domiant red, wild-type allele), the resulting zygote would be XXX (3X); it would show the dominant trait because the X-chromosome from the father contained the wild-type allele and would thus be a red-eyed female. If the nondisjunction egg (XX) were fertilized by the Y-bearing sperm, it would produce a white-eyed female which to all intents and purposes appeared normal. If the egg lacking any X-chromosome were fertilized by the X-bearing sperm (with the red allele) the resulting offspring would be a red-eye male. Now, red-eyed males resemble the father and white-eyed females the mother—the exceptions that were not to be expected under ordinary crisscross inheritance.

To verify the hypothesis of nondisjunction, Bridges set about to survey the cytology of *Drosophila* and search for examples of nondisjunctional X-chromosomes. Much to Bridges's gratification and excitement, such chromosomes did, indeed, turn up. But something else, of fundamentally greater importance to the development of the chromosome theory, emerged from these studies of nondisjunction. In his cytological investigation, Bridges found that Wilson and Stevens, in their respective papers of 1905, had considered males of most insect species as of the XO type. However, Bridges found that males of *Drosophila* did, in fact, have a second accessory chromosome in addition to the X, called the Y-chromosome.[42] The discovery of the Y-chromosome, along with the observations of nondisjunction, were seen by Bridges, and later Morgan, as the first concrete proof that Mendelian sex-linked genes were really located on the X-chromosome. This idea had been based on the assumption that what had been observed

[41] If those same eggs were fertilized by a Y-bearing sperm, however, they would not develop at all, since they would lack all of the many necessary genes located on the X-chromosome.

[42] C. B. Bridges, "Direct Proof through Non-disjunction that the Sex Linked Genes of *Drosophila* Are Borne by the X-Chromosome," *Science* 40 (1914): 107-109; and "Non-Disjunction as Proof of the Chromosome Theory of Heredity," *Genetics* 1 (1916): 1-52; 107-163.

in meiosis (i.e., separation of X's in females, and X from Y in males) also occurred in mitosis; thus the resulting association in the adult between sex of the individual and dominant or recessive sex-linked trait. However, the failure of a pair of chromosomes to separate (nondisjunction) produced a different distribution of genes.

The importance of nondisjunction in supporting the notion of a physical basis of Mendelian factors in the chromosomes was analyzed by Bridges as follows: the sex chromosomal constitution of a nondisjunctional female is represented as XXY. From breeding results Bridges concluded that approximately ninety percent of the meiotic divisions in such a female will result in the segregation of the two X-chromosomes; of that ninety percent, half the eggs will contain an X and a Y and the other half a single X. Ten percent of the meiotic divisions, however, occur in such a way that the two X-chromosomes remain together, so that half of those eggs will contain two X's, and the other half a single Y. With respect to sex-linked characters as well as with respect to chromosomal constitution, the fertilization of eggs containing the two X-chromosomes (homozygous recessive) by spermatozoa from a normal, wild-type male gives the following kinds of offspring:

1. Fertilizing of XX eggs by Y-bearing sperm gives *females* of the constitution XXY (like the original nondisjunctional parent). These females will be exact duplicates of their female parent.

2. Y-bearing eggs fertilized by X-bearing sperm give males (XY) with no extra sex chromosomes. In all sex-linked characters these males are exact duplicates of their male parent (having received their single X-chromosome from their fathers).

3. XY-bearing eggs fertilized by X-bearing sperm will give females that in terms of chromosome composition are like the original nondisjunctional female (i.e., XXY). However, in terms of the sex-linked characters they express, they will exhibit normal crisscross inheritance. XY eggs fertilized by Y-bearing sperm will be XYY males; in terms of the sex-linked traits they, too, will show normal crisscross inheritance (i.e., resemble their female parent).

4. In the normal situation, where an X-bearing egg is fertilized by an X- or Y-bearing sperm, all offspring will also show crisscross inheritance.

The crucial point Bridges noted was the occurrence of the nondisjunctional female type, as described in class 1 above. As Bridges pointed out: "These females will be exact duplicates of their mother with respect to their chromosomes, *and the breeding work showed that they are exact duplicates with respect to all sex-linked genes. This parallelism can only be explained by assuming that the X-chromosomes do in reality bear the gene for the sex-linked characters.*" (Bridges's italics.)[43] Furthermore, Bridges pointed out, this class of nondisjunctional female offspring received no X-chromosome from their fathers and therefore were observed to display no sex-linked characters from the male whatsoever. The nondisjunctional relationship showed Bridges that the association of Mendelian factors with the X-chromosome was more than mere parallelism. There *was* a physical basis for Mendel's genes on the chromosomes. More refined analyses proved the same point for autosomal nondisjunctive flies, indicating clearly to Bridges and the rest of the *Drosophila* group that the Mendelian-chromosome theory had to be viewed as more than an abstract and convenient model. It seemed to deal with real structures in the cell.

THE MORGAN GROUP, 1910-1920

A unique feature of the Morgan group was the atmosphere in which the major ideas were developed and worked out. There was a constant give and take and sense of equality among the four central members, but this spirit rubbed off to varying degrees on all those who spent any time in the fly room. Especially in the earlier years (between 1910 and 1915) there was little concern about priority or ownership of ideas. Constant conversation, infused with both humor and complete devotion to the work, characterized the best moments in the Morgan group. Sturtevant has described the mood as idyllic. He emphasized the integrity of each individual and his interrelations with the rest of the group:

> This group worked as a unit. Each carried on his own experiments, but each knew exactly what the others were doing, and each new result was freely discussed. There was little attention paid to priority or to the source of new ideas or new interpretations. What mattered was to get ahead with the work. There was much to be done; there were many new ideas to be tested, and many new experimental techniques to be developed. There can have been few times and places in

[43] Bridges, "Direct Proof through Non-Disjunction," p. 108.

> scientific laboratories with such an atmosphere of excitement and with such a record of sustained enthusiasm. This was due in part to Morgan's own attitude, compounded with enthusiasm combined with a strong critical sense, generosity, open-mindedness, and a remarkable sense of humor. No small part of the success of the undertaking was also due to Wilson's unfailing support and appreciation of the work—a matter of importance partly because he was head of the department.[44]

Elsewhere Sturtevant has spoken more explicitly of the interpersonal relations: "There was a give-and-take atmosphere in the fly room. As each new result and each new idea came along, it was discussed freely by the group. The published accounts do not always indicate the sources of ideas. It was often not only impossible to say, but was felt to be unimportant who first had an idea. . . . I think we came out somewhere near even in this give-and-take, and it certainly accelerated the work.[45]

From all available evidence, there is much truth to this picture, but in attempting to understand the social and psychological nature of science, it deserves a more critical analysis. What were the human dynamics that gave rise to this unique atmosphere? What were the personal associations and interactions among the graduate students, and with Morgan? What was the funding for this work, and did all the graduate students share in it equally? Was it all so idyllic, in fact, as Sturtevant reported? The view an outsider gets about how the group functioned depends on which insider tells the story. Sturtevant's picture is highly romantic. The idea that there might have been any concern over priorities, or any personal disagreements, for example, is reduced to one footnote in a chapter in his *History of Genetics*. H. J. Muller and Alexander Weinstein are less euphoric, and their picture illuminates human and personal details that Sturtevant tends to play down. Others who remember their experience in Morgan's lab during the period 1910 to 1915 have provided accounts that help to evaluate and balance the personal views of the major participants. Emerging from these varied reports has been a sense that the cohesiveness of the *Drosophila* group was not as strong as Sturtevant's account would suggest. In fact there seemed to develop over the

[44] A. H. Sturtevant, "Thomas Hunt Morgan," *Biographical Memoirs, National Academy of Sciences* 33 (1959): 295.

[45] Sturtevant, *A History of Genetics*, pp. 49-50.

years a rift, a split, between Muller on the one hand and Morgan and Sturtevant on the other. Bridges appears to have remained on good terms with both sides, as did the Mohrs. The possible factors entering into this split will be discussed in greater detail below.

The atmosphere of the Morgan lab was indeed lively with shop talk most of the time, and Morgan was usually in the middle of it all. One account captures this feeling very nicely. In January of 1918, Otto L. Mohr (1886-1967) and his wife Tove arrived from Norway for a year's postdoctoral work. After a harrowing experience crossing the ocean and being shadowed for several days by U.S. government officials (who thought the Mohrs, equipped with microscope slides, might be German agents), they found their way to Morgan's lab in Schermerhorn Hall. For much of that year the original group was reassembled, for Muller came back in the spring from his two-year stint at the Rice Institute in Texas, remaining at Columbia for the next two years as an instructor. Tove Mohr writes of their first impressions: "It was a crowded lab, a little bewildering to us. We had, of course, expected something wonderful in God's own country, but found it rather untidy, dusty, with milk bottles containing banana flies everywhere on tables and shelves. There was lively conversation going on, talking, joking, laughing—the professor in the midst of the group. They were like a bunch of students having a good time together. Without knowing it we had jumped right into the later so famous 'fly room' with its wonderful atmosphere. In no time Dr. Mohr was absorbed and part of the group."[46]

Sitting at their various tables around the periphery of the small room, Morgan, Bridges, Sturtevant, Metz, Lancefield, or Weinstein—whoever else was there in a particular year—would toss around new ideas and new hypotheses. Muller has succinctly described his recollection of this constant interchange: "We'd keep up a more-or-less running, often desultory, conversation as things came up. We'd speak freely to one another. 'Oh,' Bridges would say, 'Here is a strange case I just got the results from.' We'd all discuss it, and I might make a suggestion about what to do next . . . and so, we simply went along in an informal way, talking things over with one another as they came along; very seldom on

[46] Tove Mohr, "Personal Recollections of T. H. Morgan," address delivered at the Morgan centennial celebration, Lexington, Kentucky, September 1966; typescript, p. 1.

what you might call a philosophical plane."[47] Sturtevant has put it slightly differently, and with a touch of humor: "Everyone did his own experiment with little or no supervision but each new result was freely discussed by the group. There was no such thing as a coffee break or any special arrangement for discussion. Instead was discussed, planned, and argued—all day every day. I sometimes wondered how any work got done at all, with the amount of talk that went on."[48] But this talk was crucial, for through it new ideas were exchanged, brought up for analysis, and in many cases put into practice as one or another new experiment. Sturtevant has recalled a few examples of how the constant exchange of ideas facilitated the progress of the work:

> The original chromosome map made use of a value represented by the number of recombinations divided by the number of parental types as a measure of distance; it was Muller who suggested the simpler and more convenient percentage that the recombinant formed of the whole population. The idea that "cross-over reducers" might be due to inversions of sections was first suggested by Morgan, and this does not appear in my published accounts of the hypothesis. I first suggested to Muller that lethals might be used to give an objective measure of the frequency of mutation. These are isolated examples, but they represent what was going on all the time.[49]

Muller gives a more concrete (if rambling) and lively picture:

> One day Bridges said, "You know I am breeding the bar-eyed male, and there should be bar-eyed females, but one of the daughters was a non-bar-eyed female. When I bred it, it gave a two females to one male sex ratio, as if it had a lethal in it. But the bar condition was gone. It didn't show any bar in the offspring." I said to Bridges, "Well, that may be because it lost the end of its chromosome, so that both the bar was lost and the lethal appeared because of the lack of that piece of the chromosome. Why don't you breed it—two flies with recessive

[47] H. J. Muller, interview with author, March 1965, Cambridge, Mass.

[48] Quoted in Tove Mohr, "Personal Recollections of T. H. Morgan," pp. 1-2.

[49] Sturtevant, *A History of Genetics*, pp. 49-50.

> forked [bristles] which we know to be very close to bar on the chromosome, and if that part is really lost, then the daughters will show forked even though it is recessive." A couple of weeks later Bridges said, "That came out; they were forked all right." So I said, "Well, that part of the chromosome is deficient, you ought to try it with all the genes around there that you've got in our stock." He tried it, and he found that one still further toward the end of the chromosome proved to be present (contrary to expectations), as though there was a whole section of the middle of the chromosome missing. So then we discussed it further, whether it could be inactivated or lost, or what had happened, you see. Well, things went on like that in the lab.[50]

It was through such exchanges that the various interests and specialities of different members of the group could be incorporated into the total program of work. Muller, for example, was particularly adept at designing stocks of flies to cross with new mutants. Thus, when some new condition would arise, his suggestions about what stocks to breed it against would often be most useful. On the other hand, when a cytological question arose, Bridges's expertise was most valuable. But the exchange was equal from all directions and appears to have been genuine and largely unhampered by personal animosities and difficulties during the period from 1912 to 1915.

Morgan engaged in the discussions, although often he was less given to banter than some of the others. As a laboratory worker, Morgan was industrious and energetic, but sloppy and inclined to use the simplest and crudest methods. Unlike the others, he always stood at a raised table to do his reading instead of sitting at a desk. He thought that only a lazy man would sit down and constantly chided the others about their laziness.[51] As with his earlier embryological experiments, Morgan continued throughout the *Drosophila* work to use the simplest equipment possible. He examined flies with only a hand lens and said that Bridges was bringing a lot of falderal into the work by insisting on a binocular microscope (see figure 12).[52] Morgan continued using primitive methods of etherizing flies, even after Bridges had developed a simple and convenient etherizer which did not permanently harm the flies.

[50] Muller interview, 1965.
[51] Ibid.
[52] Ibid.

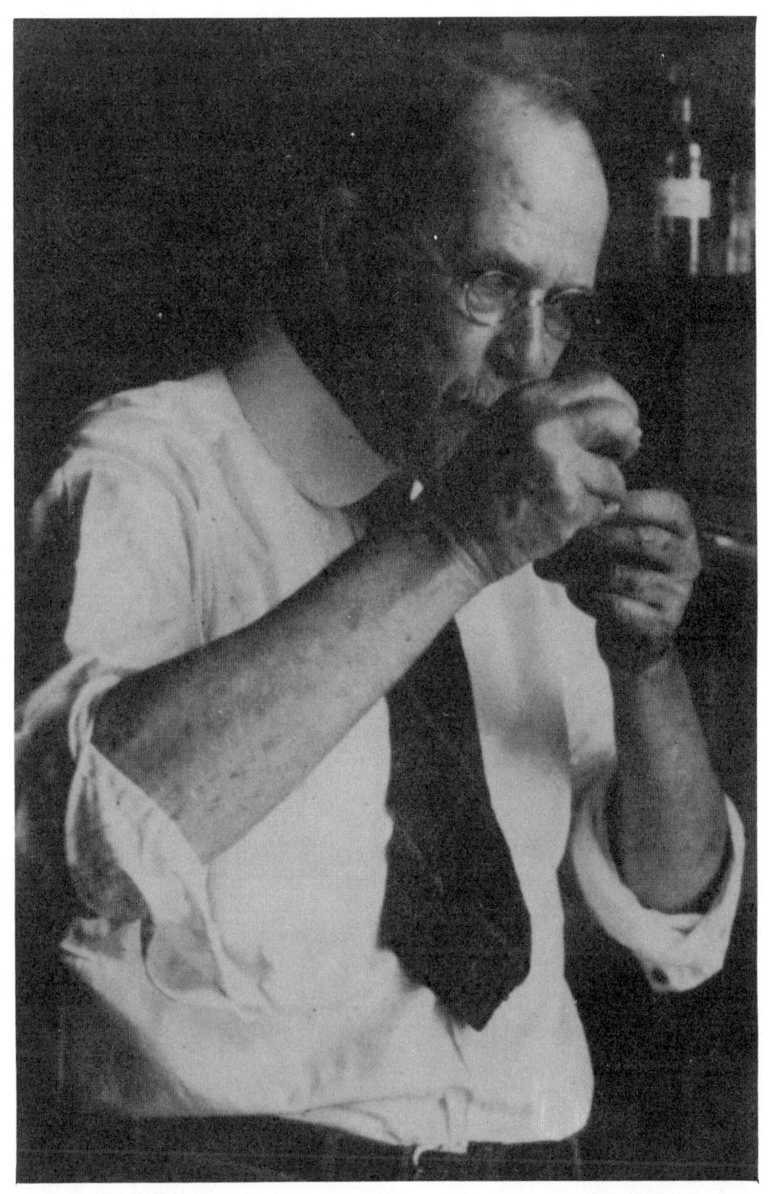

Figure 12. T. H. Morgan examining *Drosophila* in his laboratory at Columbia University, about 1915. Disdaining complex equipment, Morgan generally examined flies with a hand lens, rather than a binocular microscope. He also stood most of the time he was examining flies, frequently joking that only lazy people sit down. (Courtesy of Isabel Morgan Mountain.)

Morgan's method apparently was to dump ether on the flies, a procedure that made it difficult to breed future generations from these organisms! As Muller said, "Morgan thought fancy equipment was a sort of decadence."[53] Before Bridges developed a standard procedure for disposing of unwanted flies by placing them in a "morgue"—a bottle of ether—Morgan would get rid of those he had counted by simply squashing them into the paper or table top on which he was working. And, despite his enormously systematic mind, he had no systematic procedure for keeping data. Although he kept notebooks, much of his data was, in fact, kept on the backs of envelopes and old scraps of paper. When one of his children once commented on their father's sometimes sloppy way of keeping records by saying, "I should think he should lose them sometimes," Mrs. Morgan answered, "Well, you know he does occasionally."[54] But usually, it appears, everything that was lost was ultimately found, and seldom did Morgan seem to be without the important data he needed. There is a genius in being able to find just what you want amid apparent chaos.

At night Morgan would usually go home with his briefcase "just crammed full of papers and books, to work all evening."[55] The others, however, would return after supper. Most of the time this included Sturtevant, Muller, Weinstein, and one or two more. The only exception was Bridges, who often found other evening activities in New York City more attractive than chromosome squashes. It was often during these evening sessions that Muller, Sturtevant, and Bridges (when he was there) would plan new experiments—new crosses to make and new stocks of flies to build up. As Muller put it: "But sometimes after Morgan went home, we would get together and figure out schemes and crosses and lots of other things and what should be done. Morgan didn't really know much of what we were about. Morgan had an enormous drive and he gave us great opportunities and treated us perfectly as equals." Muller went on to say that Morgan never planned out his experiments too carefully: "He played it pretty much by ear. That was the older style, however. You poke your finger in and see what happens."[56]

While Morgan knew the broad outlines of the work that was going on, he left the members of his group alone to go their own

[53] Ibid.

[54] Isabel Morgan Mountain, "Notes on T. H. Morgan," p. 11.

[55] Muller interview, 1965. [56] Ibid.

ways and work in complete independence. It is apparent that after 1913 or 1914 most of the experiments and the plans were, in fact, laid by Sturtevant, Bridges, and Muller. Morgan had his mind on many problems, particularly such larger issues as support for the chromosome theory of heredity, the nature and origin of variation, and the relations between the new findings in *Drosophila* and the problems of evolution and embryology. The details very soon became less interesting to him than the broader picture. Sturtevant describes well Morgan's relation to the day-to-day detail work of breeding experiments: "It is true that the details of the techniques and the numerical analyses were not his forte—after about 1913 Bridges and I usually designed his *Drosophila* experiments for him—but he knew what he wanted to do and why. There was no question of his understanding the significance—it was the technical details he had to be 'educated' on."[57] Morgan provided the overall guidance and raised many new questions but left the details of working out many experiments to the eager associates who worked with him. This relationship was quickly picked up by William Bateson when he visited the Morgan laboratory (and stayed with Morgan in his home) in December of 1921. Bateson wrote home to his wife: "Saw something of Sturtevant yesterday, and thoroughly like him. He and Bridges are quite different from the type I expected. Both are quiet, self-respecting young men. Sturtevant has more width of knowledge—Bridges scarcely leaves his microscope. I wonder whether they are not the real power in the place. Morgan supplies the excitement. He is in a continual whirl—very active and inclined to be noisy."[58] Bateson continued his letter in a vein that would have shocked most members of the *Drosophila* group. Speaking of Morgan he wrote: "Nothing that he says is really interesting or original. I like him and I don't. I wish to goodness I had not been made to live in his house. I have to act more affection than I always feel. There is no meanness about him, and I recognize his sincerity—but no size above the ordinary."[59]

Once every week or so, Morgan had what amounted to Journal Club meetings at his house on Friday evenings. Beer was served,

[57] Sturtevant to T. M. Sonneborn, May 5, 1967, pp. 1-2; Sturtevant Papers, California Institute of Technology Archives; quoted with permission.

[58] William Bateson to C. Beatrice Bateson, December 26, 1921; Bateson Papers, APS.

[59] Ibid.

and there was usually discussion of some particular topic among the group. Those present included (in addition to Muller, Sturtevant, and Bridges) Lancefield, Plough, Schrader, Edgar Anderson, and anyone else who was associated with the *Drosophila* work at the time. The atmosphere was relaxed and the discussion often vivacious. As Muller described it: "We talked over problems and talked over current literature, and had some beer and crackers. That was very nice."[60] The Journal Club meetings provided an informal way in which new ideas, both related to and often distant from the *Drosophila* work, could be discussed in some detail.

All was not work with the members of the *Drosophila* group, however, for there was a considerable amount of social interaction both inside and outside the laboratory. Within the lab, especially, there was constant joking, frequently led by Morgan himself. One day, for example, Morgan received a letter from a Midwestern university asking him to recommend a teacher in biology. At the bottom of the sheet was a postscript saying "But we want a good Presbyterian." After deciding half-jokingly whom they might recommend, Morgan decided that some proof of the good religious morals being inculcated at Columbia should be forthcoming. Morgan immediately had the photograph in figure 13 taken, claiming that he was going to send a copy along with the letter of recommendation.

More dramatic is the story of Morgan's late night omelet. An ostrich egg had been delivered to the laboratory for E. B. Wilson when he was away and could not be reached. Wilson had spent considerable time and effort in locating such an egg, which he wanted for measurement of the yolk as the largest egg cell in existence. Morgan did not know what the egg was for and naturally supposed that Wilson wanted it for embryological studies. Wilson had worked on the early embryological stages of some sea animals, and as a delay might make the egg useless for such studies (that is, the egg would spoil), Morgan removed and fixed the blastoderm (the small developing embryo on the upper surface of the yolk). That evening Morgan and his group were working in the lab. Since Morgan considered the rest of the ostrich egg of no use, he made it up into an omelet on which the assembled workers in the fly room feasted. When Wilson returned he was much dismayed, since he had wanted to measure the yolk. Morgan was embarrassed and a little chagrined at his own impetuosity. However, he eventually got Wilson another ostrich egg, and Wilson was able to include

[60] Muller interview, 1965.

Figure 13. To demonstrate that "piety" reigned supreme in the Columbia laboratory, Morgan had this photograph taken in 1915. The photograph was sparked by a request from a Midwestern university that Morgan recommend one of his students in terms of religious as well as intellectual attributes. From left to right, Calvin Bridges, Alexander Weinstein, and Morgan. (Courtesy of Tove Mohr.)

the measurement in the third edition of his monumental *Cell in Development and Heredity.*[61]

More genuinely social events were occasionally held in the laboratory such as the famous "Pithecanthropus party" given to honor Sturtevant's return from military duty in 1919. At this party the featured gastronomic event was fresh squid, which Morgan had

[61] This version of the story was provided by Dr. Alexander Weinstein in a personal communication to the author, November 29, 1968.

learned to prepare while at the Naples Station many years before (see figure 14).

Especially valuable in terms of both work and social interaction were the summers at Woods Hole. In mid-June all the *Drosophila* material was packed into barrels, an enormous operation carried out by Morgan, Edith Wallace, and all the graduate students and other assistants around. The packing was arranged in such a way that several culture bottles of each strain of *Drosophila* were left

Figure 14. Morgan's group in the fly room at Columbia University in 1919. The occasion is a party for A. H. Sturtevant (front, center, leaning back in chair) who was returning from military service in World War I. The laboratory was not only a place to work, but also a location for various informal get-togethers during the heyday of the *Drosophila* work. The figure at the center of the table (back) is a dummy of the "honored guest," Pithecanthropus. Others shown in the photograph are: *back row* (left to right) E. G. Anderson, S. C. Dellinger, C. B. Bridges, Pithecanthropus, H. J. Muller, and T. H. Morgan; *front row* (left to right), Alexander Weinstein, Franz Schrader, Sturtevant, A. F. Huettner, Otto Mohr, and F. E. Lutz (barely visible on extreme right). (Courtesy of Isabel Morgan Mountain.)

behind in New York so that in case any of the specimens were damaged en route there would be ample reserves available. In addition to *Drosophila*, Morgan often took several cases of mice and a basket of roosters, pigeons, or experimental plants for some of his other breeding experiments.[62]

Morgan transported his own family to Woods Hole via boat on the Fall River Line to Fall River, Massachusetts, and from there by train to Woods Hole. Both Morgan and his children enjoyed the boat trip immensely; on approaching each of the bridges in the East River, Morgan would assemble the children on deck and pretend to get very excited about whether the boat's flagpole was going to pass under the bridge or not. Sometimes they would have picnics on deck and other times sit in the dining room where the children especially liked the "silver candlesticks and filagree trappings."[63] In addition to his own family, Morgan transported various members of the *Drosophila* group to Woods Hole each summer; Sturtevant and Bridges were regularly included, and various other graduate students and postdoctoral fellows came from time to time.

Woods Hole provided an even more informal and relaxing atmosphere for continuation of the work and discussions than the lab at Columbia. The group often had its discussions outside on the grass or at the beach. Bridges and Sturtevant particularly enjoyed horseshoes, though this was not an activity that attracted Morgan. Boating was also very popular, especially with Bridges, who was always arranging for short trips to the nearby islands, often in the company of several of his young women friends. There were, in addition, frequent outings and picnics with other biologists, usually close friends of the Morgans. These included the E. B. Wilsons, the Ross Harrisons, and the E. G. Conklins. At Woods Hole the *Drosophila* group had ample opportunity to discuss their work with biologists from a variety of other fields, including embryology, systematics, and physiology (see figure 15). It was here, too, that Morgan gave several of the famous "Friday Night Lectures" in which he publicly introduced the group's findings on *Drosophila* before the publication of the *Mechanism of Mendelian Heredity*. The emphasis on rigorous quantitative experimental work which permeated the MBL in the summer, es-

[62] Lilian V. Morgan, "Some Chronological Notes on the Life of Thomas Hunt Morgan," unpublished manuscript, Morgan Papers, Caltech, p. 3.

[63] Mountain, "Notes on T. H. Morgan," p. 23.

Figure 15. "Solving the Problems of the Universe," the title given by Morgan to this informal snapshot from Woods Hole, summer 1919. From left, clockwise around the circle. Morgan, Bridges, Franz Schrader, E. E. Just (a brilliant embryologist from Howard University who spent most summers at the MBL and maintained a continuing interest in the *Drosophila* work), Sturtevant, and an unidentified person, possibly Otto Mohr. Through sessions such as the one shown here, the *Drosophila* group maintained constant communication about the progress of the work, and exchanged or evaluated new ideas. (Courtesy of Isabel Morgan Mountain.)

pecially after 1910 or so, provided an extremely congenial atmosphere for the *Drosophila* research.

Morgan was always a part of the *Drosophila* group, yet somewhat detached from it. He was affectionately referred to as "the boss" by Bridges, Sturtevant, and others, and he referred to them as "the boys." He was very solicitous of those working with him and knew a great deal about everyone's personal and professional life. He did not encourage lengthy discussions of personal matters, but in case of problems he was always ready to listen and to help if he could. He liked to be close to his students but expected them to stand on their own; he became annoyed at those who looked to him for constant guidance, either professionally or personally. Moreover, he applied his humor at all times to make the people around him feel "special" to him. This trait was amply exhibited in his family, for each of his children appears to have felt unique and special in his eyes.

An example of Morgan's humor comes from the summer that Otto and Tove Mohr spent at Woods Hole (1919). Tove Mohr was already pregnant when she and her husband came to the United States in January of that year. Consequently, her first child was born during the stay at Woods Hole. Since Woods Hole had no hospital, it was necessary to travel around Buzzards Bay to New Bedford, Massachusetts, a distance of about forty to fifty miles. The trip was harrowing for the Mohrs, both because of the distance and because of the rough ride by auto in a strange country. They arrived in good time, however, and the baby was born satisfactorily. But the Morgans knew that the experience had been particularly exhausting for Tove. As soon as the Morgans received a telegram with the news of the birth of a baby girl (named Venke), Morgan sat down and wrote a brief note to Otto Mohr. It said in part: "A great shout of joy went up from the assembled Morgans when I read to them your telegram. I need not tell you that I felt anxious when I got your note on my arrival at the laboratory [a note saying that the Mohrs had left for New Bedford in the middle of the night] . . . and I phoned out to the hospital and heard of your safe arrival. I have no fatherly advice to offer, except not to call the little girl Drosophila. We have resisted the temptation ourselves three times."[64]

As in his relationship with members of his family, so Morgan's relationship with members of the *Drosophila* group appeared to be one of benevolent paternalism. Neither with his family nor with his laboratory group did he avoid responsibility. Yet, in both cases his role was much more that of the broad thinker and planner; the details and day-to-day operation were left to others. Because he also had a down-to-earth side to his nature, Morgan was not averse as a matter of principle to getting his hands dirty and doing the day-to-day work himself. This, in fact, he often did, especially in the laboratory. For the most part, however, he surrounded himself with people who were adept at managing practical affairs (as,

[64] Morgan to Otto Mohr, September 2, 1918; from Tove Mohr, Fredrikstad, Norway.

for example, his wife in the home or Edith Wallace in the laboratory), or in planning the details of the experiments and work to be done (as with Muller, Sturtevant, or Bridges in the laboratory). But Morgan's genius for surrounding himself with the right associates goes further than this. Those who took care of many of the practical details of his day-to-day life were themselves far-sighted thinkers, too, not mere managers or bureaucrats. Lilian Morgan was T. H. Morgan's intellectual equal and never lost sight of her interest in and devotion to her scientific or musical interests. And, as we have seen, Sturtevant, Bridges, and especially Muller were able to see both the large-scale as well as the small-scale aspects of the *Drosophila* work. It is impotrant to recognize the symbiotic relationship Morgan held with members of his group in order to understand how he as an individual, and the group as a unit, were able to function so well for so long.

There were, however, problems within the group, which increased over the years. Morgan, like all people, had his likes and dislikes, his stronger and weaker affection for those around him. Sturtevant was his closest associate, his "pet" as Muller described it. Bridges was well respected, but he was not as close personally to Morgan. As indicated earlier, Muller always stood further away from Morgan and from the group as a whole. He felt that there was always a certain coolness on Morgan's part toward him—not overt or direct, but still something present which he sensed. Several factors may have contributed to Muller's sense of "outsideness." First of all, Muller was not taken into Morgan's laboratory as early as Sturtevant or Bridges, and consequently spent his first two years in graduate school (1910-1912) studying neurophysiology. To earn his keep Muller had to do extra teaching. In addition to assisting in the Cornell physiology laboratory he also taught English to foreigners at night for the Immigration Service. As a result, he was out of direct, daily contact with the *Drosophila* work during its first two years. But there is something more to it than that. In human relations comparisons are inevitable, and Muller saw what appeared to be favored treatment to Sturtevant in particular. Sturtevant had been able to obtain a scholarship for his first semester in graduate school (the spring term of 1911-1912), thus enabling him to stay in school without having to teach.[65] In contrast, Muller's own state seemed to him to be unduly

[65] Only later was it learned that Morgan had actually provided the money from his own pocket for that one semester's scholarship. At the

hard. He felt that the heavy burden of teaching not only took time away from his studies and research, but was injurious to his health as well.[66] He felt he was not given the same opportunities as Sturtevant, thus creating a certain resentment on his part toward Morgan.

More specific causes may have contributed to Muller's feeling. One was that Morgan apparently did not particularly care for Muller's close friend, Edgar Altenburg.[67] Altenburg was not encouraged to join the *Drosophila* work but remained for his doctoral studies in the botany department. He was interested in and involved with the *Drosophila* work at various times, and no hint of a reason for Morgan's unwillingness to take Altenburg into the group can be found. Interestingly, in 1912 Muller and Altenburg started to write a textbook together to be called "Principles of Heredity." Early in 1913 Morgan proposed that Muller and Altenburg join himself, Sturtevant, and Bridges in preparing the book which later became *The Mechanism of Mendelian Heredity*. Altenburg apparently declined authorship for reasons that are unknown.[68] Since Muller was at the time (and remained throughout his life) exceptionally close to Altenburg, the coolness or actual animosity between Morgan and Altenburg probably influenced Muller's relation to Morgan.

Corroboration of this view comes from several letters Altenburg wrote to Muller after Morgan's death in 1945. Concerned that one of Sturtevant's obituary notices on Morgan was too laudatory, Altenburg suggested that Muller should prepare a notice that could "set the record straight." Altenburg advised Muller that he should state "without equivocation and without glossing over, that Morgan definitely was badly confused on practically all fundamental issues in genetics (even on the chromosome theory itself!), and it should be stated just as clearly that he did not arrive at the correct viewpoint himself after awaiting the experimental results, and then convince his 'students' along with the rest of the world, but just the opposite was true—that the experimental results pointed to the correct conclusion many years before he accepted

---

time it was assumed only that Morgan had perhaps strongly recommended Sturtevant to the fellowship committee (see chap. I).

[66] Muller interview, 1965.

[67] Ibid.

[68] H. J. Muller, *Studies in Genetics* (Bloomington, Ind.: Indiana University Press, 1962), "Explanatory note number 2," p. 6.

those conclusions and that his 'students' and Wilson were mainly responsible for setting him straight." Altenburg continued: "He undoubtedly had streaks of brilliance and went right to the point. But of course this was not a sustained brillance; he went from the sublime to the ridiculous, as when he thought he could disprove the need for natural selection because of dominance."[69] Altenburg obviously thought that Morgan was less the scientific genius than was being claimed by Sturtevant and others. Whatever sharp feelings Muller may have developed on his own were likely to have been given additional strength by Altenburg's more open disdain.

Another specific factor may have been a difference in personality and style between Morgan and Muller. Morgan was humorous and liked to tease people, as we have seen. Muller was considerably less humorous, and may have taken poorly to Morgan's propensity for bantering. Morgan did not like intellectual aggressiveness, and this may, in fact, have been a major cause for his lack of the same warmth toward Muller that he felt toward Sturtevant and Bridges. Both Muller and Sturtevant, for instance, frequently corrected mistakes that Morgan made. Sturtevant's style of correction was quiet, subtle, and humorous, and never made publicly. Muller, on the other hand, even as a student, occasionally criticized Morgan in public as well as in private, often in a direct and aggressive way. Morgan unquestionably recognized Muller's brilliance but may have felt less comfortable with him because of his style.

All these considerations affected Muller's relationship to the group when he finally joined it in 1912. As Muller explains it, he purposely chose as a subject for his graduate research an area of the *Drosophila* work that was noncompetitive with what Bridges, Sturtevant, or Morgan were already doing:

> I didn't want to butt in and step on their toes, since they already had this linkage story going; so I decided on some more out of the way things to do, such as building up stocks which had twenty mutants in a row [on the same chromosome] so that one could get the linkage of all the factors simultaneously

[69] Altenburg to Muller, March 24, 1946; Muller Papers. Reference from E. A. Carlson, "The *Drosophila* Group," *Genetics* 79, supplement (1975): 15-27; quotation, p. 23. A slightly modified version of the same article appears as "The *Drosophila* Group: The Transition from the Mendelian Unit to the Individual Gene," *Journal of the History of Biology* 7 (1974): 31-48.

> and not have the statistical error involved in different experiments. And then there were these other things I did with Altenburg, analyzing the truncate case, for example, showing that these apparently vague cases of inheritance were also strictly Mendelian. They went ahead with their mapping, and I had the fortune of finding the gene on the fourth chromosome which completed the map for them. . . . I was on the outskirts of most of that because, while I was discussing it with them, I was not actually working with it in my hands.[70]

But, of course, as we have seen, Muller did interact with the others on all sorts of problems that were strictly speaking "theirs." This produced some difficulties in Muller's eyes a few years later when the question of priority became for him an open one. Muller apparently felt that Morgan had used his and Sturtevant's work in an article published in the first volume of the *Proceedings of the National Academy of Sciences* without giving proper credit.[71] In this paper, Morgan discussed the previous three years' work with *Drosophila*, emphasizing the major tenets of the Mendelian-chromosome theory. He cited specifically the work of Bridges on non-disjunction, but failed to mention Muller's name in relation to interference, or Sturtevant's in relation to mapping. Muller seems to have felt that Morgan may have purposely slighted them both. However real this feeling may have been to Muller, it seems unlikely that Morgan's omission of specific credit in these cases was purposeful. As Sonneborn has remarked, "To me the thing that requires explanation is not why he [Morgan] left out Sturtevant and Muller, but why he put Bridges in."[72] In fact, Bridges is the only person to whom Morgan refers at all in this paper. He does not refer by name to other important investigators whose results are cited, such as Sutton, Boveri, Wilson, or Bateson. As Sonneborn has pointed out, this paper is in the nature of a general summary for non-specialists. It would have been distracting and even

[70] Muller interview, 1965.

[71] The paper in question is Morgan, "Localization of the Hereditary Material in the Germ Cells." Reference to Muller's feelings on this matter is contained in a letter from T. M. Sonneborn to Sturtevant, May 2, 1967; Sturtevant Papers. The matter is further amplified in a letter from Sonneborn to E. A. Carlson, of the same date, a copy of which is also contained in the Sturtevant Papers.

[72] Sonneborn to Carlson, May 2, 1967; Sturtevant Papers; quoted with permission.

confusing, perhaps, to have tried to cite every person whose ideas contributed to the long sequence of developments leading to the findings of the *Drosophila* group. Morgan does not claim in the paper that even the majority of the ideas and generalizations referred to are his own. Sonneborn suggests that Morgan may have referred to Bridges's work because of the outstanding significance of nondisjunction. Nondisjunction was considered to be the first direct proof that genes were actually located on the X-chromosome. Hence, in 1915 it was the single most important piece of evidence contributing to the concept of a physical basis of Mendelian heredity.

Sonneborn's analysis fits into the much larger body of evidence we have about Morgan's personality and mode of operation. Most of his books and many of his articles are, in fact, rather sparsely documented. Especially in his more popular writings, he seldom gave specific credit to every individual who contributed to a body of knowledge. Furthermore, while Morgan had his likes and dislikes, he was seldom mean or petty. It would have been highly out of character for him to purposely omit the name of an individual on the grounds of personal dislike. And, since Sturtevant was acknowledged, even by Muller, to be Morgan's pet, the omission of references to his work on mapping would also have to be explained.

Muller also appears to have felt that Morgan made it difficult for him to obtain favorable and permanent teaching positions in the United States. When Muller returned to Columbia as an instructor of zoology in 1918, he hoped to remain there permanently. In 1920, however, his contract was terminated, and he had to seek another position (this time at the University of Texas in Austin). Muller apparently thought that Morgan influenced this decision.[73] Altenburg made the point in no uncertain terms. When Muller was asked to prepare an obituary notice on Morgan in 1946, Altenburg wrote strongly: "Don't get soft all of a sudden; remember that Morgan did you incalculable damage, not only by his steals, but also by keeping you from getting a job."[74] According to Sturtevant, however, Morgan not only did not discourage Mul-

[73] Sonneborn to Sturtevant, May 2, 1967; Sturtevant Papers. References to Muller's feelings about Morgan's opposition to him are contained in Muller's short unpublished autobiography, found in the Muller Papers. I am indebted to E. A. Carlson for referring me to this autobiography.

[74] Altenburg to Muller, February 13, 1946; Muller Papers.

ler's further appointment, but actively encouraged it. Morgan was away on sabbatical leave in California at the time, but apparently urged Wilson to keep Muller on: "When Muller left Columbia to go to Texas, Morgan was in California when the question came up. At the time he [Morgan] told me that he had urged Wilson to make every effort to keep Muller at Columbia; but that Wilson, as administrative officer of the Department, had found Muller so difficult that he was unwilling to urge his advancement. I have heard (not directly from Muller) that Muller thought it was Morgan who was not enthusiastic about keeping him, and that this was one of the reasons for his somewhat anti-Morgan attitude."[75] This matter is one of those mysteries that will probably never be cleared up. However, it is both unfortunate and ironic that, if Sturtevant's statement is true, it was Wilson, rather than Morgan, who was not enthusiastic about retaining Muller at Columbia. For Wilson was one of Muller's lifelong idols.[76]

Not only did Morgan and Muller have specific intellectual and personality differences, they came from widely divergent backgrounds and social classes. Morgan never thought of himself as aristocratic or aloof, and he never acted this way overtly or consciously. Nonetheless, aggressiveness was something he neither liked nor could understand. Muller's world of the working class, immigrant New Yorker, was about as far from his own experience as anything could have been, and this distance may have contributed to his discomfort with Muller. It is difficult for a person who has always been on top to understand or be comfortable with those who have had to struggle hard to gain even a little for themselves. Such aggressive attitudes are often described as "pushiness," and Morgan must undoubtedly have felt that Muller was sometimes pushy. Morgan also may have felt uncomfortable with Muller's political attitudes. Morgan was in no way a political activist, and as noted earlier felt that it was often detrimental to their science for scientists to become involved in social causes. In addition, Muller's "radical" political leanings may have been even more difficult

[75] Sturtevant to Sonneborn, May 5, 1967; Sturtevant Papers; quoted with permission.

[76] See Muller's lengthy and laudatory essay, "Edmund B. Wilson—An Appreciation," *American Naturalist* 77 (1943): 4-37; 142-172. A similar essay appears as the introduction to the reprint edition of Wilson's *The Cell in Development and Inheritance* (New York: Johnson Reprint Corporation, 1966), pp. ix-xxxviii.

for Morgan to accept. Morgan is reported to have said one time that Tove and Otto Mohr were "in bad company" when they went with Muller and some of the others to a socialist rally one evening in downtown Manhattan.[77] Thus, on many accounts Morgan and Muller saw the world from different perspectives and as a result were less able to communicate and understand their differences.

The real problem is, of course, not simply one of differences of opinion or viewpoint. It lies, possibly, in the existence of social and economic systems that create classes of people who have little mutual understanding. In the case of Morgan and Muller, it is not to say that Morgan may have disliked Muller because the latter came from an immigrant, working-class background. However, Morgan was not used to associating with people whose behavior patterns developed out of such different cultural contexts. Thus, he may have been less able to see Muller's aggressiveness in a larger, social perspective. Muller, for his part, may have been more accustomed to seeing coolness or distance on the part of people in power as part of personal rejection. Whatever the specific causes, the more general social forces may have played a contributing role in the split between Muller and Morgan. We can never know what might have happened had Muller been as integral a part of the *Drosophila* group as Sturtevant and Bridges. Certainly Muller did a great deal on his own as a result of his more independent position. However, the research might have been furthered had Muller taken his independent position solely by choice, rather than having felt it to some extent forced on him by the feeling he perceived emanating from Morgan.

The dynamics of interpersonal relations as they influence group activity in scientific research is a complex and little-understood topic. Individual personality factors, of course, play a large role in determining the success or failure of such interactions. There is considerable need in the history of science to analyze seriously the interplay of social, psychological, and economic forces as they influence personal interactions and thus cooperative research activity.

THE MECHANISM OF MENDELIAN HEREDITY

The Morgan group consolidated its findings in 1915 with the publication of *The Mechanism of Mendelian Heredity*, a book

[77] Personal communication, Tove Mohr, April 27, 1972.

that laid out in a rigorous but nonspecialized way the fundamentals of the Mendelian-chromosome theory.[78] The book integrated cytological and Mendelian results in a manner comprehensible to the nonspecialist. It was an epoch-making work, the first major consolidation of gains in the revolution in genetics.

The Mechanism of Mendelian Heredity showed the many profound ways in which the Mendelian-chromosome theory contributed to an understanding of general biological principles. First of all, by surveying data not only from *Drosophila*, but also from other animals and many kinds of plants, the authors demonstrated clearly that Mendel's work was applicable to a large variety of species. Second, they showed that in every case that had been specifically studied, the Mendelian factor hypothesis could be related to cytological data about chromosomes. Thus, they implied strongly that Mendelian theory had a distinct physical reality. Third, they showed how the new results solved a host of old problems, including coupling and partial coupling, variations of expected Mendelian ratios, the inheritance of sex, and the distinction between continuous and discontinuous variation. Furthermore, it opened up many new avenues of research which were at the time only beginning to be explored: the actual physiological basis of gene action; the relation of Mendelian mutation and variation to the origin of species by natural selection; the differential activity of genes during embryonic development; and the origin of variability (from the change in a specific gene such as a point mutation to the more complex chromosomal rearrangements and position effects that alter gene expression).

It is significant that the term "gene" does not appear in the first edition of *The Mechanism of Mendelian Heredity*; the authors speak only of "factors." Wilhelm Johannsen coined the term "gene" in 1909, so it was well known to the authors at the time they were preparing their book. The reason they chose not to use this term, which has since become so common, reflects the group's commitment to the belief that the chromosome theory provides a material basis for Mendelian heredity. Johannsen's word "gene" was originally meant to be a completely abstract concept, conciously disassociated from any of the existing theories of hereditary particles (such as Weismann's ids, biophors, and determinants).[79] Morgan

[78] See Morgan et al., *The Mechanism of Mendelian Heredity*.

[79] Johannsen wrote in 1909: "No hypothesis concerning the nature of this 'something' shall be advanced thereby or based thereon. Therefore it ap-

and the *Drosophila* group were more committed to a material basis for Mendelian factors than Johannsen. To avoid confusion about their belief in the material basis of heredity, they chose not to use Johannsen's term in the early years.

The carefully worded introduction to *The Mechanism of Mendelian Heredity* shows the extent to which the *Drosophila* group was committed to the actual physical reality of Mendelian factors. The authors claimed that most of their coworkers would probably agree with the majority of the interpretations contained in the book, except perhaps for "the emphasis we have laid on chromosomes as the material basis of inheritance."[80] After pointing out that in fact the Mendelian theory can be treated independently of the chromosomes and still maintain its viability as a consistent model that allows prediction, they posed the question, "Why, then, we are often asked, do you drag in the chromosomes? Our answer is that since the chromosomes furnish exactly the kind of mechanism that the Mendelian laws call for; and since there is an ever-increasing body of information that points clearly to the chromosomes as the bearers of the Mendelian factors, it would be folly to close one's eyes to so patent a relation. Moreover, as biologists, we are interested in heredity not primarily as a mathematical form-

---

pears as most simple to use the last syllable 'gen' taken from Darwin's well-known word pangene since it alone is of interest to use, in order thereby to replace the poor, more ambiguous word, 'Anlage.' Thus, we will say for 'das Pangen' and 'die Pangene' simply 'Das Gen' and 'Die Gene,' The word Gen is fully free from every hyopthesis; it expresses only the safely proved fact that in any case many properties of organisms are conditioned by separable and hence independent 'Zustände,' 'Grundlagen,' 'Anlagen'—in short, what we will call 'just genes'—which occur specifically in the gametes." Wilhelm Johannsen, *Elemente der Exakten Erblichkeitslehre* (Jena: Gustav Fischer, 1909), p. 124; translation by G. E. Allen.

Johannsen's "gene" was to be used simply as a kind of accounting or calculation unit, but was not to be considered a material body. He explicitly stated that genes were not morphological structures in the same sense as Darwin's gemmules or Weismann's biophors. Furthermore, he emphasized that individual genes were not the determiners of specific adult characters. Genes were quite different in Johannsen's mind from the "living units," to use L. C. Dunn's term, of the nineteenth-century speculators. As Johannsen wrote in 1909: "The conception of the gene as an organoid, a little body with independent life and similar attributes, is no longer to be considered. Assumptions which would make such a conception necessary fail utterly." Ibid., p. 485. This passage has been translated by L. C. Dunn; see *A Short History of Genetics*, p. 93.

[80] Morgan et al., *The Mechanism of Mendelian Heredity*, p. viii.

ulation, but rather as a problem concerning the cell, the egg, and the sperm."[81]

Morgan was first and foremost a biologist and an embryologist concerned with concrete material problems of living organisms. His own initial objections to Mendelism had been on the grounds that it was too speculative and hypothetical. What won him to its acceptance was a realization that it could be tied to material structures and processes within the organism. While there was no direct, observational evidence in 1915 that chromosomes were, in fact, linear arrays of Mendelian factors, the varieties of evidence were clear enough to Morgan to lead him to this hypothesis.

Morgan's biological perspective (a perspective he transmitted to his coworkers) greatly influenced the emerging view of the Mendelian "factor" (what we shall from here on call the "gene") as a biological unit. There was little evidence in 1915 about the physiological nature of genes or the ways in which they interacted during embryological development to produce adult characters. However, Morgan and his coworkers felt a strong commitment to viewing the gene not simply as a structural, but also as a functional, unit. There had, in fact, been a long history of describing Mendelian genes in biochemical and physiological terms. Bateson's presence-and-absence hypothesis was based on a chemical analogy in which the "present" factor (a dominant) had its effect by creating a chemical substance that contributed to, or was part of, the adult character. In 1909 G. H. Shull made a strong case for Mendelian factors having their effects by altering chemical relations within the cell. He went even further to describe how this system might operate with multiple-factor heredity.[82] Muller and Altenburg, in the incomplete manuscript of their proposed text, "Principles of Heredity," discuss explicitly the possibility that genes influence the production or activity of enzymes in cellular metabolism.[83] Sturtevant was strongly committed to the idea that genes interact through biochemical processes to produce adult characters, and that it is inconsistent with the knowledge of cell chemistry to talk about one gene producing, by itself, one character. To Morgan it was the overall biological, and particularly the de-

[81] Ibid., pp. viii-ix.

[82] G. H. Shull, "The Presence and Absence Hypothesis," *American Naturalist* 43 (1909): 410-419.

[83] See Muller, *Studies in Genetics* (Bloomington, Ind.: Indiana University Press, 1962), pp. 12-16.

velopmental, phenomena rather than the specifically chemical relationships that were of interest in considering the gene as a functional unit. It was undoubtedly Morgan's viewpoint that pervades the final statement in *The Mechanism*:

> It is sometimes said that our theories of heredity must remain superficial until we know something of the reactions that transform the egg into an adult. There can be no question of the paramount importance of finding out what takes place during development. The efforts of all students of experimental embryology have been directed for several years toward this goal. It may even be true that this information, when gained, may help us to a better understanding of the factorial theory [i.e., chromosome theory]—we cannot tell; for a knowledge of the chemistry of all the pigments in an animal or plant might still be very far removed from an understanding of the chemical constitution of the hereditary factors by whose activities these pigments are ultimately produced.[84]

Morgan and all of his coworkers always regarded the gene as a functional unit. However, in 1915 as well as later, they began by focusing their attention largely on the transmissional and structural aspects of heredity (the relations of genes to chromosomes, the interactions among chromosomes, etc.). There was a reason for this approach, and in it perhaps lies Morgan's genius. Study of transmission of genes and their physical relation to chromosomes could be approached experimentally and quantitatively. Consequently, hypotheses could be developed and tested and the structure of the germ plasm and its transmissional qualities worked out in predictive ways. The study of gene function, in a biochemical or embryological sense, could not be approached experimentally in 1915. By focusing their attention on what could be studied experimentally, Morgan and his group were able to develop a highly elaborate but consistent theory. They postponed consideration of functional problems, although they realized that ultimately any theory of heredity had to account for how genes functioned and how they controlled the highly complex processes of embryonic development: "Although Mendel's law does not explain the phenomena of development, and does not pretend to explain them,

[84] Morgan et al., *The Mechanism of Mendelian Heredity*, pp. 226-227.

it stands as the scientific explanation of heredity because it fulfills all the requirements of causal explanation."[85]

The work of the Morgan group as developed by 1915 constitutes a scientific revolution by all of Kuhn's criteria. It introduced a new and consistent theory to replace a series of inconsistent and conflicting views. It provided a fundamentally new world view from which to approach the problem of heredity. It could explain better than any of its predecessors a host of old problems (dominance and recessiveness, reversion, mutation, coupling, the effects of selective breeding), and it opened up a variety of new avenues of research (mapping, gene size, the nature of mutation and recombination). Furthermore, its dissemination, largely through the writings of Morgan (*The Mechanism of Mendelian Heredity*, for example, had wide circulation), created a unified (and worldwide) school of thought, which was taken up by an increasing number of workers. In the period after 1915, the work of the Morgan group fell more and more into the pattern of what Kuhn described as "normal science." The activities of the "geneticists" were aimed at further elucidation of the details and implications of the Mendelian-chromosome theory developed between 1910 and 1915.

Characteristically, Morgan himself drifted further and further from this work. By the time he left Columbia in 1928 he had ceased being involved in much of the day-to-day activity of *Drosophila* research. He had begun to return to his first and greatest love, embryology.

[85] Ibid., p. 227.

CHAPTER VI

Politics and Biology: Development of the *Drosophila* Work (1915-1930)

In the fifteen-year period following publication of *The Mechanism of Mendelian Heredity*, Morgan's work developed in three general directions: (1) continuation of studies in genetics (which meant largely working out the details of the Mendelian-chromosome theory); (2) drawing connections between the new work in genetics and the old, but fundamental problems in evolution and embryology; and (3) publicizing the new views of heredity and their ramifications—largely through writing and lecturing. The last two areas were particularly Morgan's; the first was increasingly taken over by Bridges, Sturtevant, and various graduate students and postdoctoral research associates at Columbia. Increasingly, new facets of the *Drosophila* work itself were taken up by other laboratories throughout the United States and Europe.

By 1915 Morgan found himself on the crest of a wave of great excitement within the biological community. Although he was not a publicity seeker, he was in fact elated by the success of the *Drosophila* work. He took advantage of the general interest expressed in the work of his group to emphasize what to him were the most general and positive aspects of the revolution in genetics. He tried to apply mechanistic and experimental methods to biology in a way that put it for the first time on an equal footing with physics and chemistry. He also emphasized the use of genetic knowledge and methods to effect a deeper understanding of other biological problems such as embryonic development, evolution, and medical therapeutics.

The continued advance of Morgan's work in this period was hampered by the advent of World War I and by the growing strength of the eugenics movement, the social application, in an increasingly racist way, of genetic principles to the supposed improvement of the human race. Although he was not prone to discuss, or get involved in social and political issues, both of these phenomena profoundly disturbed Morgan. Characteristically, he did not allow either to interrupt his basic scientific research. But he did comment on both—an occurrence unusual enough to indicate that he was profoundly upset by what he saw.

WORLD WAR I AND INTERNATIONAL SCIENCE

When the war broke out in 1914, Morgan was unaware of the potential worldwide nature of the hostilities. By 1916 or 1917 he was deeply disturbed on both a social and personal level. Many of his friends in Germany, such as Hans Driesch, were caught directly in the chaos of war. To Morgan, war was useless and savage, and he could not escape thinking about it even though, as he wrote to Driesch, "We have not, of course, been seriously affected by the war so far as our general comfort was concerned."[1] Like his friend Jacques Loeb, Morgan kept himself busy with his work, which served as a diversion from thinking too much about the pain of armed conflict.[2] To Morgan, by far the most tragic aspect of the war was the personal side—the effects it had on his European friends, and especially on their scientific work. This was especially true with regard to Driesch.

Morgan sensed very deeply the pain Driesch felt during the war and was concerned both for his personal safety and for the progress of his work. In September of 1914, shortly after the outbreak of the war, Morgan wrote to Driesch:

> In these days of conflicting reports one does not know whether he is writing to the conqueror or the conquered, but for present purposes, it matters the less because I am sending only a friendly line to let you know that I sometimes think of you in the dreadful war that has overtaken Europe. Whichever way the tide may turn, the amount of suffering is fearful to contemplate. I hope you will drop me a line to let me know how it fares with you and your family and also with Herbst. I hope that he has escaped the conscription. Give him my

[1] Morgan to Driesch, November 12, 1919; Morgan-Driesch correspondence, APS.

[2] Loeb wrote to Hugo de Vries in October 1916: "I have just finished a little book on *The Organism as a Whole from a Physico-Chemical Point of View* which . . . was mainly written to divert my mind from the war." Loeb Papers, Library of Congress.

> best and most cordial greetings. If there is anything I can do for either of you I beg that you will let me know. With kindest regards to Mrs. Driesch.
>
> Sincerely your friend,
> T. H. Morgan[3]

Six months later, in early 1915, just as the *Drosophila* work was reaching its climactic point, Morgan wrote to Driesch about how insignificant laboratory work seemed compared to the international holocaust raging abroad:

> I was very glad to get your good letter a few days ago and am ashamed that I have not answered your letter that reached me in November. Not that I had forgotten to reply, for I had in mind to write very many times. But to tell the truth it is hard to know what to write about except the war. Everything else sinks into insignificance. It is true our life goes on exactly as before, but to write to you who are in the midst of this awful calamity of our little doings seems very paltry. Nevertheless you will be very glad to hear about your friends and to know that your friends often think of you and the dreadful condition of Europe. . . . Occasionally I see Harrison, Parker, Conklin, Jennings. We meet at Christmas time in Philadelphia, and while the meetings are a bore, yet they give an opportunity of renewing old friendships. Harrison's wife is, you know, from Hamburg, and their daughter Elizabeth is at present in Bonn at school. Harrison is much upset by the war and is, of course, an ardent German sympathizer.
>
> My own work and that of my students is given up entirely to *Drosophila* which continues to give off new mutant types, but so far they all (about 125 stocks) fall into four groups, that correspond to the number (four) of chromosomes. . . .
>
> As to the war, we get all the news there is—I mean the official news which comes daily from Germany, France, England, Russia, and Austria. Each one gives a different version of the same affair—or else there is nothing about it at all. So that we are left to draw our own conclusions or accept the newspapers' interpretation which is as likely to be wrong as right.

[3] Morgan to Driesch, September 16, 1914; Morgan-Driesch correspondence.

> I am filled with admiration for the efficiency and bravery of the German army, but even more at the courage of the German nation that is so undaunted with four countries attacking her on all sides.
>
> But what a horrible loss of life and energy a war entails. There ought to be some way of preventing it, for there is really nothing to gain and everything to lose. Personally I do not place the blame for this war on any one nation. They are all responsible and ourselves included in the sense that we have found no sure way of preventing such catastrophies. Everyone hopes that after this war there will be no more of it, but I do not see how it will be prevented. Still, it is better to hope than to despair. . . . Let us look forward to better times when the two families can come together and the next generation begin a new friendship. . . . We cannot stop the war, but we can always remain on the same old footing of good will and good fellowship.[4]

In 1914 Morgan felt uncertain about the causes of the war, who was at fault, or why the war was continuing. Morgan discussed the European situation frequently with Jacques Loeb, who after 1910 was at the Rockefeller Institute for Medical Research (later Rockefeller University) in New York, and had taken a house close to Morgan's on 117th Street. Loeb, a native German, was considerably more emotional about the war than Morgan, and had definite views about its origins and continuation. He wrote to Hugo de Vries in June of 1917: "To me the most ghastly part of the war is the suspicion that the immense profits which big business is making out of the war has a good deal to do with its prolongation. At present the game seems to be to conquer the trade of Russia and South America and for that other equally absurd aim the use of the world is being sacrified and the older generation driven to despair by worrying over the total collapse of reason and civilization."[5] A year earlier (1916), Loeb had expressed his suspicions even more dramatically: "About a year ago when Pierpoint Morgan returned from London, the New York Times had on its first page in big headlines, 'Pierpoint Morgan back from London. Reassures Stockbrokers Rumors of Peace Premature.' . . . Formerly professors were able to eat steak, but now

[4] Ibid., January 25, 1915.

[5] Loeb to de Vries, June 26, 1917; Loeb Papers.

we all have to buy the cheapest cuts of meat, while the bankers and munitions manufacturers, and those in league with them, are rolling in wealth."[6]

To what extent Morgan agreed with Loeb's general viewpoint in 1914 is difficult to determine. It is likely he did not share Loeb's suspicions about the role of the business community in seeking to prolong the war. Loeb was an avowed socialist, and Morgan never had leanings in that direction. However, by late 1915 Morgan was becoming convinced that Germany was the cause of the continuing hostilities.[7] Driesch was obviously more sympathetic to his country's cause, a point on which Morgan could not fail to comment. His overriding concern was still, as always, the personal and scientific well-being of his friend. He wrote to Driesch in December of 1915:

> Many times during the past year I have thought of you, and many times have I said to myself, I will write to him tomorrow, but there have been many tomorrows, much to my regret. I think I must have the feeling that to write to you about our personal doings will seem trivial to you in the midst of this gigantic struggle; and to write about the war will seem futile, because I have lost all faith in our opinions being our own. They seem rather to be made for us by the news we get and by our friends. If I were in Heidelberg I should think exactly as you do. If you were in New York, you would think exactly as I think. It is not a question of nationality, but of newspapers.
>
> We know that you are all angry with us for making ammunition, and I do not wonder that you are. Just now we are angry because our ammunition factories get blown up and I can hear you retort, "It serves you right" etc. Now I should like you to know that I and many of my friends deplore the fact that any country at peace should supply any countries at war with anything to keep the war going, but the trouble as we see it lies deeper. All that should have been agreed to when the countries were at peace and I hope it will be arranged for when the world is again quiet. Some way ought to be found

[6] Ibid., n.d., 1916.

[7] By 1918 there was no question in Morgan's mind that Germany was the originator and continual perpetrator of the war. (See letter to Ross G. Harrison, May 7, 1918; Harrison Papers, Sterling Library, Yale University.)

> to prevent war in the future, and this can only be done by an agreement to disarm both on land and on sea, and to keep only a large enough force to act as police, and treat any nations that go to war in the way that police suppress any disorder.
>
> How easy it is to write down! And how hard it will be to realize! Until we get over this mad notion that nations are of prime importance, there is not much hope but it is the only hope that I can see for the future—I mean to combine into one big human brotherhood that will be so big and so universal that its units can be kept from murdering each other.[8]

It was a source of disappointment to Morgan that Driesch supported the German cause. As the above letter shows, he tried to pass it over, but it obviously influenced their relationship. Morgan wrote to Otto Mohr after the war that he had just received a card from Driesch which gave him great pleasure. He hoped they could pick up the relationship again, but it seemed necessary from that point on not to talk about the war.[9] Morgan persisted in maintaining as friendly ties as he could with Driesch. In the postwar years, for example, when so many items, including books and journals, were in short supply in Germany, Morgan particularly (but also the Harrisons) sent food, clothing, and money orders to Driesch and his family. Morgan felt somehow that Driesch, like many other German scientists, was embittered by the hostility of world public opinion. This persistent feeling on both sides had enormous impact on the postwar development of international scientific research.[10]

An example of the growing anti-German feeling came close to Morgan in 1918. Richard Goldschmidt, recently appointed head of the genetics department at the Kaiser-Wilhelm Institute for Bi-

[8] Morgan to Driesch, December 6, 1915; Morgan-Driesch correspondence.

[9] Morgan to Otto Mohr, October 29, 1919, pp. 2-3; Morgan Papers, APS. Morgan wrote: "If we can manage to get along without talking about the war I think we can soon get back to our old friendly footing."

[10] Not only the friendship with Driesch but also a personal tragedy brought the war into the Morgans' life. Lilian's only nephew, with whom she was particularly close, was killed in France while serving with the U.S. Army. He was killed during the final days of hostilities, his family being notified only after the armistice was declared. (From Edith M. Whitaker, personal communication, October 19, 1975.)

ology in Berlin-Dahlem, was in the United States on the last leg of a return trip to Germany (from Japan) when World War I broke out. Because of the British blockade Goldschmidt was unable to get passage from the United States to Germany. For nearly two years he worked in various laboratories throughout the United States, at the invitation of some of his colleagues, principally Ross Harrison at Yale. In the spring of 1918, however, Goldschmidt was arrested and interned by the U.S. government as an enemy alien.[11] Harrison, who had received his M.D. degree in Germany and was thus sympathetic to the plight of German scientists, and Jacques Loeb, himself a German by birth, tried to help Goldschmidt by suggesting that a number of scientific men intervene with the authorities on his behalf. Among those to whom Harrison appealed was Morgan. Morgan, as his letters to Driesch indicate, did not hold German scientists individually responsible for what their government did (although many allied scientists, particularly the French, continued to do so). However, he felt that an appeal to the authorities on the basis of Goldschmidt's scientific prominence would be both unwise and dishonest. Morgan's response to Harrison is worth quoting at some length because it reveals the various factors he tried to weigh in coming to what was obviously a difficult and distasteful conclusion:

> Dear Harrison,
>
> I am very much perplexed to know what ought to be done or what can be done in Goldschmidt's case. When we saw the misstatements in the newspaper concerning him the day after his arrest one of us 'phoned down to headquarters to ask if a correct statement in regard to his scientific work before and after the war began would be of any help. We got the impression that he had been arrested on specific charges and not on general principles, so that there seemed to be nothing for us to do.
>
> I need not add that I am sorry he could not have been left at liberty, and am more sorry than I can say for Mrs. Goldschmidt and the children. For their sake at least I wish it had not happened or that something could be done to get him back to them.

[11] Curt Stern, "Richard Benedict Goldschmidt," *Biographical Memoirs, National Academy of Sciences* 39 (1967): 141-192; see p. 155.

As to the question you raised concerning an effort on our part to make a plea for him on the grounds of scientific work, I am in doubt. Our own laboratories are becoming empty in order to send men to fight Germany in a war which I personally think was instigated by Germany. I have some reason for believing that both Goldschmidt and Miss Erdmann [a distinguished German cytologist Rhoda Erdmann] believe in the German type of government as superior to our own, for instance. There are good reasons for thinking that the German government has taken every advantage of the good nature and simplicity (if you like) of other nationalities—the German embassy here, for example, appears to have been a hotbed of criminality of a very high order. You would say that *we* ought to rise superior to that kind of thing, and that neither Goldschmidt nor Miss Erdmann should be held responsible for what the German government does or has done. To the first I agree. In regard to the second, I am doubtful. If our government has specific information unknown to us I am not sure that we are called upon *under the circumstances* to press it to state what it knows or to grant special favors. I want to see fair play and am ready to do what I can to see that Goldschmidt has the best that we can offer to enemy aliens, but if he has offended, then his scientific work seems to be a matter *at present* of secondary importance. In a word, I should not urge this claim, if it be one, under the circumstances. On the other hand, if I could be shown that he has said nothing offensive or objectionable to this country, I wish very much that he could be released to continue the important work he has in hand. Until we can learn something of the charges against him, it seems to me that our hands are tied.[12]

Of course, as Morgan recognized, the whole issue hinged on whether in fact Goldschmidt had in a sense acted the part of an alien during his time in the United States. W. M. Wheeler wrote back to Morgan a few days later, agreeing with Morgan's position and adding some new information:

Of course, you know that Goldschmidt is a German reserve officer of artillery and you probably also know that when he was here [in Cambridge at the Bussey Institution where Wheeler worked] he indirectly admitted to one or two of the

[12] Morgan to Harrison, May 7, 1918; Harrison Papers.

> men about the place that while he was in Japan he acted, to a certain extent, in the capacity of information agent. This is, of course, the duty of all German reserve officers when sojourning in foreign parts. In times of peace such operations are probably admissible, and there is probably a good deal of difference between an information agent and a spy. At the present time, however, when we are engaged in a war with an absolutely ruthless and unscrupulous enemy, I am convinced that it is better to have all German aliens interned.[13]

Wheeler went on to say that the Bussey Institution faculty had decided unanimously to reject Goldschmidt's application to work at the institution, largely on the grounds that they did not wish to have a person under suspicion connected in any way with the work there. Wheeler's response to Goldschmidt's plight was considerably less sympathetic than Morgan's, for he went on to say:

> Many [German aliens] are, no doubt, quite harmless, but instead of spending thousands of dollars keeping secret agents on their track it is certainly much cheaper to simply intern them. I have a German neighbor who has four secret agents continuously on his trail. He is probably quite innocent, but these men are receiving a salary, and spend most of their time, apparently, watching his goings in and comings out of his house in the town, and this has already cost the government a great many thousand dollars. Simple internment in this case would be a great saving pecuniarily and also in other respects.[14]

Morgan believed strongly that science and politics ought not to mix. To have intervened on behalf of Goldschmidt, *primarily because of the outstanding quality of his scientific work*, would have been to confuse the two in an illogical way.

Whether or not Goldschmidt was really guilty of the charges leveled against him is difficult, if not impossible, to say. He was released at the end of the war (1919), and soon returned to his post at the Kaiser-Wilhelm Institute. The important aspect of this incident, with regard to Morgan, was the firm expression of his feeling that science and politics should be kept strictly apart.

[13] Wheeler to Morgan, May 10, 1918; from the Wheeler Papers, courtesy of Adaline Wheeler, Boston, Massachusetts.

[14] Ibid.

After the war the pace of scientific research in various countries was altered from its prewar pattern. In the United States and Europe crowded universities and heavier teaching loads greatly reduced the amount of research many investigators were able to carry out.[15] In Germany not only had many research laboratories been decimated and programs stopped, but publication of both journals and books had been seriously curtailed. Germany had been a center for scientific (and especially biological) publication during the prewar era, but with enormous inflation and social and political disorganization, publication was a matter of considerable uncertainty. For example, Otto Mohr withdrew a book from publication in Germany not out of conflicting ideology, but largely because the economic situation made any publication date so far distant and uncertain that he felt the work should be submitted elsewhere. German universities and laboratories had difficulty getting foreign journals because the exchange rate was so unfavorable (owing to inflation and the severe reparations payments demanded by the Allies) that it was impossible to pay the appropriate price. Harrison (in particular) and Morgan, among others, did much to try to get copies of journals to German laboratories on an exchange basis (a free copy of a U.S. journal would be sent to a German institute in exchange for a free copy of a German journal). Lack of journals, and thus restricted, sporadic communication, tended to isolate German scientists, many of whom had, prior to the war, been among the leaders in a variety of fields. It also necessitated the reorganization of the lines of communication and international cooperation.

There was something else more fundamental, however, than the chaos produced by the war itself. There remained in the international scientific community a pervasive anti-German attitude that for over a decade virtually excluded German scientists from intellectual intercourse with their colleagues. This affected large

[15] See Julian Huxley to Richard Goldschmidt, May 14, 1920; Goldschmidt Papers, University of California, Berkeley; and also Morgan to Otto Mohr, October 29, 1919, p. 2; Morgan Papers, APS. It is unfortunate that no full-scale study of the effects of World War I (or World War II, for that matter) on the development of twentieth-century science has been undertaken. Two short monographs by Daniel J. Kevels have made a start, however: " 'Into Hostile Political Camps': The Reorganization of International Science in World War I," *Isis* 62 (1971): 47-60; and "George Ellery Hale, the First World War, and the Advancement of Science in America," *Isis* 59 (1968): 427-429.

issues such as membership in international societies and attendance at international congresses; it also affected small questions such as what foreign authors would be allowed to publish in U.S., British, or French journals.

On the larger scale, anti-German feeling greatly changed the pattern of international scientific activity in the decade after 1918. The French dropped from foreign membership in their *Académie* all German scientists who had signed a manifesto in 1914 protesting "the calumnies and lies with which our enemies are striving to besmirch Germany's undefiled cause."[16] The Royal Society demanded, but never brought about, the removal of the signers of the manifesto from its foreign register.[17] The United States National Academy of Sciences did not go that far, but anti-German hysteria among the scientific community began to grow by 1915. By the end of the war the U.S. populace as a whole, along with many scientific workers, had developed a strong hostility to everything German. For example, George Ellery Hale (1868-1938), an American astrophysicist, director of the Mount Wilson Observatory (Pasadena), and architect of the National Research Council[18] worked long and hard during the course of the war itself to build an international research council—excluding Germany, of course.

International congresses invariably ran afoul of the question of whether or not to extend invitations to member German societies. Generally, the French, Belgians, and British took a strong stand against such invitations; some of the more neutral European powers, such as Holland or Italy, were more lenient. United States scientists generally tried to sidestep the issue. For example, when a proposal was made in the 1920s that the International Genetics Congress meet jointly with the Eugenics Congress in the United States in 1921, a double specter was raised, especially to geneticists such as Morgan. First of all, Morgan wanted to separate clearly genetics from eugenics, which he considered to be largely a pseudoscience. Second, he attempted to skirt the issue of a con-

[16] See George F. Nicolai, *The Biology of War* (New York: Century, 1918), pp. xi-xiii; quoted in Kevels, "'Into Hostile Political Camps,'" p. 48.

[17] Kevels, "'Into Hostile Political Camps.'"

[18] The National Research Council (NRC) was created by the National Academy of Sciences as a joint organization of experts from the universities, industry, government, and the military, designed to promote cooperative efforts in science in general, and coordinate research and technical studies for defense in particular (ibid., p. 49).

frontation with representatives of other nations over invitations to Germany. As he wrote to William Bateson:

> It turns out that they are going to try to have a Eugenics Congress here in 1921. . . . I was personally reluctant to see the attempt to call an international congress next year, because it will undoubtedly raise in acute form the question of whether the Germans and Austrians are to be invited to the Congress. My own opinion is that no congress can be international without extending to them an invitation, and that if the Congress is held this must be the necessary consequence. On the other hand, knowing something of the bitterness of personal feeling, I am afraid that more harm than good will be done in attempting to bring together men of the warring nations, and that it would be wiser to wait a little longer, until things have gotten either better or worse, before we attempt to have such a congress which after all has little scientific value in my opinion and is for social purposes rather than scientific.[19]

Bateson's response was clear and in agreement:

> I agree with all you say and moreover am very unwilling that our congress, whenever it meets, should be mixed up with the Eugenics Meeting. We might do ourselves great harm. . . . I am desirous of getting international relations back to normality as soon as possible, but a move made so soon would lead to bitterness and trouble, I feel sure. . . . But I have been disquieted with the way our scientific world has behaved. I was at the Congress at Brussels last July, hoping to promote moderation, but I got no chance. . . . The Congress in Brussels, acting in sympathy with France, did all that it could to make the breach definite and impassable. I declined to be an officer of the "Union" there constituted. . . . One of the functions of science is to promote international amity. But we can only regard this as in suspense.[20]

Individual and personal relationships may have persisted, but normal exchange between scientists in the Allied and German countries was greatly diminished. Morgan hoped this situation would not persist, but he was realistic enough to realize that to force

[19] Morgan to Bateson, April 17, 1920; Bateson Papers, APS.

[20] Bateson to Morgan, May 19, 1920; Bateson Papers.

amity is to ask for further bitterness and recrimination. Typically, Morgan felt it was better to let things ride until passions cooled.

On the less general, more day-to-day level, the persistence of anti-German feeling continued to leave its mark. For example, in 1920 Goldschmidt submitted a paper to the *Journal of General Physiology*. Loeb, who was then editor, wrote back rejecting the paper because "both Osterhout and Flexner [coeditors] insisted that the appearance of your paper would result in a boycott of the journal in the United States on account of the intense feelings against the Germans."[21] Morgan deplored the loss of open communication but felt there was little he personally could do to cool the passions generated by four years of war.

One effect of the war and the international economic, political, and scientific relations ensuing from it was that German hegemony in science in general, and in biology in particular, declined in the postwar years. As we saw in earlier chapters, German predominance in the biological sciences had been unquestioned in the later years of the nineteenth and even the early years of the twentieth century. Though German dominance was declining before the war, the effects of the conflict hastened that decline considerably. More and more in the postwar years biology became dominated by research emerging in the United States.[22] Students began to come for graduate education to America rather than going to European universities. Germany was wounded economically and politically by the war and was unable to support the kind of research and educational opportunities that had made it a dominant educational force in the last half of the nineteenth century. To varying degrees both Britain and France suffered the same problems, and their educational and research programs likewise declined. As a nation of foremost scientific activity, the United States, however, began to come into its own. Similarly, the economic fortunes of the United States blossomed after the war, and, despite the depression of the 1930s, the country grew continually more prosperous through the decades following World War II. It took several dec-

[21] Jacques Loeb to Richard Goldschmidt, February 16, 1920; Goldschmidt Papers.

[22] For example, in Norway before the war most textbooks in science and medicine were written in German; Norwegian students and scholars went to Germany to increase their background. After the war the books were written in English (most frequently by American authors), and students went to the United States for advanced work (personal communication, Tove Mohr, 1974).

ades, of course, before the United States gained the kind of international ascendancy in scientific research that Germany had held in the latter part of the nineteenth century. But that expansion of scientific activity was only made possible by the strong economic base the United States could muster, especially after World War II.

Largely through the work of the Morgan group, the United States became the center of genetic research. Students came to this country from all over the world to learn the theoretical and practical aspects of Mendelian-chromosome heredity. The applications of Mendelian principles to plant and animal breeding, developed in the United States through the work of R. A. Emerson at Cornell and E. M. East at Harvard's Bussey Institution, were only a few examples of the practical implications that attracted agriculturalists and biological research workers from every corner of the globe. Thus, in the aftermath of the war, the work of Morgan and his group, despite the decrease in communication with European colleagues, could continue to grow and develop.

MORGAN, EUGENICS, AND SOCIAL BIOLOGY, 1910-1930

Less dramatic, but in the long run no less disruptive to the pursuit of genetics than the World War was the growth of the worldwide eugenics movement between 1910 and 1930. "Eugenics" is a term first used by Francis Galton to refer to that branch of the study of human heredity specifically applied to the "improvement" of the human germ plasm. This meant not only elimination of known genetic diseases, but also the selection of favorable traits for encouraged breeding. The eugenics movement began, historically, as a branch of heredity in England. It took lively root in the United States under the leadership of Charles B. Davenport, founder of the Carnegie Institution's Laboratory for the Experimental Study of Evolution at Cold Spring Harbor, and a few years later (1907) of the Eugenics Records Office, also at Cold Spring Harbor. Although there were many eugenics societies throughout the United States, the Eugenics Records Office stood as the central institution for the pursuit of research about human heredity and the improvement of the human race.

The eugenics movement in the United States gained considerable momentum in the early years of the century among biologists and lay people alike. Prior to 1915 Morgan supported the movement, and was a member of the Committee on Animal Breeding of the American Genetics Association, the scientific society that fostered work in practical breeding and eugenics. While there were some positive advances made in the study of human heredity in this early period, eugenicists increasingly claimed that personality traits, intelligence, and behavior patterns were genetically determined—claims most geneticists realized had no basis in fact. Even worse were the increasingly racist overtones the eugenics movement (both in the United States and abroad) took after 1910. Not only were certain traits (of individuals) claimed to be inferior and others superior, whole races and ethnic groups were rated on a scale from "best" to "worst." The eugenicists were instrumental in the passage of the Immigration Restriction Act (the Johnson Act) in 1924, legislation that was based on the belief that American "blood" was becoming degenerate because of the influx of southern Europeans and other "biologically degenerate" stock. They also promoted the passage of compulsory sterilization and antimiscegenation laws in over thirty states.[23]

Although Morgan supported the aims of the eugenics movement in the early years of the century, by 1915 he had lost sympathy with it. In January of that year he wrote to C. B. Davenport resigning from the Committee on Animal Breeding of the American Genetics Association because of the "unsubstantiated" and "reckless" use of genetics to support social and political conclusions:

> Dear Davenport;
>
> I have just written to Mr. Popenoe [Paul Popenoe, editor of the *Journal of Heredity*] resigning from the Committee on Animal Breeding. I am sending you just a line to give a further explanation of why I have done so. For some time I have been entirely out of sympathy with their method of procedure. The pretentious title, for one thing [I presume here Morgan is

[23] Kenneth Ludmerer, *Genetics and American Society* (Baltimore: Johns Hopkins Press, 1971). For more historical information about the eugenics movement, see the following: Mark Haller, *Eugenics: Hereditarian Attitudes in American Thought* (New Brunswick, N.J.: Rutgers University Press, 1963), which deals with the early history of the movement in the United States, especially its sociological basis; Donald Pickens, *Eugenics and the Progressives* (Nashville, Tenn.: Vanderbilt University Press, 1968), which covers the movement as social history; and G. E. Allen, "Genetics, Eugenics and Class Struggle," *Genetics* 79, supplement (1975): 29-45, which analyzes the political, social, and economic forces behind the movement.

> referring to the eugenics committee, which had been named "The Committee to Study and Report on the Best Practical Means of Cutting off the Defective Germ-plasm in the American Population"], the reckless statements and the unreliability of a good deal that is said in the Journal, are perhaps sufficient reasons for not wishing to appear as an active member of their proceedings by having one's name appear on the journal.

But then Morgan went on to say:

> If they want to do this sort of thing, well and good; I have no objection. It may be they reach the kind of people they want to in this way, but I think it is just as well for some of us to set a better standard, and not appear as participators in the show. I have no desire to make any fuss, or to discuss the matter; but personally I would rather be out of it and remain a simple member of the Association, for the sake of the Journal.[24]

Morgan took a principled stand on this issue; he saw the prejudice and lack of scientific evidence rampant in the eugenics movement, and he disassociated himself from it. It was consistent with his desire to avoid unnecessary controversy and to separate science from politics that he took the stand he did. In general, Morgan's priorities dictated that he select carefully the battles he would wage. A public battle against a social or political movement was foreign to his manner of operating. Two other incidents, in which Morgan came up against aspects of the eugenics movement, indicate his reticence about attacking eugenic ideas in a public way. In 1916, Jacques Loeb, as one of the editors of *The Biological Bulletin*, received a paper by Professor J. E. Wodsedalek, a biologist formerly of the University of Wisconsin, but at the time at the University of Idaho. Wodsedalek proposed a comparison of the mulatto (a cross between a Negro and a Caucasian) to the mule (a cross between a horse and a donkey), to show that such hybridization produced an inferior product. Loeb wrote a fiery denunciation of Wodsedalek's paper, which he sent to Morgan with the following request: "Will you do me the favor of looking over the enclosed manuscript which I have thought of sending to Lillie

[24] T. H. Morgan to C. B. Davenport, January 18, 1915; Davenport Papers, American Philosophical Society.

to controvert some foolish and rather vicious statements in a paper by Wodsedalek, of whom, by the way, I have never heard. . . . I am sure that expressions of the kind that I quote should not be published in the *Biological Bulletin* because they surely will be picked up by fanatics like Professor Ross and then utilized to justify the further maltreatment of the Negro."[25] Morgan wrote back the following day:

> I am afraid that you will give our friend Wodsedalek's paper too much advertisement if you call attention to his contribution. . . . On the whole, I am inclined to think that he did not intend to compare the Negro with the ass, and that the suggestion is merely a piece of stupidity on his part, and am even inclined to think it possible that when he said "unfortunately the mulatto is fertile," he meant unfortunately for the comparison. . . . My feeling is that we had better give him the benefit of the doubt. Personally, as I have already gone for Geyer, who is his teacher, in regard to the same matter, I would rather not take up the question again in this connection.[26]

Loeb wrote back on February 5, agreeing to withdraw his attack on Wodsedalek. Morgan's desire to avoid an unpleasant, and what he thought to be unnecessary, controversy carried the day.

In 1918, Loeb brought another eugenics matter to Morgan's attention. In that year Scribner's put out a second (revised) edition of Madison Grant's *The Passing of the Great Race.* The central thesis of this book is that heredity and race are everything in the development of human history, and that the United States, specifically, is on the decline because it is allowing the mixing of racial elements. Grant was blatant in his assertion of Aryan superiority, even claiming at one point that New York was becoming a "cloaca gentium" of Jews, Poles, Irish, Italians, and others of inferior genetic quality. America was on the decline, Grant claimed, because it allowed the superior qualities of Aryan blood to be swamped by polluted immigrants. When J. McKeen Cattell published a moderately favorable review of Grant's book in *Science,*

[25] Loeb to Morgan, February 3, 1916; Loeb Papers. (E. A. Ross was a controversial sociologist at the University of Wisconsin, a leader of the progressive movement, and a staunch eugenicist who claimed that the black race was distinctly inferior to all others.)

[26] Morgan to Loeb, February 4, 1916; Loeb Papers.

Loeb was aroused to indignation. He wrote to Morgan on October 29, 1930 that some one ought to write a counterreview showing the bias and lack of scientific basis for Grant's conclusions. Morgan wrote back:

> I quite agree with you that Cattell ought not to have published [the] review of Madison Grant's book or any other review of that book that did not condemn it. However, you fellows when you become editors have to look out for copy and sooner or later everything is grist that comes to your mill. Cattell knows what I think about Madison Grant . . . but he thinks I take a very extreme position. . . .
>
> The first edition of Grant's book was reviewed by Boas in *The New Republic*, and Muller tells me that it was adversely reviewed elsewhere. . . . I suppose it ought to be reviewed by somebody who knows something of history and anthropology, but unfortunately these are just the kinds of men that indulge *themselves* in that kind of thing. All the zoologists could do is point out the insufficiency and the inaccuracies of his statements wherever it was possible to check them up. As my colleague and your warm, personal friend has fathered the book with an introductory statement, our hands are completely tied.[27]

Morgan's views on Grant's book are perfectly clear. Again, however, he felt that it was appropriate to take no action.

Given the subsequent history of the eugenics movement, with its anti-immigration bias at home and its effect on Nazi race the-

[27] Ibid., October 30, 1918. The introduction to Grant's book was written by Henry Fairfield Osborn, at the time president of the American Museum of Natural History, head of the museum's department of mammalian paleontology, and Da Costa Professor of Zoology at Columbia. Morgan and Osborn were not close personal friends, but because they were colleagues in Columbia's zoology department (which Osborn had established in 1891), Morgan felt he could not criticize the (to him) socially unacceptable parts of Grant's book. Such criticism among colleagues he believed should be reserved for technical details or matters of scientific interpretation. For example, Morgan strongly criticized as vague and indefinite Osborn's use of the term "energy" and other terms in the latter's book, *The Origin and Evolution of Life* (New York: Charles Scribner's Sons, 1917)—but not in print, only in a personal letter to Osborn on December 26, 1917. (See G. E. Allen, "T. H. Morgan and the Emergence of a New American Biology," *Quarterly Review of Biology* 44 [1969]: 168-188.)

ories in Europe, in retrospect one might wish that Morgan and many other geneticists had actively fought in the public arena to stop the spread of eugenic ideas. It is clear that the constant propaganda campaign the eugenicists waged for thirty years contributed to the deepening gulf between whites and other racial or ethnic groups in the United States. The social application of eugenic theories led to specific, detrimental effects on the lives of scores of immigrant families in the United States and to the genocide against Jews in Germany. There is no doubt that geneticists alone could not have stopped the spread of eugenics, no matter how much they spoke out in public. But there were other groups fighting the eugenics movement (immigrant organizations, social workers' groups, etc.), and a strong, public stand by eminent geneticists would undoubtedly have been some help. But scientists in Morgan's day had had less experience with the social and political misuse of science than their contemporaries in the post–atomic-bomb period. Like most of his contemporaries in genetics, Morgan appears to have thought the eugenics movement was a lunatic-fringe development of no long-range consequence.

However, the rapid development of the eugenics movement after 1915 began to rouse Morgan's alarm, and in some of his later writings he tried to suggest a more moderate view of race and heredity. In the second edition of his book *Evolution and Genetics* (1925) Morgan added a phrase that was not present in the first edition (1916); he complained of the "acrimonious discussion taking place at the present time concerning racial differences in man."[28] In the last chapter, which had no counterpart in the first edition, Morgan examined the issue of racial differences in man. He pointed out that the various human races have far more genes in common than they have differences, and that compared to the magnitude and quality of the similarities, the differences are of considerably less importance.[29] Morgan went on to say, as if chiding the eugenicists, "A little good will might seem more fitting in these complicated questions than the attitude adopted by some of the modern race propagandists."[30]

On the whole, Morgan felt that many of the social ills the eugenicists hoped to cure by a whole host of genetic procedures (including sterilization, selective breeding, and antimiscegenation

[28] T. H. Morgan, *Evolution and Genetics*, 2d ed. (Princeton, N.J.: Princeton University Press, 1925), p. vi.

[29] Ibid., p. 184.

[30] Ibid., p. 207.

laws) could be better served by ameliorating environmental conditions. In *Embryology and Genetics* (1934) Morgan pointed out that while biological evidence suggests a possible role for heredity in determining some human behavioral traits it also suggests a stronger role for environment.[31] Modern embryology and genetics, he went on to say, point out that all adult traits are the result of an interaction between genetic and environmental factors. Some may be more the result of one than of the other, but the more complex the behavioral trait the more complex the interaction. The lesson from these facts for social reform, Morgan pointed out, was that the environmental causes of social ills should be remedied first, before any attempt was made to correct genetic causes (if indeed they exist to any measurable extent). He chose his Nobel Lecture in 1934 to emphasize this point more strenuously than he had in the past:

> Geneticists can now produce, by suitable breeding, strains of populations of animals and plants that are free from certain hereditary defects; and they can also produce, by breeding, plant populations that are resistant or immune to certain diseases. In man it is not desirable, in practice, to attempt to do this, except in so far as here and there a hereditary defect may be discouraged from breeding. The same end is accomplished by the discovery and removal of the external causes of the disease (as in the case of yellow fever and malaria) rather than by attempting to breed an immune race. . . . The claims of a few enthusiasts that the human race can be entirely purified or renovated at this later date, by proper breeding, have I think been greatly exaggerated. Rather, must we look to medical research to discover remedial measures to insure better health and more happiness for mankind.
>
> While it is true, as I have said, some little amelioration can be brought about by discouraging or preventing from propagating some well-recognized hereditary defects (as has been done for a long time by confinement of the insane), nevertheless it is, I think, through public hygiene and protective measures of various kinds that we can more successfully cope with some of the evils that human flesh is heir to.[32]

[31] T. H. Morgan, *Embryology and Genetics* (New York: Columbia Uniton University Press, 1934), p. 109.

[32] T. H. Morgan, "The Relation of Genetics to Physiology and Medicine," *Scientific Monthly* 40 (1935): 5-18; quotation, p. 18.

Morgan believed strongly that biology, and particularly genetics, could be of direct benefit to man. His Nobel Lecture, in fact, is aimed at bringing out that relationship. He hoped that the findings of the *Drosophila* school, along with the work of other geneticists, would be useful in alleviating some of the problems of the human race. But he felt that the proper social uses of these findings could be accomplished by scientists sticking to the laboratory to make sure the ideas were correct, rather than themselves becoming propagandists for one or another social cause. Propagandizing for the social uses of science was totally foreign to Morgan's personality.

THE *DROSOPHILA* WORK, 1915-1930

Although it is impossible to do full justice to the many lines of work emerging after 1915 from *Drosophila*, several areas deserve particular attention. Increasingly, it should be noted, after 1915 contributions to genetic theory came from laboratories other than Morgan's at Columbia. The methodology and viewpoint of the *Drosophila* group gradually spread to numerous laboratories in the United States and abroad.

Within the Columbia group, work developed generally along the broad lines outlined in the previous chapter: (1) the complex and lengthy task of mapping the chromosomes of *Drosophila melanogaster* for all known mutants, and its expansion to include comparative studies of other species; (2) further modification of the original Mendelian theory as a result of the concept of "position effect"; (3) from the studies of nondisjunction, the development (largely by Bridges) of a form of what was called the "balance theory" of sex determination.

Continuation and Refinement of the Mapping Work

Between 1913 and 1925 a considerable part of Bridges's and Sturtevant's time went into mapping the four chromosome groups of *Drosophila melanogaster*. In one way, the task was a never-ending one, since each new mutant provided another gene to be mapped. While both Sturtevant and Bridges participated in this activity (as did Muller until he left in 1916), it was largely Bridges who did the most fundamental collection and organization of data. Muller designed a number of stocks with specific chromosome markers, as well as the more traditional stocks bearing a number of recessive mutants on the same chromosome.[33]

To produce accurate maps required an enormous amount of data, since the number of recombinants was always small to begin with; calculations based on a small number of offspring would be seriously jeopardized by sampling error. As Morgan wrote to E. B. Babcock at Berkeley in 1918: "It is surprising how many controls must be made before one can feel secure that the location of a factor expressed numerically is correct, or fairly so."[34] Bridges's persistence in counting flies under his binocular microscope and his apparently inexhaustible energy made his observations particularly valuable. Not only did he continually count offspring and keep accurate data on them, but he could also observe new mutant types or anomalies in phenotypes at the same time. It was tedious work, by any criterion. It was reported that one day in the laboratory, somewhere around 1919, Bridges looked up from his microscope after a particularly concentrated and uninterrupted period of observation, and simply said, "God, what a life." After which he returned to the microscope.[35] New ideas and discoveries, however, constantly intervened, and sometimes the more prosaic work of constructing maps was laid aside. In the spring of 1921, for instance, Morgan wrote to Otto Mohr of the slowness of the mapping project: "But the combinations papers on the IV and III chromosomes are dragging along, and with so much excitement [Bridges's discovery of the triploid *Drosophila* and its implications for sex determination] I haven't the heart to put on the screws and get that long overdue piece of drudgery out of the way. The longer it is deferred, the worse it gets to me."[36]

[33] Markers are visible structures on individual chromosomes which make it possible to follow the inheritance of the chromosome, or even segments of a chromosome (e.g., after crossing-over), from one generation to the next. Using mapping techniques, the location of one or more genes can be pinpointed to a marker region. Hence, markers allow the correlation of breeding data (the appearance of particular phenotypes in offspring) with cytological data (the inheritance of a particular chromosome or part of a chromosome). Such correlations added weight to the chromosome theory of heredity. See A. H. Sturtevant, *A History of Genetics* (New York: Harper & Row, 1965), pp. 53-54.

[34] Morgan to Babcock, May 1, 1918; Papers, Department of Genetics, University of California, Berkeley; housed at the American Philosophical Society.

[35] Tove Mohr, personal communication, 1973.

[36] Morgan to Mohr, May 4, 1921; Morgan Papers, APS.

Bridges's role in the mapping project was crucial in three ways. First, he counted large numbers of flies, providing the primary data on which recombination values were calculated. Second, and simultaneously, he identified new mutants when they occurred. Third, he provided the cytological analysis of mutant strains. This was extremely important. While it was not possible at that time to observe changes in fine structure of individual chromosomes, the existence of gross markers on certain chromosomes made it possible to correlate certain phenotypic changes with physical changes in particular chromosomes. For example, with appropriate markers it was possible to tell when one section from the middle of a single chromosome had been broken away and reinserted in an inverted position. Inverted regions show considerably less crossing-over than noninverted chromosomes as observed in the occurrence of fewer recombinants among offspring. With such "inversion" strains map distances appear different from those for noninversion strains. Bridges's skill in making such cytological observations was essential to the construction of accurate maps.

An anecdote from one summer at Woods Hole indicates how valuable Morgan considered Bridges for the work on this project. Bridges, who was very fond of swimming, took an unusually long and deep dive from the end of the pier in front of the MBL. When Bridges did not reemerge in what seemed like an inordinately long time, Morgan, growing somewhat agitated, is reported to have remarked, "Oh, there goes my fourth chromosome![37]

Sturtevant's role was largely to construct the maps from the data Bridges collected. By the mid-1920s the chromosome maps of all four groups of *Drosophila melanogaster* were well developed; with the exception of new mutations being fitted in, the maps stand quite accurate today. Sturtevant and Bridges continued work on a number of related species, for comparison, yielding dramatic support and extension of the theory. They found that when another species had five chromosome pairs (as in *Drosophila obscura*) or six chromosome pairs (as in *Drosophila virilis*), there were, respectively, five and six linkage groups.[38] Another important finding was that the genes falling into any particular linkage group varied considerably from species to species. It thus seemed that there was nothing magical about the location of any gene, or group

[37] Quotation from Tove Mohr, personal communication, 1972.

[38] T. H. Morgan, "On the Mechanism of Heredity," *Proceedings of the Royal Society* 94 (1922): 187. This is Morgan's Croonian Lecture for 1921, which summarized the state of the *Drosophila* work up to that time.

of genes, on any particular chromosome. This did not mean that their distribution within the chromosomes had no effect on phenotypic expression, but rather it indicated that a number of possible arrangements could exist, and that each produces its particular set of effects in a particular and unique way. The whole genome within the genus *Drosophila* had considerable flexibility with regard to the architecture of the germ plasm itself.

Bar Eye and New Concepts of Gene Expression

In 1913 S. P. Tice discovered in a stock bottle of *Drosophila* one male whose eyes instead of being the normal oval shape (the wild-type condition), were reduced to a pigmented bar across the middle (see figure 16). The condition was called Bar eye; the gene producing it was located at map position 57 on the X-chromosome, and hence it was a sex-linked characteristic. Since Bar eye was dominant to its allele, the wild-type oval eye, all males would be expected to show the condition. Among those who worked on Bar eye was Charles Zeleny, formerly a student of Morgan's and at the time at the University of Illinois. Zeleny observed that (a) Bar mutants reverted to normal more frequently than most mutants; (b) the condition arose almost exclusively in females—i.e., males seemed to inherit their mutant Bar gene from their mothers; and (c) homozygous Bar-eyed females showed a more extreme reduction in eye shape than females heterozygous for Bar, or males with only one Bar gene. Various explanations were offered for the Bar-eye condition. Some workers suggested incomplete dominance, others, chromosome deficiency, and still others the action of nonchromosomal particles such as "episomes."[39]

The *Drosophila* group, however, was intent on determining whether the Bar-eye condition was consistent with the Mendelian-chromosome theory. They thus concentrated their efforts on trying to correlate phenotypic variability in Bar-eye character with observable chromosome changes. In keeping with their strong Mendelian biases, Morgan and his group were especially concerned to show that Bar-eye condition was not the result of qualitative variability in the gene itself.

In studying the Bar-eyed stock, Sturtevant noted that not only did the overall size of the eye vary considerably in different ge-

[39] For a summary see E. A. Carlson, *The Gene: A Critical History* (Philadelphia: W. B. Saunders, 1966), pp. 109-110.

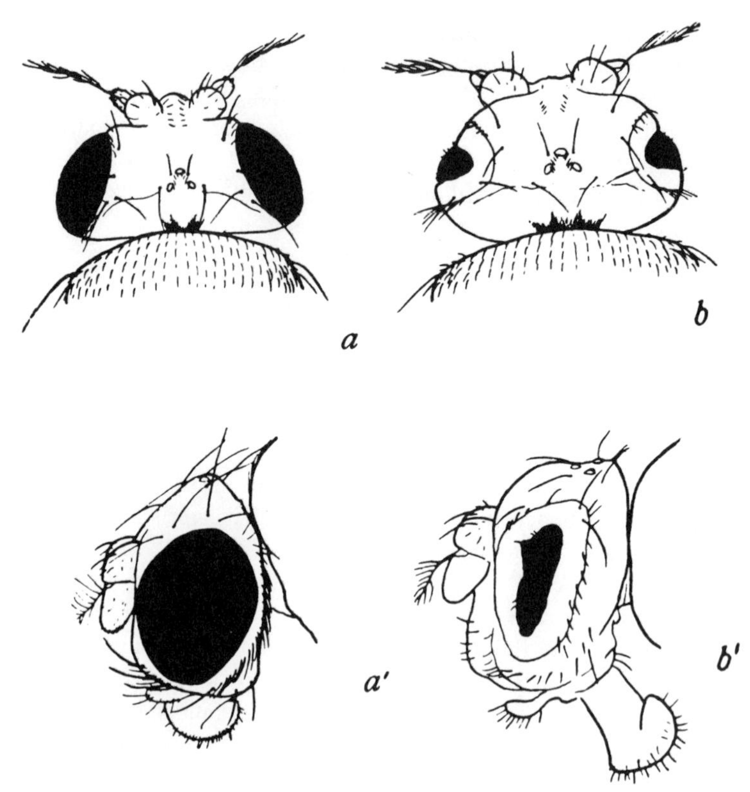

Figure 16. Drawings of normal eye (left, a,a′) and Bar eye (right, b,b′) in *Drosophila.* The mutant phenotype is characterized by reduction in the number of pigmented facets to a narrow bar running vertically through the eye. (From Morgan, Sturtevant, Muller, and Bridges, *The Mechanism of Mendelian Heredity* [New York; Henry Holt & Co., 1915], figure 15, p. 29.)

netic crosses, but the average number of pigmented segments (called ommatidia) also varied. Since ommatidia in an insect eye are discrete structures, they could be counted, thus providing an easily quantifiable character. With varying known Bar phenotypes, Sturtevant found that the average facet number bore a specific relationship to locations of the Bar gene on the X-chromosome, as shown below in table 1.

To explain these results, Sturtevant invoked two hypotheses. The first, which he and Morgan originally suggested in 1923, held that Bar was not a typical point mutation, but resulted from unequal crossing-over between two X-chromosomes (where one chromosome ends up with two identical regions, while the other ends up with no copy of this region; see figure 17). According to this scheme, the chromosome with the two regions produced the Bar phenotype. Morgan and Sturtevant came to these conclusions by

TABLE 1

Genotype	Symbols	Average No. of Ommatidia
Homozygous Bar	B/B	68.1
Heterozygous double Bar/Wild	BB/+	45.4

SOURCE: Summarized from A. H. Sturtevant, "The Effects of Unequal Crossing-Over at the Bar Locus in *Drosophila,*" *Genetics* 10 (1925): 117-147.

NOTE: B/B means that one Bar gene was found on each member of the homologous chromosomes, while BB/+ means that two Bar genes were located side by side on one of the X-chromosomes, while a single wild-type allele was located on the other homolog.

analyzing the reverse "mutation" of Bar to normal, showing that it was generally associated with another unequal crossover which thus counteracted the effects of the first unequal crossing-over.[40] The idea of unequal crossing-over was Morgan's, but the theory did not explain completely the perplexing observation that different phenotypes were obtained with the same number of Bar genes, located differently within the chromosome pairs. Here Sturtevant invoked a second (and his own) hypothesis, what he called "position effect." This was the idea, suggested in a 1925 paper, that the position a gene occupied on a chromosome—that is, what genes were directly next to it or even some distance away—could influence the phenotypic effects that that gene itself displayed.[41]

[40] A. H. Sturtevant and T. H. Morgan, "Reverse Mutation of the Bar Gene Correlated with Crossing-Over," *Science* 57 (1923): 746-747.

[41] A. H. Sturtevant, "The Effects of Unequal Crossing-Over at the Bar Locus in *Drosophila,*" *Genetics* 10 (1925): 143. Later cytological work (after 1934) showed that the original Bar gene did, in fact, arise as a result of unequal crossing-over producing a tandem duplication of the segment 16A of the X-chromosome (the region determining wild-type eye). Reversion (back to the wild type) or further extension of the Bar gene (known as ultra-Bar or infra-Bar) resulted from the mechanism of further unequal

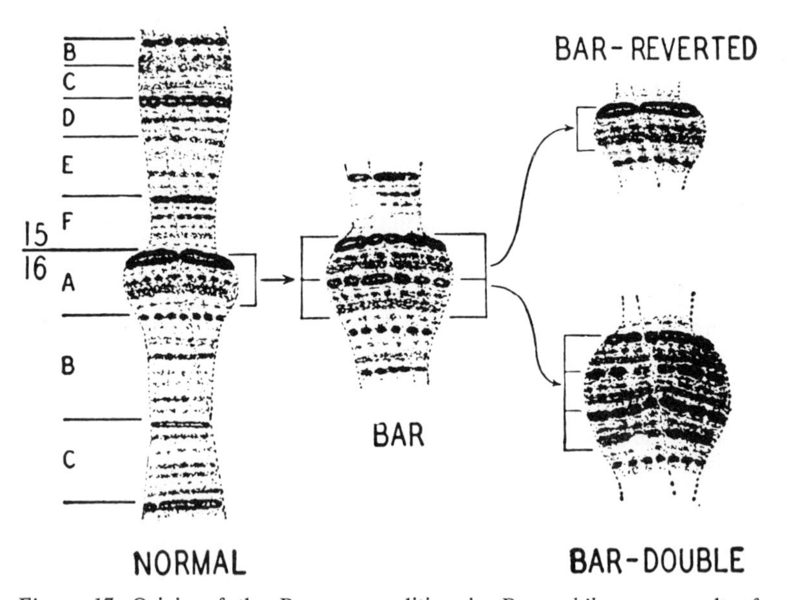

Figure 17. Origin of the Bar-eye condition in *Drosophila* as a result of duplication of a specific region of chromosome I. For comparison, the normal chromosome I structure is shown to the far left. The typical Bar phenotype is produced by a single duplication of the region shown in brackets (center). Double Bar is produced by still another duplication within the same region (lower right), and Bar reverted is produced by loss of the duplicated region (upper right) with a corresponding return of eye phenotype to the wild-type condition. Details of the genetic mechanism producing the various Bar-eye phenotypes were worked out by a number of investigators, including Bridges, Muller, and Prokofyeva (the last two in the Soviet Union) in the late 1920s and early 1930s. (After Bridges, from E. Sinnott, L. C. Dunn, and T. Dobzhansky, *Principles of Genetics*, 4th ed. [New York: McGraw-Hill, 1950], p. 451.)

crossing-over in that same, already duplicated, region. The actual details of the origin of the Bar-eye condition could not have been determined accurately with the cytological materials and techniques available in the early and mid-1920s. The definitive studies on Bar eye had to wait for the discovery of the giant salivary-gland chromosome of *Drosophila* by Painter in the late 1920s and early 1930s (see below), and for improved techniques of staining chromosomes. These studies were carried out largely by Muller and A. H. Prokofyeva, during Muller's three-year stay in the Soviet Union. (H. J. Muller, A. A. Prokofyeva-Belgovskaya, and A. V. Kossikov, "Unequal Crossing-Over in the Bar Mutant as a Result of Duplication of a Minute Chromosome Section," *Doklady Akademii Nauk SSR New Series*

The idea of position effect was a notable departure from the pure Mendelian view that genes are independent units and thus not influenced by other genes associated with them. It also marks a notable departure from the old idea of the Morgan group itself, known as the "beads-on-a-string" chromosome model. In this model, the chromosome was thought of as a long, linear array of individual, atomistic genes, arranged side by side like beads on a string. As with the older Mendelian view, the beads-on-a-string model claimed that the expression of a particular unit factor was independent of the other factors around it. The concept of position effect, like that of modifier genes, introduced into the hereditary process a greater degree of complexity than the older notion, which derived from some of the early Mendelians, that saw genes as rigidly determined "characterlets." Now, even *position* was seen as affecting the phenotypic expression of a gene. Between 1910 and 1925 the work of the Morgan school increasingly emphasized the interactions between genes, and between genes and the environment, rather than the simple, purely morphological, view represented by the original beads-on-a-string analogy. The theory of position effect was a further step toward understanding the molecular and chemical processes of gene action.

The Balance Theory of Sex Determination

The third area in which the Columbia laboratory pioneered during the period 1915 to 1930 was the balance theory of sex, rising out of Bridges's work on nondisjunction. When Bridges first noted the nondisjunction flies in 1913, both he and Morgan had seen the obvious implications for understanding the mechanism of sex determination.[42] Between 1913 and 1915 Bridges and C. W. Metz, one of Morgan's students at Columbia, had shown that while the Y-chromosome was in fact present in *Drosophila*, it was

1 [no. 10, 1935]: 87-88.) Independently, Bridges had also observed that the region of the X-chromosome concerned with the Bar mutation was, in fact, a duplication with two identical regions side by side. (C. B. Bridges, "The Bar 'Gene,' a Duplication," *Science* 83 [1936]: 210-211.) Once again cytology eventually showed what had already been deduced from theoretical considerations of breeding data.

[42] See T. H. Morgan, "Mosaics and Gynandromorphs in *Drosophila*," *Proceedings of the Society for Experimental Biology in Medicine* 2 (1914): 171-172.

not necessary for determining maleness.[43] Thus, from *Drosophila* alone, one might conclude that sex was determined by the number of X-chromosomes. However, several lines of evidence from other species argued against this simplified view. For example, in the plant *Melandrium* (in which the male and female flowers are on different plants) the Y-chromosome is in fact the sex-determining element. Individuals with the Y are male, and those without it are female.[44] While it might have appeared in 1915 that *Melandrium* was a special case, as time went on, more and more species appeared to have a similar sex-determining mechanism. At times, geneticists despaired that despite all the new advances in chromosome theory, sex determination was as elusive a problem as ever.

But then in 1921 Bridges observed a number of unusual looking flies in one strain with which he was working. These flies appeared to be intermediate in many ways between males and females, and he described them as "intersexes." As he wrote: "The intersexes, which were easily distinguished from males and females, were large-bodied, coarse-bristled flies, with large roughish eyes, and scalloped wings. Sex-combs (a male character) were present on the tarsi of the forelegs. The abdomen was intermediate between male and female in most characteristics. The external genitalia were preponderantly female. The gonads were typically rudimentary ovaries; and spermathecae were present. Not infrequently one gonad was an ovary and the other a testis; or the same gonad might be mainly ovary, with a testis budding from its side."[45] Bridges then went on to make an even more interesting observation: "The intersexes showed considerable variation, apparently forming a bimodal group—on the one hand a more 'female-type' the extreme individuals of which might even lack sex combs, and on the other hand a more 'male-type,' many of the individuals having large testes and normal genitalia. All intersexes proved sterile."[46] From previous studies Bridges suspected that these flies might have

[43] C. W. Metz, "Chromosome Studies in the *Diptera*. I," *Journal of Experimental Zoology* 17 (1914): 45-58. The Y-chromosome was, however, necessary for fertility of the strains being studied at the time (i.e., males without a Y were infertile).

[44] See Sturtevant, *A History of Genetics*, p. 83. The same situation seems to apply in a number of other species of plants and animals, including man.

[45] C. B. Bridges, "Triploid Intersexes in *Drosophila melanogaster*," *Science* 54 (1921): 252-254; quotation, p. 252.

[46] Ibid.

an extra set of chromosomes or at least chromosome parts. Upon making a cytological survey of the cells of these intersexes he made the dramatic discovery that they were almost all triploid (that is, containing *three* members, as opposed to *two*, of each homologous chromosome pair, including the sex chromosomes). Triploidy had been observed for isolated segments of chromosomes before, or for single chromosome pairs, but this case was the most extensive degree of triploidy Bridges had encountered. To him it immediately suggested a possible means of analyzing the sex-determining mechanism in *Drosophila.*

The original parents of the intersexes were found to be triploid, yet they were fertile. Thus, they could be bred with other triploid flies, or with normal diploid flies, to produce a variety of chromosome combinations in their offspring. Bridges carried out a number of such crosses and discovered that he could produce a range of morphological types, from abnormal males through intersexes, to abnormal females. Cytological observation of these different types demonstrated an interesting relationship between chromosome count and degree of maleness or femaleness. The relationship, Bridges noted, was not in the total number of chromosomes, but in the ratio of X-chromosomes to number of members in each set of autosomes, i.e., all the chromosomes except the sex chromosomes. Normal, diploid individuals have two members of each set of autosomes; triploid individuals, three members; tetraploid individuals, four; and haploid individuals, one. Normal males have one X-chromosome to two members of each autosomal pair and would thus be written as 1X/2A, and expressed as a ratio of 1/2 or reduced numerically to 0.50. A normal female has two X-chromosomes to two members of each autosomal pair, and would thus be written 2X/2A, or 2/2 or 1.00. Within the offspring of his triploid crosses, however, Bridges observed a whole range of X/A ratios as summarized in table 2. Bridges now made a generalization which was the basis for his balance theory of sex determination: "A significant new conclusion proved by the intersexes is that sex in *D. melanogaster* is determined by a balance between the genes contained in the X-chromosome and those contained in the autosomes. It is not the simple possession of two X-chromosomes that makes the female, and one that makes the male. A preponderance of genes that are in the autosome tend toward the production of male characters; and the net effect of genes in the X is a tendency to the production of female characters."[47] Bridges noted that the ratio of 1.00 represented a normal female, whatever the total number of chromosomes. A ratio of 0.50 represented a normal male. A ratio of less than 0.50, as for example, 0.33 (1X/3A), produced an abnormal male in which the male characteristics were more distinctive, more exaggerated in size. These males he called "supermales." A ratio higher than 1.0, as in 1.33 or 1.50 (4X/3A, or 3X/2A, respectively), produced abnormal females in which the female characteristics were

TABLE 2

Actual Chromosome Numbers	Numerical Ratios		Morphological Type
IX/3A	0.33	increasing maleness ↑	Abnormal male
IX/2A, 2X/4A	0.50		Normal male
2X/3A	0.67		Intersex
3X/4A	0.75		Intersex
2X/2A, 3X/3A	1.00	increasing femaleness ↓	Normal female
4X/3A	1.33		Abnormal female
3X/2A	1.50		Abnormal female

SOURCE: C. B. Bridges, "The Genetics of Sex in *Drosophila*," in *Sex and Internal Secretion*, ed. Edgar Allen (Baltimore: Williams & Wilkins, 1934), p. 82.

exaggerated. These forms Bridges called "super females."[48] Moreover, Bridges saw the balance operating not between individual chromosomes, but between the total complement of genes involved in the various chromosome groupings.

To Bridges the success of the balance theory lay not simply in its intellectual integrity, but in the fact that it led to such accurate

[47] Ibid., p. 253.

[48] Ibid.

predictions. For example, Bridges's earlier work with what were called haplo-IV flies had shown that a certain strain of flies normally had only one member of the fourth chromosome pair (and was thus haploid for that chromosome); haplo-IV organisms had a distinct set of observable phenotypic traits. Bridges predicted that a triplo-IV fly, when it occurred, would show the opposite development of those same characters. When such an organism finally appeared, the prediction turned out to be true.[49] Even more elegant to Bridges was the fact that two gaps existed in the range of observed chromosome ratios. In 1921, when Bridges originally proposed his theory, neither completely haploid individuals (1X/1A) nor tetraploid individuals (4X/4A) had been observed. Lilian Morgan, who had by that time returned to the laboratory and was studying the cytology of certain *Drosophila* strains, found a tetraploid fly, which was indeed female.[50] Bridges, meanwhile, found an interesting case of haploidy. Completely haploid individuals do not develop. However, in a special strain Bridges found a few mosaics in which parts of the body were haploid, while the rest was diploid. The haploid parts showed the female traits, as the theory predicted.[51]

While Bridges found the idea a compelling one, Morgan never fully endorsed it. In a letter to Otto Mohr in 1923, Morgan wrote explicitly of his skepticism: "I entirely agree with the criticism of the inadequacy of the theory of balancing, of which Bridges is so inordinately fond. It is, and must remain, a fiction so long as we cannot attach any objective values to the unit or units involved. It may be quite true, and probably is, in a way, but it tells us nothing more than the facts do themselves. This I have urged on Bridges from the beginning, but without any effect."[52]

In his 1926 book *The Theory of the Gene* Morgan explained Bridges's idea briefly in two pages, falling short of endorsing it.[53] Morgan's objection to the balance theory was twofold. First, it

[49] C. B. Bridges, "The Genetics of Sex in *Drosophila*," in *Sex and Internal Secretion*, ed. Edgar Allen (Baltimore: Williams & Wilkins, 1934), pp. 80-81.

[50] L. V. Morgan, "Polyploidy in *Drosophila melanogaster* with Two Attached X-Chromosomes," *Genetics* 10 (1925): 148.

[51] C. B. Bridges, "Haploidy in *Drosophila melanogaster*," *Proceedings of the National Academy of Sciences* 11 (1925): 706; see also Bridges, "Haploidy in *Drosophila* and the Theory of Genic Balance," *Science* 72 (1930): 405.

[52] T. H. Morgan to O. L. Mohr, March 1, 1923; Morgan Papers, APS.

[53] T. H. Morgan, *The Theory of the Gene* (New Haven: Yale University Press, 1926), pp. 241-243.

seemed to only restate the facts; it was just another way of saying that there was a correlation between sex phenotype and the ratio of X-chromosomes to autosomes. Second, Morgan could not accept Bridges's conclusion that the bulk of the genes determining maleness were on the autosomes, and those determining femaleness on the X. As he wrote: "This evidence from triploids gives no specific information as to the occurrence of genes for sex-determination. If we think of the chromosomes only in terms of genes, it follows that genes are involved, but the evidence does not show what they are like. Even if genes are involved, we cannot state whether there is one gene in the X that stands for femaleness, or hundreds of such genes. Similarly for the ordinary chromosomes —the evidence does not tell us whether the genes for maleness, if there be such, are on all the chromosomes or on only one pair."[54] Morgan also objected to the numerological aspect of Bridges's theory. The existence of ratios might in itself provide a quantitative description of certain otherwise qualitative observations. But there was nothing magical about ratios or "balances." Morgan always held a strong skepticism for theories that attempted to explain something in terms of numerical consistency. He strongly favored the use of quantitative data, but he did not favor the inference from such relationships to physical reality. It was on this same ground that he initially objected to Mendelian theory (see chapter IV), and on which he also objected to the work of the English school of biometricians.

Both the balance theory of sex determination and the concept of position effect contributed to the gradual moving away of the *Drosophila* school from their earlier, atomistic view of genes. The balance theory as well as the idea of position effect were based on a concept of gene interaction. Genes were no longer seen as isolated elements, each acting out a predetermined role, but rather as a complex set of interacting chemical agents intimately involved in the metabolic processes of the cell. Clearly Sturtevant, Bridges, and Muller all espoused such a view.

It appears that Morgan was the least willing to make explicit statements about the physiological nature of genic action. While mirroring in many cases the point of view enunciated by Sturtevant, Bridges, and Muller that genes were molecules interacting as parts of biochemical processes in the cell, since these processes were not subject to experimental analysis at the time, Morgan did

[54] Ibid., p. 242.

not wish to waste his time with hypotheses that could not be tested. Hence, his main focus remained on the problem of transmission of the genic material, whatever its molecular basis and physiological role.

Work outside of Morgan's laboratory proceeded in a number of directions, only three of which bear discussion at this time. One was Muller's work on the X-ray induction of mutation; another was the extension of *Drosophila*-type analyses, i.e., the correlation of breeding results and cytological observations, to plants; and a third was the development of special cytological techniques and materials, such as Painter's work with the giant salivary-gland chromosomes in *Drosophila.*

Of all the above lines of work, Muller's studies on mutation are undoubtedly the most brilliant and far reaching. In a 1927 paper, Muller reported a number of experiments in which, with specially designed strains of *Drosophila,* he showed consistently that, within certain limits, the frequency of mutations was directly proportional to the dosage of radiation.[55] Muller's work provided some of the most basic structural as well as functional information for extending the concept of the gene from its original 1915 form.

Muller's work rested on what is called the "target theory." If a target has distinct boundaries, then all other factors being equal, the number of hits will be proportional to the number of volleys fired at it. Thus, a proportionality between dosage and number of mutations (i.e., number of hits) validates the target-theory assumption. X-ray studies thus confirmed the idea that genes were, in fact, discrete units which had definable and even measurable dimensions.

Extensions of the Mendelian-chromosome theory to plants occurred partly through R. A. Emerson at Cornell and his school investigating the genetics of maize (corn). Emerson, along with E. M. East at Harvard's Bussey Institution and Donald F. Jones at the Connecticut Agricultural Station, applied Mendelian and chromosomal analyses to a variety of plant types. They elucidated linkage groups, showed a correlation between the number of linkage groups and number of chromosome pairs, and explored the mechanisms of sex determination in a variety of plants. Emerson's group, in particular, developed many strains of corn with specific chromosomal markers, analogous to the marked strains developed

[55] H. J. Muller, "Artificial Transmutation of the Gene," *Science* 66 (1927): 242.

by the *Drosophila* group at Columbia. In addition, they mapped the chromosomes of many strains of corn and showed that plant chromosomes, like those of animals, could be viewed as composed of a linear array of genes.

Studies in mutagenesis in plants comparable to Muller's studies in animals were carried out by L. J. Stadler at the University of Missouri. Although Stadler's results did not lead him to exactly the same conclusions as Muller, they nonetheless aided in the formulation of a gene theory in plants that was wholly compatible with that in animals.

Last of all were the cytological developments, occurring mostly around 1930 and 1931, which showed that the chromosome theory as developed by the Morgan group was not based solely on consistent deductions. Until this time the mechanism of crossing-over had never been observed directly but only inferred from stained and prepared material. Aside from Bridges's original work with nondisjunction, there had been no direct observational evidence that chromosomes actually carry the factors that are inherited in Mendelian fashion. In 1931 two groups of workers, Curt Stern with *Drosophila* and Harriett Creighton and Barbara McClintock with corn showed that crossover was an observable phenomenon. Using particular marked chromosomes, Stern was able to observe with live material the actual process of exchange of parts among homologous chromosomes.[56]

Probably the most important single piece of evidence for the chromosome theory was the discovery of the giant salivary-gland chromosomes in *Drosophila* by T. S. Painter (1889-1969) in the late 1920s. Painter had been a student of Ross Harrison at Yale, and afterward of Theodore Boveri at Würzburg. From Boveri, in particular, he had acquired a sense of the importance of cytology as a means for understanding hereditary and developmental problems. In the late 1920s Painter discovered that the salivary-gland cells of *Drosophila* have inordinately large chromosomes (produced by repeated replication of the original chromatid strand and

[56] Curt Stern, "Zytologisch-genetische Untersuchungen als Beweise für die Morgansche Theorie des Faktorenaustausch," *Biologische Zentralblatt* 51 (1931): 547-587; and Harriett Creighton and Barbara McClintock, "A Correlation of Cytological and Genetical Crossing-Over in *Zea mays*," *Proceedings of the National Academy of Sciences* 17 (1931): 485-497. Creighton McClintock did not work with live material, but using chromosome markers demonstrated in fixed material that crossing-over did occur.

the subsequent failure of these strands to separate). Because of their size these giant chromosomes offered excellent observational material. Painter, Bridges, and others were now able to observe many small changes in chromosome structure, which until that time had only been inferred from genetic analysis. For example, the idea that the Bar gene was the result of a duplication could not be observed in the early 1920s when it was first proposed by Morgan and Sturtevant. However, subsequent analysis using the giant salivary chromosome showed that such a duplication was, in fact, present (see figure 16).

Another striking example is the case of chromosomal inversions. In 1917 Sturtevant observed some unusual genetic patterns in a certain geographic strain of *Drosophila melanogaster* which he interpreted as resulting from a block of genes in one chromosome having been rotated by 180 degrees (i.e., the middle segment of the chromosome somehow became dislodged and reinserted back into the same chromosome upside down). The evidence for the existence of such inversions was almost exclusively genetic, based on analysis of peculiar phenotypic ratios in offspring of these crosses. It was almost two decades later when, through use of the giant salivary chromosome material, such cytological patterns of inversion as Sturtevant and Morgan had predicted were actually observed (see figure 16). It was a credit to the brilliance of Sturtevant and the methods of the Morgan group as a whole that such chromosomal aberrations could actually be inferred from the genetic data before they could be seen cytologically. Yet for the Mendelian-chromosome theory to acquire a more fundamental basis in material reality, the cytological observation of such aberrations was required. It was through the discovery of the giant salivary chromosome that such observations became possible. By 1931 there was little doubt, except in the minds of the most intractable, that the Mendelian-chromosome theory was a reality. Genes were discrete units borne on the chromosomes.

FUNDING THE *DROSOPHILA* WORK

In an age of "big science," it is difficult to realize the simple and modest financial support on which much work in Morgan's era was based. In the first three decades of the twentieth century there were no large foundation- or government-supported research grants. Funds for scientific research were difficult to come by. In some cases, universities provided meager resources, but except for a few large and comparatively wealthy schools, this was not a widespread practice. For certain areas of biology, state agricultural stations sometimes provided funds. And, except for a few entrepreneurs such as Louis and Alexander Agassiz, private funding for scientific research was not available. Morgan, like many of his contemporaries, had adapted to this condition by carrying out research on relatively inexpensive topics. In his early embryological work he relied only on a few petri dishes, magnifying glasses, dissecting tools, a small supply of chemicals, and sometimes a microscope. The specimens Morgan collected for most of his embryological research were easily obtained, and often renting a table at Woods Hole or the Naples Station for a few months was his major expense.

Similarly, the *Drosophila* work could be carried out, at least initially, on a relatively modest basis. The flies were obtained at no cost in the wild. They were grown in milk bottles which Morgan collected from his friends and associates. The major expense was for *Drosophila* food, which by 1915 was costing anywhere from $100 to $200 per year. In fact, late in 1929, Morgan found a new and cheaper method for preparing food which reduced the cost to $50 per year! After about 1915, however, additional expenses loomed on the horizon. It was necessary to obtain more binocular microscopes as the number of people working in the fly room increased. For Bridges, it was necessary to obtain a high quality microtome (apparatus for slicing thin sections of tissue to be observed under the microscope). And it was increasingly necessary to hire assistants, at first to help in the day-to-day preparation of food and washing of bottles in the lab. Later it became necessary to provide steady salaries for people like Sturtevant (research associates not paid by Columbia University), and for artists and administrative assistants like Edith Wallace.

In the early years (between 1910 and 1915) the *Drosophila* work was largely financed by Morgan personally, with small contributions from Columbia University. Graduate student support, of course, came from the university, mostly as teaching assistantships, but in a few cases as scholarships. The university also provided basic equipment such as magnifying glasses, binocular microscopes, and glassware. Most of the rest, however, came directly from Morgan's own funds. For example, in addition to furnishing a scholarship for Sturtevant for a semester (1912-1913), Morgan also paid Bridges's salary as bottle washer (1912-1913). This is something Morgan could afford largely because he and his wife had independent resources—a course not open to every biologist at that time.

By 1915, however, the work had expanded in so many directions that it was necessary to seek more money if the new genetics was to fulfill its promise. Morgan thus turned to the Carnegie Institution of Washington, whose willingness to support basic research in heredity and evolution had been well known since Davenport established his laboratory at Cold Spring Harbor in 1904.[57] The Carnegie Institution of Washington agreed to support the *Drosophila* work for an initial five-year period. Between 1915 and 1919 the Carnegie Foundation provided Morgan with an annual grant of $3,600, which was used solely to pay the salaries of Bridges and Sturtevant as full-time Carnegie Institution research associates. The remainder of the expenses, such as the salary of Edith Wallace, was still paid for by Morgan from his own funds.

It was thus understandable that Morgan was "annoyed and flabbergasted" when he learned in 1919 that his former student Charles W. Metz was proposing to set up a large-scale *Drosophila* research program at the Carnegie laboratories in Cold Spring Harbor and was asking for considerably more financial support than Morgan had. Even more disturbing was Metz's suggestion that the Morgan group limit itself only to *Drosophila melanogaster*, leaving the

[57] For Davenport's role in the founding of the Cold Spring Harbor Laboratory, see E. Carlton MacDowell, "Charles Benedict Davenport, 1866-1944, A Study in Conflicting Influences," *Bios* 17 (1946): 3-50; especially pp. 15-24. See also Charles E. Rosenberg, "Charles Benedict Davenport and the Beginning of Human Genetics," *Bulletin of the History of Medicine* 35 (1961): 266-276. As reflected in their support of the eugenics movement, the wealthy "robber barons" of the early 1900s had much to gain by supporting research into the question of heredity (especially human heredity). The conditions of deprivation and poverty found among the poor immigrant and working-class elements of society could be explained in one of two ways: such conditions were caused by unequal distribution of income, or degradation and poverty were innate in the hereditary material of the lower classes. Obviously the financial elite favored the second alternative, since it provided not only a justification for their own present position at the top of the social and economic pyramid, but an argument against welfare programs and other social reforms that might cut into their profits. The economic basis of support for such theories of biological determinism has been extensively discussed by Richard Hofstadter in *Social Darwinism in America* (Boston: Beacon Press, 1964).

Cold Spring Harbor group to study all other species, and his inquiring whether Bridges might not be able to work better at Cold Spring Harbor than at Columbia. Metz's letter outlining his plan has not been found, but Morgan's reply has and deserves to be quoted at some length. It shows not only Morgan's hardheadedness in the face of a possible loss of the work and group he had built up to that time; it also shows his forthrightness in telling Metz exactly what he thought:

> My dear Metz:
>
> I am glad that you wrote fully and in a friendly spirit about the plans you are making "to take up *Drosophila* work on a life size scale"—a scale that has been beyond our reach, but not beyond our dreams. I am heartily in favor of any work and all work that will promote the interest of heredity in *Drosophila*, but some of the questions which you raise are so serious for us, that I feel sure you will want to look at the matter from all sides before entering upon such a far-reaching undertaking. There are three points that are more important than the rest and in this letter I will do little more than touch on these.
>
> *First.* If you are prepared to offer any of the men working with me better opportunities for work, or greater freedom, or time for research than they can get in this laboratory (where, so far as I know, not a single restriction has been placed on them) I shall gladly help you to give them a lift and wish them Godspeed. But if the difference is only that you can get more money for them from the Carnegie than I have been able to get, then, since it is paid by the Carnegie in either case, it would only be fair, I think, to offer them such an amount and let them decide which opening gives better opportunity for their future, and for their immediate work.
>
> *Second.* If I have been too slow or lazy—I should like to think too modest but I dare not—to urge our necessities on the Carnegie (because I thought they were carrying all the load they could stand), and if you have been more successful in presenting the claims of this field or research, go ahead, and I am ready to help your plans in every way I can. You know, of course, that the $3600 given to me by the Carnegie for the work is being spent entirely on the salaries of Bridges

and Sturtevant; that I have had to pay out of my own pocket for an assistant to keep the stock going and make the necessary illustrations for our work. This has cost already somewhere between $6000 and $8000, and it has been done at a personal sacrifice since we haven't really the money to spare from ourselves and from our children, but it seemed worthwhile to do it in order to carry on the work. Perhaps it has been my own fault that I have not been able to get more support, but heaven knows that is not because I have not felt the need of it, and so you will understand that I cannot but envy you your good fortune in coming into the same field and getting more of the good things than have fallen to our lot.

Third. You intimate that your group would work on all the other Drosophilas except the particular one, melanogaster, with which we are chiefly concerned. I am pretty sure you do not intend to imply that we should have an understanding to discontinue our work on any of these other forms. Of course we could make no such arrangement! And from your point of view also, such a division would not be desirable, for any work that did not take the fullest advantage of the rich material that we have accumulated would handicap itself unnecessarily. For example, the problems on which you have made the most progress, namely, in parallel mutations, crossing over, etc., are only significant when compared with melanogaster data.

I need not point out to you that even before you came to Columbia, we have had other species under observation and tried crosses with them, and later obtained mutations from them. Sturtevant has, as you know, recently catalogued the wild species of Drosophila of North America, and he and others working in this laboratory have found mutations, one of which recently obtained has proven by all odds the most interesting one for comparison with melanogaster yet obtained. Only this summer and autumn Dr. O. L. Mohr and Sturtevant have discovered the cause of a high male ratio that appeared in funebris, a solution that we had failed to obtain in melanogaster although we had three times at least found high male ratios. Why then should we narrow our researches when other flies may help us out with melanogaster? Why in

> short is not this laboratory the best place for such work, if funds for the work can be found? We have the ambition, the energy, the room, the material, and I hope the brains.
>
> Concerning the undesirability of such limitations as you suggest, I feel sure that on further thought you will agree with me; for, if not, the import of thc teaching in the Columbia laboratory has failed to take root, viz. that the best work can be carried out by constant cooperation between individuals working on the best materials obtainable, and that the moment they begin to appropriate special fields of influence, the work will suffer.
>
> I beg that you will take all these things into consideration, for, I feel quite sure that you have the same general purposes and ideals that we have in carrying on the work, and such being the case we can undoubtedly find a way by means of which we can see the Drosophila business through.
>
> Sincerely,
> T. H. Morgan[58]

Morgan's letter is the essence of sound reasoning, candor, and diplomacy. He expressed his disappointment at the prospect of losing support for the same work to someone else, but he phrases it in terms of envy, and even willingness to aid Metz because, after all, the most important thing is to get on with the work. At the same time, he argued against Metz's presumption that a division of labor as regards different species of *Drosophila* might be effected between the Cold Spring Harbor and Columbia laboratories. As Morgan points out, even Metz's own most significant work (the parellelism in mutation and crossover rates among different species of *Drosophila*) is only meaningful when compared to *Drosophila melanogaster* data, which by that time were so extensive.

It is at the end of the letter that Morgan makes what is probably his most significant point. Scientific work should not be carried out in an atmosphere of competition between groups, each vying for the greater share of the resources or the results. The best work can be carried out by constant cooperation and *not* by dividing up a field into isolated topics. The success of the *Drosophila* group, as has been pointed out, was rooted in the constant inter-

[58] Morgan to Metz, February 26, 1919; Morgan Papers, APS.

action among its members and in the fact that no one specialized so heavily in one area of the work that he did not at all times know what was going on in others. What Metz was proposing seemed to Morgan to be scientific empire building, which he felt would be in the long run detrimental to the work.

Possibly as a result of Metz's proposal, Morgan entered into new negotiations with the Carnegie, and their support more than tripled in the following year (beginning in 1920). Furthermore, Metz's group was incorporated into Morgan's instead of working separately, under a funding arrangement from the Carnegie. Morgan wrote with some satisfaction to Otto Mohr two weeks after writing to Metz: "We have had quite a turn-over in our relationship with the Carnegie. We shall be much better off than heretofore. The salaries of the boys will be considerably increased. I am to have an artist and a technical assistant, with a fund for running expenses and another for apparatus. Metz also is to have a similar group (if he can get as good a one) and next winter, at any rate, the two groups will work here together in this laboratory. So if you both will come back again, you will find the flies humming."[59]

For 1920 the Carnegie's total bequest was $11,000 and this was gradually increased over the next decade, decreasing slightly after 1931 as a result of the depression. Far and away the greatest portion of each year's grant went to salaries for Sturtevant, Bridges, and Wallace. Salaries were not large in those days, Sturtevant and Bridges each receiving $2,750 in 1920, increasing to $3,500 in 1921 and $4,000 in 1926. Wallace received $1,100 in 1920, increasing to $1,800 by 1928.[60] A more precise breakdown of the Carnegie grant by year is given in table 3. The Carnegie Institution supported the *Drosophila* work for the rest of Morgan's life.

In obtaining and allocating funds for his research, Morgan was always known to be tightfisted with institutional funds and generous with his own. As Sturtevant has said: "When his own pocket was involved, he was very generous, and many a student was helped by him (though he did not like this to be known); but he was very saving—sometimes it seemed almost miserly—when the source was institutional. He always tried to stay as far as he could under any budget that he administered, and he was reluctant to ask for

[59] Morgan to Otto and Tove Mohr, March 6, 1919; Morgan Papers, APS.

[60] Wallace was handicapped by epilepsy and frequently did not work full time.

TABLE 3

CARNEGIE FOUNDATION GRANT BUDGETS, 1920-1934

Year	Total	Salaries			Apparatus		Travel
		AHS	CBB	EMW	Equip.	Food	
1920	$11,000*	2,750	2,750	1,100	500	200	1,000
1921	12,500*	3,500	3,500	1,500	160	250	
1922	11,200	3,500	3,500	1,500	1,400#	250	
1923	11,285	3,500	3,500	1,500	300	250	
1924	11,374	3,500	3,500	1,700		250	
1925	12,374	3,500	3,500	1,700		250	
1926	12,475	4,000	4,000	1,700		250	
1927	Data not available						
1928	12,575	4,000	4,000	1,800		250	
1929	Data not available						
1930	11,775	3,500*	4,500	1,800		50**	
1931	11,775	3,500*	4,500	1,800		50**	
1932	10,000						
1933	10,000						
1934	10,000						

SOURCE: Morgan Papers, Cal Tech.

* Actual amounts appropriated by Carnegie (according to notification letters from Carnegie). All other amounts taken from Morgan's letters of request submitted to Carnegie each year.

For section cutter ($1,200) and microscope ($200).

** New method of food prep.

NOTE: Breakdown figures do not add up to full budgeted amounts. This indicates only that full breakdown was not available from Morgan's books. The difference may have included a general fund for unexpected expenses, or it may have been designated in categories Morgan failed to record (he kept most of the records himself in a series of soft-back notebooks).

an increased budget."[61] Examination of Morgan's extensive correspondence with and annual financial report to the Carnegie Institution reveals that Morgan repeatedly wrote with pride about the amount of money he could turn back to the institution at the end of each year. He was particularly proud of his new method of preparing the food which reduced the cost by a factor of four. To him, the best science asked simple questions of nature for

[61] A. H. Sturtevant, "Thomas Hunt Morgan," *Biographical Memoirs, National Academy of Sciences* 33 (1959): 298.

which a simple answer could be expected. Although Morgan did not always deplore the introduction of expensive or elaborate equipment into biology (increasingly, he recognized the importance of such "hardware"), he did feel there was a strong tendency among research workers to depend on fancy gadgetry. To him, equipment removed the investigator one step further from the organism. It was the organism, and its visible behavior, that always excited Morgan the most.

INFLUENCE OF THE *DROSOPHILA* WORK, 1915-1930

Science proceeds not only by innovations in ideas and techniques, but also by the influence of people and schools of thought. The widespread acceptance of the Mendelian-chromosome theory by the middle or late 1920s was in no small part due to the direct influence of Morgan and his group. That influence came from two sources. Morgan himself played a key role in the dissemination of the new genetics. Through his voluminous writing, his influential position in various societies and on the editorial boards of important journals, and through lecturing in the United States and Europe, the new discoveries of the *Drosophila* group became known. A second source of influence came from the work of many investigators trained in the Morgan lab over an eighteen-year period. Through the graduate students and postdoctoral fellows who worked in the fly room the wonders of the new genetics were transmitted to students all over the world.

One of the most direct means Morgan used to get before the scientific, and to some extent the general, public was by writing. A quick survey of Morgan's bibliography shows that between 1910 and 1915, he published two books dealing solely with the new work in genetics[62] and fifty journal articles. In addition, he published articles on a number of other subjects, including the cytology of chromosomes in parthenogenetic forms and embryology. Between 1915 and 1930 he published two more books devoted solely to the new genetics[63] and one on the relationship between

[62] T. H. Morgan, *Heredity and Sex* (New York: Columbia University Press, 1914); and *The Mechanism of Mendelian Heredity* (New York: Henry Holt & Co., 1915, written with Sturtevant, Muller, and Bridges).

[63] *The Physical Basis of Heredity* (Philadelphia: Lippincott, 1919); and *The Theory of the Gene.*

the new genetics and evolution.[64] He also wrote, as author or co-author, another fifty articles dealing either with genetics alone or with the relationship between genetics and evolution or embryology. The *Drosophila* group as a whole (Morgan, Muller, Sturtevant, and Bridges) published a laboratory manual on *Drosophila* experimentation for college genetics courses.[65] Morgan's unusually lucid writing style made the new results with *Drosophila* eminently comprehensible to specialists and nonspecialists alike. Furthermore, he wrote for a wide variety of journals whose audiences included a large percentage of the biological and scientific community. His articles appeared in highly specialized journals such as *Genetics*, *Proceedings of the National Academy of Sciences*, and *The American Naturalist*; in less specialized scientific journals, such as *Popular Science Monthly* and *Science*; and, for the general reader, in *Harper's* or *Atlantic Monthly*. He also wrote the article on "The Theory of Organic Evolution," for the thirteenth edition of the *Encyclopedia Britannica* (1926).

Through his influential position in various professional societies and on the editorial boards of several journals, Morgan made the work of his group well known from the very first. Morgan's influence in this regard was especially important in the early years before his younger associates had established their own reputations and developed their own channels for quick and easy publication. For example, when the National Academy decided to begin producing a *Proceedings* in 1915, Morgan made certain that a report of the group's findings was published in the first issue. As a founder and member of the editorial board of two very important journals, *The Journal of Experimental Zoology* (founded 1904) and *Genetics* (founded 1916), Morgan saw to it that the *Drosophila* work received its fair share of space. Morgan's personal connections also helped in this regard. He was a close friend of J. McKeen Cattell, long a professor of psychology at Columbia and throughout the 1920s and 1930s the editor of *Scientific Monthly*, *The American Naturalist*, and *Science*, in all of which Morgan and his students published frequently.[66]

[64] *A Critique of the Theory of Evolution* (Princeton, N.J.: Princeton University Press, 1916), revised later as *Evolution and Genetics* (1925).

[65] T. H. Morgan, et al., *Laboratory Directions for an Elementary Course in Genetics* (New York: Henry Holt & Co., 1923).

[66] For a discussion of Cattell's life, see Michael Sokal, "The Unpublished Autobiography of James McKeen Cattell," *American Psychologist* 26 (1971): 626-635.

Morgan was invited to give a number of special lectures throughout the United States and Europe between 1915 and 1930. His personal influence was thus an additional factor in emphasizing the many important aspects of the new genetics. Morgan was not a campaigner; he did not like to engage himself in anything that smacked of publicity mongering and fanfare. During his lifetime he turned down a number of invitations to give talks on subjects he did not feel competent to discuss, or which he felt played on sensationalism or superstition, such as eugenics or "biology and religion." At the same time, however, he accepted those invitations that were important enough to gain for the work the recognition he felt it deserved.[67]

Morgan was also invited to give a large number of papers at symposiums and general scientific meetings. Furthermore, his position as an elected officer in a number of professional societies gave him considerable leeway in arranging programs for annual meetings. During the period of the *Drosophila* group's most rapid development, roughly between 1910 and the early 1930s, Morgan was president of the Society for Experimental Biology and Medicine (1910-1912), the National Academy of Sciences (1927-1931), the American Association for the Advancement of Science (1930), and the Sixth International Congress of Genetics (1932). He was basically modest and did not use his personal position or prestige to further work that could not stand equally well on its own. However, considering the positions of influence Morgan held,

[67] Among these were the Lewis Clark Vauxeum Lectures at Princeton (on the relation of genetics to Darwinian natural selection, 1916); the Hitchcock Foundation Lectures at the University of California, Berkeley (on the same subject, also delivered in 1916); the Middleton Goldsmith Lectures (on genetics and pathology, 1922), published as *Some Possible Bearings of Genetics on Pathology* (Lancaster, Pa.: New Era Printing Co., 1922); the Croonian Lecture, before the Royal Society, London (on the basic mechanism of the Mendelian-chromosome theory, 1922), published as "On the Mechanism of Heredity," *Proceedings of the Royal Society* B 94 (1922): 162-197; the Mellon Lecture at the University of Pittsburgh Medical School (on human inheritance, 1924), published as "Human Inheritance," *American Naturalist* 58 (1924): 385-409; the Silliman Lectures at Yale (a series of public lectures outlining the theory of the gene, 1926), published as *The Theory of the Gene*; and, ultimately the Nobel Lecture, presented in Stockholm (on the relation of genetics to medicine, 1934), published as "The Relation of Genetics to Physiology and Medicine," *Scientific Monthly* 41 (1935): 5-18.

it would be unrealistic to deny their role in drawing attention to the work of his group.

The influence of Morgan's students, subsequent to leaving the Columbia laboratory, was also an important factor in making known the new work in genetics.[68] It should be pointed out that the influence that graduate students and postdoctoral investigators carried away from Columbia was not only Morgan's; it was also E. B. Wilson's. So similar were Morgan's and Wilson's ideas on many subjects, and so intertwined their day-to-day association, that both were strong influences, in different ways, on the younger workers in the laboratory. Morgan's influence was of the "romantic" kind: a flair for new ideas, overt and visible excitement about the work, boundless energy, and a feel for the connection that could be made between breeding data and the biological phenomena of heredity in living organisms. Wilson's influence was in many ways less directly visible. It was of the "classic" kind: less overt, patient, moving more slowly, in a more calculated direction, fiery and intense in its own way, and above all, committed to the idea that the fundamental principles of biology can be discovered by analysis of details at the cellular level. A basic aspect of the *Drosophila* work, described earlier as the unification of the breeding tradition (concerned with whole organisms) and the cytological tradition (concerned with cellular events), was personified in the work of Morgan and Wilson. Through their many students, the newly unified Mendelian-chromosome theory of heredity and the excitement surrounding a young field was communicated throughout the world.

Those who worked in the fly room subsequently took positions in diverse regions of the United States and Europe, thus ensuring that a wide range of new students were brought into contact with the *Drosophila* work.[69] Muller, as pointed out earlier, was probably the most widely traveled; he taught at Rice University (Houston) and the University of Texas (Austin), held a Guggenheim Fellowship at the Kaiser-Wilhelm Institute (Berlin), spent three

[68] Sturtevant has drawn up a "scientific genealogy" for the Morgan group, as shown in appendix 2 (from *A History of Genetics*, pp. 140, 142). This genealogy indicates to some extent the pathway of influences affecting Morgan, and through which he affected others.

[69] Morgan did not, of course, seek to place students in jobs in order to achieve a wide geographic distribution. Many of the fly-room students took jobs where they were available, which also happened to take them far and wide.

years in the Soviet Union, worked at the University of Edinburg, taught at Amherst, and finally at the University of Indiana. In each place he built up a group of students devoted to the *Drosophila* work. Fernandes Payne (Ph.D., 1909) spent his entire teaching career at Indiana University, where he was dean of the faculty and, in 1945, instrumental in bringing Muller to Bloomington. Charles W. Metz (Ph.D., 1917) spent fifteen years as a research worker at Cold Spring Harbor, and subsequently taught for a time at Johns Hopkins. H. H. Plough (Ph.D., 1917) spent his entire career at Amherst; during the summers, however, he taught at the Marine Biological Laboratory and thus transmitted the *Drosophila* work to many graduate students. Donald E. Lancefield (Ph.D., 1921) taught at Columbia from 1922 until 1938 and at Queens College from 1938 to 1963. Others were strongly influenced by the *Drosophila* work, although not technically graduate students of Morgan's. Franz Schrader (Ph.D. under Wilson, 1919) was a cytologist who taught subsequently at Bryn Mawr and in 1930 returned to Columbia on Wilson's retirement. Charles Zeleny (Ph.D., Chicago, 1904) had worked with Wilson at Naples, and through Wilson met Morgan. Zeleny became a champion of the *Drosophila* work at the University of Illinois from 1911 on. Leslie C. Dunn wanted to study with Morgan in 1914, but Morgan's facilities were already crowded, so Dunn went to Harvard instead and took his degree with Castle studying mammalian genetics.[70] In 1928, however, Dunn returned to Columbia to replace Morgan and carry on the work in genetics there.

Among others who worked in the fly room were three of international importance: Otto L. Mohr from Norway, Theodosius Dobzhansky from the Soviet Union, and Curt Stern from Germany. All came to Morgan as postdoctoral investigators. Mohr and Stern returned to their native countries and carried the new genetics along with them. Dobzhansky remained in the United

[70] Dunn recalled in 1965: "I stood in the doorway of that room [the fly room] in the spring of 1914, having read Morgan's *Heredity and Sex* [published 1913] which had inspired in me the hope that I might be allowed to take graduate work with the group in the following year. But as Morgan explained, and was obvious from the doorway, there was little room." (L. C. Dunn, *A Short History of Genetics*, New York: McGraw-Hill, 1965, p. 140.) Dunn was always an enthusiastic proponent of the work of the Morgan group. His entrance into the fly room was finally achieved in 1928, with the onerous task of cleaning out Morgan's lab after the group had departed for California Institute of Technology.

States and carried the *Drosophila* work in new and imaginative directions. In Norway, Mohr extended Mendelian-chromosome principles not only to animal and plant breeding, but also to human genetics. Stern came to Morgan in the late 1920s from Germany and was an important member of the group for two years. After returning to Germany, he developed his significant work on the Mendelian and chromosomal nature of genetic mosaics, work that he continued after coming permanently to the United States (University of California, Berkeley) in 1933. Dobzhansky came to the United States as a Rockefeller Fellow of the International Education Board in 1927; he worked with Morgan's group at Columbia for one year, moving subsequently to Cal Tech, Columbia, Rockefeller Institute, Santa Cruz, and finally the University of California at Davis. Dobzhansky brought to the *Drosophila* group a strong background in the Russian school of population genetics. He combined the study of Mendelian heredity and cytological analysis of chromosomes in work on evolutionary problems with populations of *Drosophila* in the wild. It was Dobzhansky, more perhaps than anyone else, who exploited the evolutionary possibilities the *Drosophila* work had opened up.

The new genetics was also propagated among biologists through the influence of Morgan and his group at Woods Hole. He and his group spent every summer there from 1910 through 1919 and from 1922 through the early 1940s. Because of his long association with the MBL, Morgan knew most of the prominent investigators who worked there regularly during the summers. In the early years (1910-1920) they occupied only one or two rooms on the top floor of one building. However, by the early 1920s Morgan could write to Otto Mohr that the *Drosophila* work had expanded to cover almost the entire third floor. Morgan also gave lectures in various MBL courses as well as occasionally in the famous Friday Night Lectures in which he discussed various aspects of the continuing genetic work. Because so many prominent investigators, graduate students, and young teachers came to Woods Hole in the summers from all over the country, this was a particularly important way in which the new ideas gained ground.

Through these various channels the *Drosophila* work was well and even enthusiastically accepted throughout most of the United States by 1920. Yet acceptance was not complete. A variety of criticisms emerged during the late teens and early twenties which attacked some of the most basic developments of the Morgan school.

OPPOSITION TO THE MENDELIAN-CHROMOSOME THEORY, 1915-1930

From the carly days of the *Drosophila* work, a number of criticisms were launched against the chromosome interpretation of Mendelism. Many of these came from outside the field of heredity. Morgan took note of the objections by nongeneticists in 1917 when he wrote:

> The objections have taken various forms. It has been said, for instance, that the factorial interpretation is not physiological but only "static" whereas all really scientific interpretations are really dynamic. It has been said that since the hypothesis does not deal with known chemical substances, it has no future before it, that it is merely a kind of symbolism. It has been said that it is not a real scientific hypothesis for it merely restates its facts as factors, and then by juggling with numbers pretends that it has explained something. It has been said that the organism is a whole, and that to treat it as made up of little pieces is to miss the entire problem of "organization." It has been seriously argued that Mendelian phenomena are "unnatural," and that they have nothing to do with the normal process of heredity and evolution as exhibited by the bones of defunct mammals. It has been said that the hypothesis rests on discontinuous variation of characters, which does not exist. It is objected that the hypothesis assumes the genetic factors are fixed and stable in the same sense that atoms are stable, and that even a slight familiarity with living things shows that no such hard and fast lines exist in the organic world.[71]

Most of these objections are similar, if not identical, to those brought prior to 1910 against the Mendelian theory alone. Ironically, they are also the objections Morgan himself made against Mendelism in his earlier days!

In addition, a whole range of more technical problems with the chromosome theory were put forth by professional geneticists. Some, such as W. E. Castle argued that the Morgan school had made too many modifications in the basic Mendelian hypothesis (such as gametic purity), or that the whole phenomenon of crossing-over on which the chromosome theory rested was based on ques-

[71] T. H. Morgan, "The Theory of the Gene," *American Naturalist* 51 (1917): 513-534.

tionable evidence. Others, such as Goldschmidt, Bateson, and W. M. Wheeler, argued that there was no concrete evidence that genes had any material relation to chromosomes. At best, they claimed, Morgan's argument rested on mere parallelism. Some of these criticisms were due largely to misinterpretations of the *Drosophila* work. Others, however, were based on insight into very real problems that the Mendelian-chromosome theory had encountered. Three examples will serve to illustrate the most prominent criticisms.

Between 1907 and 1914, W. E. Castle at Harvard with his colleagues Hansford MacCurdy and J. C. Phillips, carried out a series of selection experiments on the hooded rat. Hooded rats are white with a darkly pigmented area on the head and usually a narrow line of pigmentation down the back. A hooded rat crossed with a wild-type rat produced all wild-type F_1. In the F_2 the ratio of wild to hooded was 2:1, which Castle interpreted as meaning hooded was recessive to wild-type as a single Mendelian factor. Castle observed further that the hooded individuals showed a much greater variability in the F_2 than the original hooded grandparents. Castle interpreted this observation to mean that the recessive hooded gene was altered by existing in the hybrid along with genes for the normal, wild pigmentation. If this hypothesis were correct, Castle reasoned, then selecting one or the other extreme of this variability in the F_2 ought to produce further modifiability that remained stable. In one line, called "plus," Castle and his colleagues selected for an increase in the extent of the hooded pattern. Selection was effective and ultimately yielded individuals far beyond the limits of variability of the original parents, or the F_2 strain. After thirteen generations, selection in the minus strain produced almost solid white rats, while selection in the plus strain produced almost totally pigmented individuals.

To determine whether the altered genes were stable, Castle carried out an equivalent number of generations of backward selection. Using the nearly white (minus) strain rats, Castle found that thirteen generations of backward selection yielded rats that were not nearly as highly pigmented as the original parents from which he had started. In an important paper of 1914, "Piebald Rats and Selection," Castle pointed out that selection seemed to permanently modify genes by bringing them into contact in the germ cells with other genes that contaminated them.[72] This concept of "genetic

[72] W. E. Castle and J. C. Phillips, "Piebald Rats and Selection: An Experimental Test of the Effectiveness of Selection and of the Theory of

contamination" went against the basic Mendelian assumption of purity of the gametes, but Castle found no objection to this. As he wrote in 1916: "Many students of genetics at present regard unit characters as unchangeable. . . . For several years I have been investigating this question and the general conclusion to which I have arrived is this, that unit characters are modifiable as well as recombinable. Many Mendelians think otherwise, but this is, I think, because they have not studied the question closely enough. The fact is unmistakable that unit factors are subject to quantitative variation."[73] The fact was unmistakable that selection did produce new phenotypes. What was at stake was the fundamental question of whether genes produced permanent effects on other genes simply by being present together with them in the same germ cells. Castle's work raised a serious question as to the validity of some basic Mendelian assumptions.

Morgan and his colleagues took exception to Castle's conclusions. They, too, were beginning to question the basic Mendelian conception of nonmodifiability and noninteraction of genes, but their approach came from a different direction. It was particularly Sturtevant and Muller who attacked Castle. Muller argued first that Castle's assumption that hoodedness was due to a single recessive gene was highly questionable. If many factors were involved, instead of a single pair of alleles, and if the original set of parents were heterozygotes (as might be expected), then, Muller argued, the effects of continued selection could be explained on the basis of sorting out of modifying factors (either increasing or decreasing the number) rather than as a process of continual contamination of the hooded gene.[74] While Muller was not committed to any concept of immutability of the gene, he held that genes *are* known to be relatively constant and that contamination implied a kind of modifiability for which there was relatively little evidence. So far, most of the mutants the *Drosophila* group had encountered appeared to have occurred randomly rather than in a fixed direction or in any relation to the other genes with which they were combined.

Gametic Purity in Mendelian Crosses," *Carnegie Institution of Washington Publication* 195 (1914): 1-54.

[73] W. E. Castle, "Can Selection Cause Genetic Change?" *American Naturalist* 50 (1916): 248-256.

[74] H. J. Muller, "The Bearing of the Selection Experiments of Castle and Phillips on the Variability of Genes," *American Naturalist* 48 (1914): 567-576.

Castle responded to Muller's and Sturtevant's criticisms, and there was a considerable exchange, especially in *The American Naturalist*, between the Harvard and Columbia groups over a period of several years. Some of that discussion, according to Sturtevant, was "rather heated."[75] Muller and Sturtevant had used Johannsen's earlier experiments of 1903 on selection for bean size as an argument against Castle's interpretation. Johannsen's work had shown that there was a limit to the modifications selection could produce in a given direction; the Morgan group interpreted that limit to be the result of the presence, in the initial population, of a fixed number of modifier genes. Castle opposed Muller's use of Johannsen's work as wholly inapplicable to the present case. He also seriously questioned Muller's belief in the fundamental principle that genes are not contaminated by other genes. As he wrote, "Now, when, I should like to inquire, did these principles become 'fundamental,' by whom were they established, and on what evidence do they rest?"[76] Castle rejected Muller's hypothesis of modifier genes, claiming that it, too, assumed the modifiability of genes, and therefore could not represent any real improvement on his hypothesis in that regard. The Morgan group, however, maintained its stand, and Morgan himself criticized Castle's results in his Vauxeum Lectures at Princeton in 1916: "Selection has accomplished this result [i.e., Castle's rats] not by changing factors, but by picking up modifying factors. The demonstration of the presence of these factors has already been made in some cases [i.e., for *Drosophila*]. Their study promises to be one of the most instructive fields for further work bearing on the selection hypothesis."[77]

However stubborn he was, Castle did not rest assured about his conclusions. Beginning in 1916 he designed a new series of experiments to test his genetic contamination hypothesis. He reasoned that if genes were permanently altered by selection, then outcrossing a highly selected plus or minus strain with wild-type rats should not bring about the loss of the newly altered gene. Using a plus strain he found that when highly pigmented rats were crossed with wild rats, the hooded character was lowered no more than ¾ of a grade in three successive generations. This seemed to Castle to

[75] Sturtevant, *A History of Genetics*, p. 61.

[76] W. E. Castle, "Mr. Muller on the Constancy of Mendelian Factors," *American Naturalist* 49 (1915): 37-42.

[77] Morgan, *A Critique of the Theory of Evolution*, p. 165.

bear out his original hypothesis that the gene had been permanently modified. However, when he performed the same experiment with a highly selected minus strain, he found that in three generations the effects were lost; i.e., pigmentation returned rapidly so that the third generation resembled wild-type rats. Castle first announced the findings of his experiments to a seminar of his colleagues and students at Harvard, many of whom (including his friend at the Bussey Institution, E. M. East) had criticized the contamination theory for years. Castle concluded, "These findings harmonize with the idea that the residual heredity in question consists of several modifying genes independent of the hooded gene proper."[78] Castle's public announcement of his error gave further support to the modifier-gene hypothesis championed by the Morgan school.

No sooner had Castle extricated himself from this debate, than he locked horns again with the *Drosophila* group in another controversy. In the same year that he retracted his contamination theory, he put forth a fundamental criticism of the chromosome theory itself. He wrote: "That the arrangement of genes within a linkage system is strictly linear seems for a variety of reasons doubtful. . . . In reality it has been found that the distances experimentally determined between genes remote from each other are in general less than the distances calculated by summation of supposedly intermediate systems. . . . To account for this discrepancy, Morgan had adopted certain subsidiary hypotheses of 'interference,' 'double crossing-over,' etc."[79] Castle developed his own theory, known as the "rat-trap" model for chromosomal structure. He constructed a three-dimensional model based on simple recombination frequencies, ignoring interference or double crossing-over. Castle conceived of the chromosome as a single molecule; although it was sausagelike in overall shape, the fine structure was assumed to contain many three-dimensional shifts in position which gave the overall structure considerable complexity. One of the reasons Castle developed the rat-trap model was his skepticism over the actual physical process of crossing-over. Castle claimed that there was

[78] W. E. Castle, "Piebald Rats and the Theory of Genes," *Proceedings of the National Academy of Sciences* 5 (1919): 126-130.

[79] W. E. Castle, "Is the Arrangement of the Genes in the Chromosome Linear?" *Proceedings of the National Academy of Sciences* 5 (1919): 25-32; quotation, p. 26.

no direct evidence that such a process of exchange actually occurred. Therefore, the assumption on which the whole mapping process rested could be called into question.[80]

Again, it was Muller who replied in print to Castle's objections. Muller severely criticized Castle's arguments and claimed that Castle misunderstood the fundamental point of double crossing-over.[81] Castle had predicted, for example, that on a three-dimensional model, genes more distant from each other should appear in a less straight line than genes close together. If, Muller argued, the chromosome were viewed as a sausage, as Castle agreed, and the genes were spread throughout it like peppercorns, then any two peppercorns at opposite ends of the sausage would *more likely* appear in a straight line than two peppercorns close together somewhere in the middle. Muller did not argue for strict linearity in the sense of beads on a string. He did, however, argue for the relative linearity of genes along the length of the chromosome with little variation in three-dimensional space. Castle's rat-trap model was not only more complex than the linear model developed by the *Drosophila* group, but it also could not predict the inheritance pattern of newly discovered mutants in relation to established genes as effectively as the linear model could.

Castle's objections to both the modifier-gene concept and the linear chromosomal model were based on his personal unwillingness to accept any idea that could not be derived directly from observation. More fundamentally, however, they represented an attempt to see the work of the *Drosophila* school in terms of the simplified Mendelian concepts developed largely by the Bateson school in the early 1900s. They also demonstrated Castle's lack of familiarity with the cytological evidence on which the chromosome interpretation was based. Without cytology he could not appreciate the basic simplicity and importance of the linear chromosome model.

Another persistent critic of the chromosome theory was the German biologist, Richard Goldschmidt. Goldschmidt was always a Mendelian, and from his exposure to cytological studies as a student, did not question the reality of chromosomes or their relation to the hereditary process. He was, however, especially after the

[80] Ibid.

[81] H. J. Muller, "Are the Factors of Heredity Arranged in a Line?" *American Naturalist* 54 (1920): 97-121.

mid-1920s, a persistent opponent of the idea that Mendelian genes are arranged in a linear fashion on the chromosome. Goldschmidt's criticisms were by and large less specific than either Bateson's or Castle's, and involved a barrage of attacks on the work of the Morgan school for over twenty years.[82]

Like Castle, Goldschmidt argued that the *Drosophila* group invoked too many subsidiary hypotheses in order to save the Mendelian concept of discrete, immutable genes.[83] The idea of multiple alleles and modifier genes meant to Goldschmidt a complete reevaluation of the doctrine of genetic stability which, in turn, raised the question of the kinds of factors that might influence genetic changes.

What really brought Goldschmidt to launch his severest attacks on the chromosome interpretation of Mendelism developed out of the Bar-eye case worked out by the Morgan group. Goldschmidt felt that the concept of "position effect" was the ultimate absurdity devised by the Columbia group to save the theory of the discrete gene. To Goldschmidt, the fundamental question of genetics was always how the individual units of heredity, whether they were discrete genes or whole chromosomes, functioned in a physiological way to produce adult characters. The hypothesis of position effect seemed to have no physiological relevance: how could the position of a gene actually affect its chemical functioning? In place of a linear array of separate hereditary units, Goldschmidt proposed his "continuum model" of the chromosome. In this view, Goldschmidt dispensed with discrete genes entirely. The continuum model proposed that the chromosome was a single molecule (again reminiscent of Castle's ideas) with genetic information arranged along its entire length. No one segment of the chromosome could be identified with any single character, although a change in one or another part would be reflected in some visible phenotypic change. Different parts of the chromosome could be active at different times during development, or even during adult life, with the interaction of parts ultimately determining the nature of the

[82] A detailed analysis of Goldschmidt's arguments against the Mendelian-chromosome theory can be found in G. E. Allen, "Opposition to the Mendelian-Chromosome Theory: The Physiological and Developmental Genetics of Richard Goldschmidt," *Journal of the History of Biology* 7 (1974): 49-92.

[83] See, for example, Richard Goldschmidt, "A Preliminary Report on Some Genetic Experiments Concerning Evolution," *American Naturalist* 52 (1918): 28-50.

adult characters.[84] To Goldschmidt, the value of the continuum model was that it seemed more in harmony with physiological facts.

The response of the Morgan group to Goldschmidt's criticism was probably the most damning they could give: they simply ignored what he said. Only occasionally did either Sturtevant or Morgan refer to Goldschmidt's continuum model. Even though Goldschmidt stepped up his attack on the chromosome interpretation when he arrived in the United States in 1936 as a political refugee from Germany, the Morgan group still refused to take up the challenge. It is ironic that Goldschmidt criticized Morgan and his coworkers for producing essentially a morphological and static theory. If there is anything Morgan would have opposed it was being called a morphologist. It is true that the Morgan group had concentrated on transmission genetics—which was amenable to investigation. But they had always taken into account the physiological (translational) dimensions of genetics. As Morgan said in his Nobel Lecture:

> Certain students of genetics inferred that the Mendelian units responsible for the selected character were genes producing only a single effect. This was careless logic. It took a good deal of hammering to get rid of this erroneous idea. As facts accumulated it became evident that each gene produces not a single effect, but in some cases a multitude of effects on the characters of the individual. It is true that in most genetic work only one of these character-effects is selected for study—the one that is most sharply defined and separable from its contrasted character—but in most cases minor differences also are recognizable that are just as much the product of the same gene as is the major effect.[85]

Goldschmidt's criticisms were of value in the development of genetics because he persistently raised the issue of genes as functional units. At a time when, despite Morgan's warnings, the biological community was in danger of becoming wholly captivated with the transmission problem alone, Goldschmidt reminded en-

[84] Richard Goldschmidt, "Position Effect and the Theory of the Corpuscular Gene," *Experientia* 2 (1946): 197-203; 250-256; especially pp. 201-202.

[85] T. H. Morgan, "The Relation of Genetics to Physiology and Medicine," *Scientific Monthly* 41 (1935): 5-18, quotation, p. 10.

thusiasts that any theory of heredity must ultimately be compatible with the facts of embryology and cell physiology.

A third major critic of the chromosome theory was William Bateson. Bateson, who had done so much to promote the acceptance of Mendelian theory between 1900 and 1910, developed an early skepticism for the chromosome idea and never fundamentally accepted its relationship to Mendelism. Bateson's skepticism began almost as soon as any suggestion was made that chromosomes might have anything to do with Mendelian factors. One of Bateson's students wrote much later in his life: "I was one of the first students of Bateson in 1903-04. . . . One of my most vivid recollections during these days was his . . . antagonism to the chromosome theory of the mechanism of heredity. I remember coming across Sutton's papers in the Library, and being very excited by them. I took these to Bateson and asked him what he thought of them. He would have nothing of it—and I remember his going over the figures and remarking that chromosomes didn't really have anything to them."[86] Bateson had written with a certain enthusiasm to E. B. Wilson in 1905 about the latter's work on chromosomes and sex: "Your letter announcing the discovery of the significance of the X and Y chromosomes for sex determination has put me into great excitement. Of course, in a way we have been waiting for some such announcement, but to have actually brought the thing off is splendid. Of all the discoveries I should like to have made, this is just about the best, and I do most heartily congratulate you. It is indeed a new beginning."[87]

Nonetheless Bateson could not accept a chromosome interpretation of the general hereditary process. He raised several arguments against the Morgan group's interpretations. First, like Castle and Goldschmidt, he felt that the work with *Drosophila* had turned up too many exceptions to classical Mendelism. The concept of linkage of genes on chromosomes and a number of subsidiary hypotheses invoked by the *Drosophila* group seemed to Bateson to

[86] Agar to E. B. Babcock, 1947; Papers, Department of Genetics, University of California, Berkeley; housed at the American Philosophical Society.

[87] Letter in the possession of Nancy Wilson Lobb, South Hadley, Massachusetts. Bateson's lifelong opposition to chromosomes is analyzed at great length in William Coleman's, "Bateson and Chromosomes: Conservative Thought in Science," *Centaurus* 15 (1970): 228-314.

call for a complete revision of the theory rather than for its continual modification. As he wrote in 1921: "What frightens me off is the number of 'bolt-notes.' Lethal factors, 'genes,' modifying linkage, and similar expedients may all be sound, but to get *proof*, each hypothesis must stand independently on its own bottom."[88] Second, Bateson also pointed out that by every chemical and cytological test available, the chromosomes within the nucleus of any cell all appear to be identical—that is, they all stain alike, look very much alike, etc. If the chromosomes really contained discrete regions which produced wholly different types of characters, it would be expected that more visible heterogeneity could be observed.[89] Third, Bateson noted that the chromosomes remained constant in shape and morphology throughout dramatic changes in the growth and development of organisms (for example, through the life cycle of an insect). If chromosomes were involved in development, as they must be if they carried genes, then we should observe some visible changes during different stages in development. Since this was not observed, it appeared to Bateson that chromosomes did not contain discrete Mendelian genes.[90] Fourth, on the surface, Bateson's most persuasive argument against the chromosome interpretation, as advanced by the Morgan group, was that it was based exclusively on the parallelism between cytological events and breeding results, that it did not have any independent

[88] William Bateson to Charles Benedict Davenport, February 11, 1921, p. 2; Bateson Papers.

[89] One of Bateson's shortcomings was lack of familiarity with cytological evidence. Even by 1905 it was clear to most cytologists that not all chromosomes in any organism were identical; each had its special shape and size. Later, fine-structure analysis with giant salivary-gland chromosomes further confirmed these early observations.

[90] In these arguments Bateson perched himself on an illogical limb. He assumed that for regions of the chromosome to be qualitatively different, or for different chromosomes to be active at different times, there must be some related, observable effect. As we know today, different regions of the chromosome (or, in some cases, whole chromosomes) are active or inactive at different periods during development. Yet often it is impossible to notice any visible difference between those that are active and those that are inactive. The observation that chromosome "puffing" may be associated with differential regional activity of specific chromosomes has been put forward since the late 1950s. (See Adrian M. Srb, R. D. Owen, and Robert F. Edgar, *General Genetics*, 2d ed. [San Francisco: W. H. Freeman, 1965], pp. 362-364.)

evidence of its own. This argument was advanced in the period prior to 1921, before the evidence from deletions and other chromosomal abnormalities or nondisjunction began to fortify the parellelism.

Behind all Bateson's arguments lay an aspect of his own philosophy of which he was probably unaware: a philosophical idealism and distrust of materialistic theories in science. Coleman has shown in his detailed study of Bateson's opposition to chromosomes[91] that Bateson was strongly influenced by the idealistic physics of James Jeans and others of the Cambridge school of physics in the period between 1910 and 1925. The Cambridge school attempted to derive basic postulates of physics without reference to any concrete theory of matter, i.e., without reference to atoms. To the idealistic physicist, among whom many of the early quantum theorists may be grouped, the peculiar nature of the universe was unknowable. Most important to them was that predictions of observable phenomena could be made on the basis of mathematical theories without commitment to any notion of material reality. It appears that this same line of thinking affected Bateson's view of heredity. He wrote in his review of *The Mechanism of Mendelian Heredity* in 1916: "It is inconceivable that particles of chromatin or of any other substance, however complex, can possess those powers which must be assigned to our factors or gens [*sic*]. The supposition that particles of chromatin, indistinguishable from each other and indeed almost homogeneous under any known test, can by their material nature confer all the properties of life surpasses the range of even the most convinced materialism."[92] Bateson did not deprecate the value of the work of the *Drosophila* group, for at the end of the above review he wrote: "not even the most skeptical of readers can go through the *Drosophila* work unmoved by a sense of admiration for the zeal and penetration with which it has been conducted, and for the great extension of genetic knowledge to which it has led—greater far than has been made in any one line of work since Mendel's own experiments."[93] And in a letter to Davenport in 1921 Bateson wrote, "But don't mistake—I do admire Morgan and the character of his work, and I am quite

[91] Coleman, "Bateson and Chromosomes."

[92] William Bateson, "The Mechanism of Mendelian Heredity (A Review)," *Science* 44 (1916): 536-543; quotation, p. 542.

[93] Ibid., p. 543.

prepared to believe if and when I must. Some days I am much drawn that way, at other times I see the difficulties more."[94] What Bateson apparently failed to see was the necessity of combining the abstract and idealized Mendelian theory with the material theory of chromosomes.

Bateson was not alone in questioning the material reality or necessity of individual genes. Goldschmidt shared these views to some degree, though he at least admitted that chromosomes had something to do with heredity. William Morton Wheeler, on the other hand, argued similarly to Bateson. What is interesting and suggestive historically, is that all three men appear to have been influenced by the philosophical issues arising from the advent of quantum theory in the first two decades of the century. Wheeler, for example, was greatly influenced by his colleague in the philosophy department at Harvard, Alfred North Whitehead (1861-1947), who saw genes as purely arbitrary and abstract units, like quanta, with only mathematical reality.[95] Similarly, Goldschmidt was influenced by Einstein's abstract views of the relation between time and space.[96] It is important that a more detailed study be made

[94] Bateson to Davenport, February 11, 1921; Bateson Papers.

[95] See Whitehead's book, *Nature and Life* (Chicago: University of Chicago Press, 1934), pp. 13-14. Wheeler's copy of this book is heavily underscored where Whitehead makes the following statement: "The fundamental fact, according to the physics of the present day, is that . . . the notion of the self-contained particle of matter, self-sufficient within its local habitation, is an abstraction. . . . This general deduction from the modern doctrine of physics vitiates many conclusions drawn from the application of physics to other sciences, such as physiology, or even such as physics itself. For example, when geneticists conceive genes as the determinants of heredity. The analogy of the old concept of matter sometimes leads them to ignore the influence of the particular animal body in which they are functioning. They presuppose that appellative matter remains in all respects self-identical whatever be its changes of environment. So far as modern physics is concerned, such characteristics may, or may not, affect changes in the genes, changes which are important in certain respects, though not in others. Thus, no *a priori* arguments as to the inheritance of characters can be drawn from the mere doctrine of genes." Wheeler knew Whitehead well, often joining him and other philosophically minded Harvard faculty at dinner-club meetings. See Mary Alice and Howard Evans, *William Morton Wheeler, Biologist* (Cambridge, Mass.: Harvard University Press, 1970), pp. 11; 296.

[96] See, for example, Goldschmidt's "Vorläufige Mitteilung über Weitere Versuche zur Vererbung und Bestimmung des Geschlechts," *Biologische Zentralblatt* 35 (1915): 565-570; see also Curt Stern, *Genetic Mosaics and Other Essays* (Cambridge, Mass.: Harvard University Press, 1968), p. 93.

of the influence of the growing trend toward nonmaterial theory in physics in the 1920s on biological—and especially genetic—thought.

Bateson struggled with the chromosome theory for a number of years. Ultimately, he made at least one public statement in which he accepted the reality of chromosomes. This was in 1922, after he had visited Morgan's laboratory during a stay of several months in the United States. Bateson came to the North American continent in December of 1921 not only to visit laboratories and universities here, but also to deliver a vice-presidential address to the American Association for the Advancement of Science which was meeting in Toronto. In the weeks preceding the meeting, Bateson stayed in New York City with Morgan at his home, and spent several intensive days in the Columbia laboratory examining the work of Bridges and Sturtevant. As we have seen, Bateson *was* impressed with the work of both Bridges and Sturtevant. What about his view of Morgan?

Personally, Bateson found Morgan difficult to deal with. After his first day at the Morgan house, Bateson wrote home to his wife:

> The evening and the morning have been the first day. We had a long discursive talk last night, avoiding anything like an actual clash. I fancy more or less that is to be the order of our proceedings. Morgan has a rough good nature that attracts, but I have just the same impression that I got 14 years ago, that he is of no considerable account. His range is so dreadfully small. Off the edge of a very narrow track, he is not merely puzzled, but lost utterly. . . . He is totally free from pretense—he is almost without shame in his ignorance—I mean of things scientific.[97]

In another letter just a few days later, Bateson revealed to his wife: "I wish I liked Morgan better. I think he has made a great discovery, but I can't see in him any quality of greatness."[98] Professionally, however, Bateson found Morgan's work difficult to refute. The impressions the *Drosophila* work made on him ultimately forced him to reconsider his position on chromosomes. Of his visit to the Morgan group, Bateson wrote:

[97] William Bateson to C. Beatrice Bateson, December 20, 1921; Bateson Papers.

[98] Ibid., December 24, 1921.

> Yet, there is no denying the fact that by intensive methods they [Morgan and his group] have got a long way.[99]

> Part of each morning is devoted to chromosomes. I can see no escape from capitulation on the main point. The chromosomes must in some way be connected with our transferrable characters. About linkage and the great extensions, I see little further than I did. . . . Cytology here is such a common-place that everyone is familiar with it. I wish it were so with us. Bridges inspires me with complete confidence.[100]

And a few days later he wrote: "I am heartily glad I came. I was drifting into [an] untenable position which would soon have become ridiculous. The details of the linkage theory still strike me as improbable. Cytology, however, is a real thing—far more important and interesting than I had supposed. We must try to get a cytologist."[101]

Bateson's public capitulation came in his AAAS address a few weeks after visiting Morgan:

> We have turned still another bend in the track and behind the egg and sperm cells we see the chromosomes. For the doubt—which I trust may be pardoned in one who has never seen the marvels of cytology—cannot, as regards the main thesis of the *Drosophila* workers, be any longer maintained. The arguments of Morgan and his colleagues, and especially the demonstration of Bridges, must allay all skepticism as to the direct association of particular chromosomes with particular features of the zygote. The transferable characters borne by the gametes have been successfully referred to the visible details of nuclear configuration.[102]

Yet Bateson's capitulation was not complete. He still retained serious doubts about the many claims of the Morgan group, and still questioned whether all the hypotheses about crossing-over and interference were really supported by the cytological facts. Unfortunately, Bateson did not live to see the more convincing proof that emerged with the study of the giant salivary chromosome; he died in 1926, just before that era in cytogenetics began.

[99] Ibid., December 20, 1921.

[100] Ibid., December 24, 1921.

[101] Ibid., December 26, 1921.

[102] William Bateson, "Evolutionary Faith and Modern Doubts," *Science* 55 (1922): 55-61; quotation, p. 57.

The objections of Castle, Goldschmidt, and Bateson reflect to some degree the important role cytology played in the development of the chromosome theory. Of the three, Castle and Bateson had had least general contact with cytology as a science. Neither had any direct contact with the newer techniques of cytology and only passing familiarity with newer findings in this field. Although Goldschmidt had carried out some early cytological studies, his efforts in this area had decreased considerably by 1910. Lack of first-hand experience may well have been a major factor in the continued skepticism that Bateson, Castle, and Goldschmidt manifested toward the chromosome theory of heredity.

The objections of Goldschmidt and Bateson indicate something of the role an investigator's philosophical assumptions can play in evaluating a new idea. Holding basically antimaterialist positions, both men had a strong disposition against postulating morphological units as the determiners of complex developmental and physiological processes. Somehow both men seemed to feel it ought to be possible to understand the *process* of heredity without recourse to some type of material unit. Furthermore, all three men were trained as old-style morphologists, with a special emphasis on embryology.[103] As embryologists, all three tended to constantly mix vertical with horizontal concepts of heredity.[104] A vertical concept of heredity is concerned with the transmission of hereditary elements between generations [i.e., vertically]; a horizontal concept is concerned with the translation of those hereditary elements into adult characters. Embryologists tend to be interested largely in horizontal concepts, since their whole area of concern is the development of the fertilized egg into differentiated adult characters. Those concerned with breeding experiments, on the other hand, tend to be more interested in vertical concepts. Failure to distinguish between the demands of these two very different concepts of heredity led workers such as Bateson, Goldschmidt, and Castle to expect too much from Mendelism. They wanted all research into Mendelism, a vertical theory, to yield information about embryonic development, a horizontal issue. Like the old-

[103] Castle under E. L. Mark; Bateson under Adam Sedgwick, W.F.R. Weldon, and W. K. Brooks; and Goldschmidt under Otto Bütschli and Richard Hertwig.

[104] This point is closely related to the distinction, respectively, between genotype and phenotype, which has been carefully discussed by Frederick Churchill in "Wilhelm Johannsen and the Genotype Concept," *Journal of the History of Biology* 7 (1974): 5-30.

style morphologists, they hoped that any theory of heredity would also explain development and evolution. Ultimately, of course, any theory of heredity must at least be consistent with other explanations of development and evolution. But Goldschmidt, Bateson, and Castle wanted too much too soon. They were skeptical of the chromosome theory because it seemed too limited. It was merely a materialistic interpretation of the vertical concept of heredity. It did not contribute anything to the horizontal aspect of heredity.

Ironically, Morgan was also trained as an embryologist in the old morphological tradition. And for a time, as seen in chapter 4, he objected to the Mendelian theory on the grounds that it explained nothing about developmental processes. But it was Morgan's genius that he could ultimately distinguish between the horizontal and vertical aspects of heredity and focus his attention on the one that was more amenable to experimental analysis. Thus the Morgan group brought together at one time and place a unique combination of viewpoints which enabled them to exploit the *Drosophila* material so fully. They rigorously distinguished between vertical and horizontal components of heredity and focused on the former. Initially through Wilson, and later through Bridges, they were in direct contact with the newest results of cytology—especially of the relation between chromosomes and heredity. Finally they had a down-to-earth materialist philosophy which saw no difficulty in associating hereditary determiners with discrete material components of the cell.

NATIONAL INFLUENCES ON RECEPTION OF THE *DROSOPHILA* WORK

The varying acceptance in different countries of the Mendelian-chromosome theory, as enunciated by the Morgan group after 1915, shows some interesting patterns. In science, as in any other aspect of culture, national traditions may contribute strongly to the acceptance or rejection of particular ideas.

Between 1915 and 1930 the Mendelian-chromosome theory gained acceptance not only in the United States, but also in Norway, Sweden, Denmark, the Soviet Union and, to a lesser extent, in Germany. It did not take hold strongly in England or France. In all cases acceptance seemed to depend on two elements: the existence of a well-developed cytological tradition and the presence of someone with direct experience in the Morgan lab who could communicate the excitement and novelty of the fly work. In countries where the chromosome theory was not well accepted, three conditions appeared to exist: lack of a well-developed cytological tradition, the presence of one or more persons in a prominent position who opposed the chromosome theory, and, probably because of this, the failure of students in these countries to visit Morgan's laboratory.

The spread of Mendelian ideas through Europe was accomplished by many people: especially Otto Mohr in Norway and Muller in England and Germany. Mohr was originally an M.D. who also received a Ph.D. under Brachet (Brussels) in cytology (1918). Mohr had been induced to come to the United States to work with Morgan's group after reading the *Mechanism of Mendelian Heredity*. The year with Morgan's group had converted Mohr from an anatomist and cytologist to a geneticist. After returning to Oslo in 1919, Mohr pioneered in areas of human genetics, and organized the first institute for the study of human genetics in Norway.[105] An enthusiastic proponent of the chromosome theory, in 1923 he published a book for the general reader incorporating all the latest *Drosophila* findings.[106] He was by that time, according to Sturtevant, "the recognized European representative of the *Drosophila* school."[107] It was through personal and professional contacts that Mohr contributed much to the spread of the chromosome theory. For example, less than a month after returning to Europe from Morgan's group he presented the newest *Drosophila* findings to Bateson's group at the John Innes Horticultural School at Merton College. Mohr was in direct contact with Hans Nachtsheim, Goldschmidt's assistant at the Kaiser-Wilhelm Institute in Berlin-Dahlem. He encouraged the rather shy Nachtsheim, who was in complete sympathy with the *Drosophila* work, to spread the new ideas within the skeptical Goldschmidt camp.[108] Through Nachtsheim, Mohr was invited to give a major address before the European Congress of Animal Breeders in

[105] For details of Mohr's career see A. M. Dalcq "Notice sur la vie et l'oeuvre de M.O.L. Mohr, correspondent étranger," *Bulletin de l'Académie Royale de Médicine de Belgique* 7 (1967): 691-698.

[106] O. L. Mohr, *Arvelaerens grundtraek* (Det Norske Studentersamfund Flakeskrifter, no. 6, 1923).

[107] A. H. Sturtevant to Edgar Sturtevant (brother), May 28, 1922; Sturtevant Papers, California Institute of Technology Archives; quoted with permission.

[108] Nachtsheim was responsible for the German translation of *The Physical Basis of Heredity*, which appeared in 1921 as *Die Stoffliche Grundlage der Vererbung* (Berlin: Gebrüder Borntraeger, 1921).

Vienna in 1919. The original lecture, delivered in English, made such an impression that Mohr was asked to repeat it the same evening in German![109] Equally successful was his appearance at a major biological conference in Berlin a year and a half later.[110] Throughout Scandinavia Mohr's influence was more direct and persistent. His student Gert Bonnier (1853-1922) went to Morgan's lab (at Mohr's suggestion) and returned to his native Sweden where he exerted considerable influence both in agricultural and scientific circles.[111] Mohr was also a friend of the Swedish breeder Hermann Nilsson-Ehle (1873-1949), and through him encouraged the spread of the ideas of the Morgan school at the influential experimental breeding station at Malmo. And through his association with the Danish professor Otto Winge (1886-1964), Mohr was invited, shortly after his return from the United States, to discuss the *Drosophila* work in the Biological Society of the University of Copenhagen. Despite Wilhelm Johannsen's early skepticism about the chromosome theory (he was a Mendelian, however), Winge's influence prevailed; by 1926 Johannsen had "come around" and admitted the possible reality of chromosomes as the physical bearers of Mendelian genes.

Muller was also influential in the spread of the *Drosophila* work throughout Europe. In 1920 he gave "one of the most brilliant speeches of his career"[112] at the Royal Society (London). Much later, in 1933, he went to work at the Kaiser-Wilhelm Institute on a Guggenheim Fellowship—though by that time the *Drosophila* work was well known in Germany, and despite Goldschmidt's skepticism, had gained numerous supporters.

Of course, the planting of ideas in new ground does not automatically mean they take root. There is a notable lack of even the most rudimentary studies of the reception of the chromosome theory in different countries (except England, where Bateson's negative influence, and the Soviet Union, where Muller's positive influence, as discussed below, have been partially documented).

[109] Tove Mohr, personal communication, August 1974.

[110] Ibid. Exact name and date of the conference unknown.

[111] Bonnier came from a prominent Jewish family (publishers) in Stockholm. After more than ten years as head of the government's Animal Breeding Institute, he returned to academic circles, where at the University of Stockholm he had many students pursuing *Drosophila* work.

[112] Tove Mohr, personal communication, August 1974. Mrs. Mohr was present with her husband at Muller's lecture.

A beginning has been made with regard to Norway[113] and Dunn, among others, has discussed the spread of the new ideas in general terms. However, while we can judge that the influence of enthusiastic disciples such as Mohr and Muller was considerable in making the *Drosophila* work known and accepted, more evidence of the direct lines of influence is needed.

In the Soviet Union, the *Drosophila* work received an enthusiastic welcome when Muller made his first trip there in 1923. At that time he took with him a large number of bottles of flies of special stocks, which could be used to test wild species for the presence of innate mutations. It was largely Muller's initial influence that stimulated workers in the USSR to undertake their own genetic work with *Drosophila.* There was no cytological tradition to speak of in Russia at the time, but there were several factors that made that country ripe for encouraging the new work in genetics. One was the persistence of food shortages caused by the attempts to equalize distribution amid drought and the chaos of reorganizing agricultural production.[114] Lenin had remarked during the famine of 1921 that the time to prevent the next famine was "now." The result was the founding of the Institute of Applied Botany and priority for support for biological (especially breeding) research.[115] A Russian school of *Drosophila* geneticists arose in the next decade which made outstanding contributions in two areas: cytological work on chromosome structure, and population genetics using wild populations of *Drosophila* to study factors affecting gene frequency.[116] It was Muller's actual presence and his transporting of the "clean" and systematically developed *Drosophila* stocks to the Soviet Union that provided the impetus for the rapid acceptance of the chromosome theories. It is thus par-

[113] Per Slagsvold and Tore Fremstad, "Mendelian Influence on Norwegian Biological Thinking in the '30's of the 20th Century," *Folia Mendeliana* 7 (1972): 43-49.

[114] For example, many large landowners had slaughtered their cattle and destroyed their seeds rather than allow them to be turned over to peasants during the early days of land reform. See David Joravsky, *The Lysenko Affair* (Cambridge, Mass.: Harvard University Press, 1970), chaps. 1-3.

[115] L. C. Dunn, "Science in the USSR: Soviet Biology," *Science* 99 (1944): 65-67; see also Mark Adams's excellent paper, "The Founding of Population Genetics: Contributions of the Chetverikov School, 1924-1934," *Journal of the History of Biology* 1 (1968): 23-39, especially pp. 24-25.

[116] Adams, "The Founding of Population Genetics."

ticularly lamentable that with such an auspicious beginning, Russian genetics ran afoul of the Lysenkoist movement, beginning in the early 1930s. This movement set itself specifically against the genetics of the Morgan school, labeling the theory of the gene as "Mendelism-Morganism-Idealism."[117] Lysenko's movement thwarted both the professional geneticists and the practicing farmer by promoting methods for improving strains that produced no long-term beneficial results. The Lysenko movement simply drove *Drosophila* genetics underground, thus greatly retarding its development for several decades.

Two countries where the *Drosophila* work failed to gain a foothold during the years 1915 to 1930 were England and France. The reasons were different for each, yet tell us something about the sociological factors involved in the spread of scientific ideas. In England, the result was largely due to the previously discussed antichromosomal bias of William Bateson. Bateson was the dean of English genetics, and his refusal to accept the results of cytology and the chromosome theory in particular doomed the study of heredity in England to a slow death.[118] The many advances Bateson and his coworkers had made in illuminating aspects of the Mendelian theory prior to 1910 proved less and less useful in the abstract, that is, dissociated from the cytological evidence. In addition, Bateson's stubborn adherence to the presence-and-absence theory made it less and less possible for his work to include such well-established phenomena as multiple alleles and modifier genes. Bateson's recalcitrance may have been largely responsible for the failure of a strong cytological school of geneticists to develop in England.

In France, although Lucien Cuénot in the early 1900s had done much to establish the applicability of Mendelian laws to animals, he worked alone and did not develop a school or a group of students to carry on genetics work. In some respects the reasons for

[117] For a full history of this movement the best source is Joravsky, *The Lysenko Affair*. It is particularly interesting that Morgan's work would have been linked with philosophical idealism, since the great merit of the chromosome theory, in Morgan's eyes, was that it gave a physical, material basis to the hypothetical "units" of Mendel (see chapter VIII).

[118] W. E. Agar wrote: "I think it was Bateson's and Punnett's stubbornness in this matter that caused the decline of importance of English genetics for so many years after such a promising start." (Agar to E. B. Babcock, 1947; from Papers, Department of Genetics, University of California, Berkeley; housed at American Philosophical Society Library.

this may lie in features of Cuénot's own personality. In addition, however, there was a strong tradition of neo-Lamarckism in French biology, which was a considerable deterrent to seeing any significance in the new genetics. It was a common view in France that the most important problems to study were related to evolution and origin of variations. Going on the Lamarckian principle of inheritance of acquired characters, many workers in France thought that the characters of evolutionary importance were the *adaptive* ones, produced by the inherited effects of use and disuse. Mendelism, on the other hand, appeared to deal largely with non-adaptive characters, such as bristle number or eye color. Thus Mendelian inheritance had little or nothing to do with evolution. As Felix Le Dantec, professor of the Faculty of Sciences at the Sorbonne stated: "There are in every individual, two kinds of characters: first, mechanical essential or adaptive characters necessary to life which result from a slow evolution and do not in any way confirm Weismann's theory. The other category of characters comprises ornamental characters, peculiarities of form, which may be ruled by different laws. The latter are without any importance as regards the evolution of the species. It is in this category that all the Mendelian characters belong and Mendelian cases are not universal, but exceptional."[119] In Delage and Goldsmith's *The Theories of Evolution* (1912), Mendel's laws are relegated to a few pages as a minor component of contemporary biological theory. This preoccupation with evolution, particularly by Lamarckian inheritance, may well have stifled interest in the new genetics in France. In addition, there was no cytological tradition in France, and hence no concern for or knowledge of the details of cell structure and chromosomes.

Between 1915 and 1930, the Mendelian-chromosome theory began to spread throughout the world. Although there were critics, by the mid-1920s their number had considerably diminished (some, such as Bateson, had died), and the force of the new theories began to attract increasing numbers of workers. Younger geneticists in England, such as C. D. Darlington, took up the chromosome theory seriously. Others, such as Johannsen and Castle, admitted the errors of their earlier objections. Still others, such as Goldschmidt, kept up their opposition, but to an increasingly

[119] Le Dantec is quoted in Yves Delage and Marie Goldsmith, *The Theories of Evolution,* trans. André Tridon (New York: B. W. Huebsch, 1912), p. 184.

uninterested audience. Indeed, it can be said that by 1930 the revolution in genetics was complete: the Mendelian-chromosome theory was well established throughout most of the scientific world.

Because of its focus on the Morgan group, the preceding chapter might suggest that no other groups contributed in any significant way to the growth of Mendelian theory in the early twentieth century. This, of course, is not the case. Active centers for genetic research existed throughout the world during the period 1915 to 1930, many long antedating the appearance of Morgan and *Drosophila* on the scene in 1910. In the United States alone, much important work in heredity was carried out using organisms other than *Drosophila* and methods other than those developed by the Morgan group. Harvard had an active group of workers: W. E. Castle at the biological laboratories in Cambridge and E. M. East at the Bussey Institution in Jamaica Plains. Castle made significant contributions to the understanding of heredity in mammals, East in crop plants such as potatoes, corn, and tobacco. At Cornell, in Ithaca, New York, R. A. Emerson developed corn (maize) to the same refined position in the study of plant genetics that *Drosophila* held in animal genetics. Using maize, Emerson and his students led the way in developing hybrid corn, and in demonstrating the cytological basis for crossing-over. At the Connecticut Agricultural Station in Storrs, Donald Jones (along with E. M. East, who worked at the station before going to the Bussey) developed ideas on inbreeding and outbreeding, and coined the term "hybrid vigor." These and other centers contributed significantly to the growing body of genetic information between 1915 and 1930. Morgan and his students gained much from interaction with these groups; in turn, work with *Drosophila*, partly because of the rapidity with which the Columbia group achieved results, contributed significantly to the development of work along these different, but related, lines.

CHAPTER VII

Genetics, Embryology, and Evolution (1915-1930)

After 1915 Morgan saw the value of the work of the *Drosophila* group as stretching far beyond the boundaries of genetics itself. The expansion of the Mendelian-chromosome theory made new interpretations of both heredity and variation possible, and Morgan became an open champion of interpreting Darwinian theory in terms of the new genetics. Through the increasing evidence for gene interaction, he tried to show how the problems of embryonic development—notably differentiation—could be approached in genetic terms. Finally, he saw the *Drosophila* work as representing a new methodology—experimentalism combined with mechanistic materialism—in biology. With Jacques Loeb, Morgan became an outspoken advocate of "the new biology,"—an extension to *all* fields of the methods and philosophy of *Entwicklungsmechanik.*

EVENTS, 1915-1928

The period 1915 to 1928 was framed on the earlier side by the publication of *The Mechanism of Mendelian Heredity* and on the later side by Morgan's decision to leave Columbia University for the California Institute of Technology. Within Morgan's scientific career it represents a coherent period, characterized by the development and dominance of certain ideas. It was, in fact, the last period of intellectual upheaval in his career. His ideas on natural selection, genetics, embryology, and scientific methodology were well formulated and remained more or less fixed for the remainder of his life. By 1928, he was certain of his position on critical biological issues, and on the best methods for pursuing biological research. In that year he brought his enormous experience to bear on the founding of the biology division of the California Institute of Technology.

During the years 1915 to 1928 Morgan engaged in many activities in addition to the work with *Drosophila.* He was invited to give a large number of lectures throughout the country and abroad. In collaboration with Loeb and W.J.V. Osterhout, he set up a monograph series on experimental biology for the Lippincott Company. He took his one and only sabbatical leave in the academic year 1920-1921, working on the West Coast at the Hopkins Marine Station, at Stanford, and at Berkeley. It was the period of ascendancy of his professional reputation. Already recognized as a leader of American embryology by 1915, after the publication of *The Mechanism of Mendelian Heredity*, Morgan was considered to be the most important biologist in America, and one of the most important in the world.[1]

Yet despite the honors and recognition that came to Morgan for the work of his group, he busied himself continually with practical matters relating to friends, organizations, and professional work. Morgan was never too busy to help organize some new effort or meet a crisis. In 1917, for example, James McKeen Cattell, one of Morgan's close friends at Columbia and since 1907 the editor of *The American Naturalist*, found himself wishing to dispose of the journal. Cattell had been fired from Columbia after twenty-six years as a professor of psychology, officially over his public stand against United States involvement in World War I, but also partly as a result of his constant quarrels with Columbia's president, Nicholas Murray Butler.[2] Cattell found that his affairs had become quite involved, and he wanted to put the *Naturalist* in other hands.

He turned at once to Morgan. After discussing the matter with Cattell for some time, Morgan wrote to W. M. Wheeler at Harvard:

> Yesterday Cattell came in to talk over the management of *The American Naturalist*. . . . His affairs are now in such a

[1] With the exception of the Nobel Prize (1933) Morgan received his most prestigious awards during this period. He was elected to the National Academy of Sciences in 1909 and became president and vice-chairman of its executive board in 1927 (to 1931). He was elected to the American Philosophical Society in 1915 and to foreign membership in the Royal Society in 1919; he received the Darwin Medal from the Royal Society in 1924. He also received a number of honorary degrees during this period: from Johns Hopkins (LL.D., 1915), McGill (LL.D., 1921), Edinburgh (Sc.D., 1922), Michigan (Sc.D., 1924), University of California, Berkeley (LL.D., 1930), and in the early thirties an honorary Ph.D. from Heidelberg (1931) and an honorary M.D. from the University of Zurich (1933).

[2] See Michael Sokal, "The Unpublished Autobiography of James McKeen Cattell," *American Psychologist* 26 (1971): 626-635, especially p. 625.

> condition that he feels the necessity of making some change. Stopping *The Naturalist* is one possibility, but this I think most of us would very much regret, as it would be cutting off a means of publication at a time when we are likely to need every possible outlet. A second suggestion is that it might be merged with the *Popular Science Monthly*, but that, of course, would put it entirely outside the needs of scientific men. A third plan . . . is that a committee be appointed, composed of men in different fields, such as zoology, botany, entomology, agriculture, paleontology, psychology, anthropology, etc., who will support the Journal in the sense that they will act as advisors to the managing editor, who is to be some young man selected by the committee who will have charge of the management of the journal.[3]

Cattell was the legal owner of the journal and was willing to retain ownership, since it seemed unlikely that any group of academics could raise the money to buy it. Morgan used his influence and position, partly through the National Academy of Sciences, to form a group that would act as editorial supervisors: "I promised Cattell that I would bring the matter before a group of men who would be likely to take a real interest in the *Naturalist*. My opinion is that if a few men could be gotten together at the meeting of the National Academy in Philadelphia on the 20th we can discuss the matter informally and then look around for the proper man to manage the journal and consider the advisability of forming a larger committee into which we could later become merged."[4] In addition to Wheeler, Morgan wrote to D. T. MacDougal, of the New York Botanical Garden, E. G. Conklin of Princeton, Raymond Pearl of Johns Hopkins, and George Parker of Harvard. The meeting took place in Philadelphia as Morgan had promised, but Cattell decided that he could remain editor of the *Naturalist* after all, so matters rested as they had before. The *Naturalist* incident is illustrative, however, of Morgan's constant willingness to act not only in behalf of friends and associates, but in behalf of professional activities in which he believed. He was always willing to try to devise a plan to circumvent a difficulty. He was never one to sit by and let matters slide when something directly involved him.

[3] Morgan to W. M. Wheeler, November 8, 1917; Wheeler Papers, courtesy of Adaline Wheeler, Boston, Massachusetts.

[4] Ibid.

A second example comes from a few years later. Harking back to his earlier experiences collecting marine specimens in Jamaica (1891) and Bimini (1892), in 1923 Morgan fathered a scheme to establish a tropical marine station in the Western Hemisphere. As he wrote to Ross Harrison in 1924:

> Last summer at Woods Hole a few of us who are interested in the establishment of a tropical or subtropical marine laboratory, came to the conclusion that the time was propitious to discuss unofficially the desirability of such a station. A list of the men who have worked in marine laboratories in one or another part of the world was made out, and I agreed to send out letters before the Christmas meetings in Washington, to ask whether an informal discussion could take place concerning the various places at which a marine station, or stations, could be located to the best advantage.
>
> We have no ulterior motives, or plans, or backing, and no particular station in mind, but if a general consensus of opinion could be reached as to where a station could be most profitably placed, it is quite possible that in the next few years the opinion of those zoologists who have been interested in marine stations might be useful to anyone considering the possibility of establishing a tropical marine laboratory.[5]

Morgan sent such a letter to thirty-one zoologists, of whom several, including Harrison, were enthusiastic. Others, such as W. M. Wheeler, were skeptical. Wheeler argued that existing facilities in Cuba, such as the Cienfuegos Station near Harvard's Soledad Botanical Gardens and Laboratory, or perhaps a station in Panama in the Canal Zone might be more appropriate. The matter was dropped at this point, as there did not appear to be sufficient interest in establishing a single site to engender further action. It is illustrative of Morgan's love for marine organisms and natural history that, at the height of his work with *Drosophila*, which represented highly mathematical, analytical, and laboratory-oriented research, he was still working to further the pursuit of marine biology.

During this same period the older Morgan children were all in school. It was Lilian's conviction, however, that the school system

[5] Morgan to Harrison, December 15, 1924; Harrison Papers, Yale University.

did not allow enough time out of doors for young children, so she undertook to teach first Edith, then Lilian and Isabel, at home. Consequently the three girls did not enter school until the third grade. Home teaching, especially among intellectuals, was not uncommon in the early twentieth century.[6] When all the children were in school, around 1919, Lilian Morgan decided it was time for her to go back to her own work. She gave serious consideration to devoting herself full time to music, as she was at the time a member (leader of the second-violin section) of the Columbia University Orchestra. However, her original career choice as a biologist won out, and she returned to the laboratory. Morgan gave her no preferential treatment in the lab, but she became part of the group in a quiet and efficient way. Morgan apparently did not push her toward returning to science as opposed to music. However, it is likely that his own interest, as well as the enthusiasm emanating from his laboratory, weighted the decision heavily in that direction. Joining the *Drosophila* group, she applied her considerable cytological skills to analyzing chromosome complements of various fly strains.[7] Lilian Morgan remained part of the laboratory group until her husband's death in 1945, and indeed retained a workroom in the laboratory until the end of her life (see Figure 18). She published a number of papers and, as will be discussed shortly, made several important contributions to the progress of the fly work.

Repeatedly in his fifteen-year stay at Columbia, as well as his prior thirteen years at Bryn Mawr, Morgan had passed up sabbatical leaves when they came due. At Columbia, his first sabbatical came up in 1910 or 1911 when he was just beginning the *Drosophila* studies, and his family was too young to make travel easy. The second sabbatical opportunity came in 1918, and Morgan began to think about it seriously. He finally made plans to take a full year sabbatical in 1920-1921, at a time when the *Drosophila* work could benefit from relocating for a while and when his children could enjoy the adventure. He felt that the group as a whole had developed a considerable routine by 1919—working at Columbia in the winter and Woods Hole in the summer—and that a variation might prove stimulating. His thoughts turned

[6] Isabel Morgan Mountain, personal communication, August 1974. Other intellectuals who chose to educate their children at home included Pierre and Marie Curie and Bertrand Russell.

[7] See Isabel Morgan Mountain, "Notes on T. H. Morgan," pp. 14-15.

Figure 18. Lilian V. Morgan counting fruit flies in her laboratory at California Institute of Technology, about 1930. After returning to the laboratory part-time in the 1920s, L. V. Morgan made several significant contributions to genetic studies in *Drosophila* (see text for details). (Courtesy of Isabel Morgan Mountain.)

toward California, to which he had been attached since his summer trip in 1904. The two weeks spent at Berkeley in 1918 when he gave the Hitchcock Lectures had renewed his interest in California.

Professionally, Morgan saw several advantages to relocating the entire laboratory on the West Coast for a year. First, there was some important new work in plant genetics being carried out at

Berkeley, largely under the direction of E. B. Babcock, and Morgan thought that there would be some advantage to the zoological group through "cross-fertilizing" with the botanists. More concretely, he wanted to be able to compare the species of *Drosophila* from California with the East Coast species which formed the largest bulk of those with which his group had worked. Morgan's interest was directed particularly to a comparison between types of mutations, as well as chromosomal abnormalities.

To Morgan, a sabbatical meant freedom from teaching and the technical details involved with the Columbia department—duties that had fallen increasingly on him because of E. B. Wilson's poor health. A sabbatical also meant a chance to complete a number of papers and projects that had been sketched out, but not fully written. It also meant a chance for Morgan, who had always maintained a strong love for nature, to take advantage of the California climate and spend an entire year more or less in the out of doors investigating many new species in the tide pools along the Pacific coast. To effect these plans, Morgan arranged with W. K. Fisher, administrator of the Hopkins Marine Laboratory at Pacific Grove (operated under the auspices of Stanford University), to work at the station for the summer of 1920.[8] The following winter would be spent at Stanford, and somewhat later arrangements were made to spend the summer of 1921 working with Babcock's group at Berkeley. The aim was to continue the fly work at each location—something that was possible because of the simple requirements of the work. As Morgan indicated to Babcock, "The work can be done almost pretty well anywhere so long as we have a table, and an electric bulb."[9]

With this plan in mind, of course, Morgan could not think of leaving the *Drosophila* group behind. So, he made arrangements for Sturtevant, Bridges, Wallace, and Phoebe Reed (who was also working with the group as an artist and assistant) to accompany the Morgan family west.[10] Arrangements were made to leave on June 1, 1920. As the time approached, excitement reigned in the Morgan household and in the lab. Morgan and his three

[8] Morgan to Otto Mohr, March 1, 1920; Morgan Papers, APS.

[9] Morgan to Babcock, June 15, 1920; Papers, Department of Genetics, University of California, Berkeley; housed at the American Philosophical Society.

[10] Morgan to Otto Mohr, March 1, 1920; Morgan Papers, APS.

daughters went to the bank and carried home $1,000 in cash to purchase the tickets to California![11] The house at Woods Hole and in New York City were rented, and on the day of departure the group assembled at Pennsylvania Station where they were to take the train to Chicago. Some of the more valuable *Drosophila* cultures were not entrusted to the baggage car with the bulk of the culture material, but were carried in the Pullman car, where the family had a drawing room and a seat-and-bed section. Morgan's son Howard kept notes of the train ride which give an interesting insight into the jolly and novel experience which a trip across the country represented in 1920:

> *June 1.* Board train after having picture taken of the whole party in the Pennsylvania station. Train pulls out at 2:08. I eat a little and then sit and read. Boy across the way in berth 11 plays on the violin. Supper we use lots of powdered milk. Then sit in smoker with daddy until 9:30. Go to bed and read until 10:20. At 10:30 father comes to bed. Watch the mountains out the window until 10:45 . . . after awhile I go to sleep.
>
> *June 2.* . . . I wake up at 6:00 and read. At 7:30 I get dressed. At 8:00 father and I get up and go to the diner. We order strawberries and cream and also coffee and cocoa. We come back and have breakfast in the compartment. . . . At lunch we all go to the diner. After diner [*sic*] I sit with Miss Reed and Wallace who are fathers assistance [*sic*]. We arrived in Chicago on time and went to the other station on a funny horse driven cart.
>
> *June 3.* After breakfast we got off at Omaha after crossing a very muddy river. Here we were ten minutes late because they had to put new brake shoes on the car. . . . We passed a large sand shovel that was way down in a hole and was digging sand. It was a great contrast to the limestone quarries that were in Ohio. We are still going through prairie. At dinner we went to the diner. Then we had afternoon lemonade with Miss Reed and Wallace. I mean they had it with us. We are now in the foothills of the Rockies.
>
> *June 4.* I wake up in the morning and look back to see the mountains that we passed through. We had breakfast in the

[11] From Mountain, "Notes on T. H. Morgan," pp. 1-2.

compartment and then went to the observation where I took some pictures. Then we had some candy with Miss Reed and Wallace. . . . We passed Devil's Slide in Echo Canyon, but I did not see it. . . . We passed the great Salt Lake and it is very hot. . . . We came to a great salt desert that was the bottom of the old lake and it was 4-16 feet deep and extended 4 miles to each side. We can now see the snow capped peaked tops of the Sierra Nevadas. . . . We ate dinner in the drawing room and then Daddy and I went to the smoker until it was time for bed.

June 5. When we reached summit of the Sierras the downgrade started. Today we have been through 41 miles of snowshed. We ran through the foothills and ate dinner in the diner while stopping in Sacramento. When we got out of Sacramento we came to the flat country where our engine broke down. Now we are one hour and 20 minutes late. With the help of a small engine we got to the ferry. There our train was broken up in sections and put on the "Solano," the largest ferry boat in the world. When we got started, we made up five minutes with the crippled engine. We reached Oakland 1 hour and 20 minutes late. At Oakland we meet Mr. Sturtevant, an assistant, and he told us that the auto we were going to buy was in Yosemite but would be back soon.[12]

While Howard busied himself with these observations, the three young girls vied with each other to see who could count the largest number of cows as the train steamed through the Midwest plains.

From San Francisco the group went south to Pacific Grove where Morgan had arranged for a small house for the summer. The family loved California from the very first. For the children, especially, it was an exhilarating experience. Howard liked the short hikes and camping trips that he and friends were able to make into the surrounding country. For the girls, the beaches were particularly attractive as were trips with their mother into the small towns nearby. For Tom and Lilian it was a beautiful countryside with much opportunity to be outside and observe nature. As Morgan wrote ecstatically to Otto Mohr: "New York has no charms for us at this time, for the country here is simply superb. But for that, so is it

[12] Excerpted from Howard Morgan's account of the California trip, a copy of which was sent to the Mohr family in Norway, June 1920. Courtesy of Tove Mohr.

everywhere at this time. Only here we live out in it all from morning to night. When the fruit trees were in bloom, we took some pictures to sent [*sic*] to you. I will try and get them off in a few days."[13]

Scientific work during the academic year at Stanford centered around *Drosophila*. During the summer of 1920 and the ensuing winter, it was largely Morgan, Bridges, Sturtevant, Mrs. Morgan, Wallace, and Reed, who comprised the group. In June of 1921, however, when the group moved to Berkeley, Morgan tried to induce several of his special friends to come and join them for a month or two. He invited Metz, then at Cold Spring Harbor (any differences growing out of the potential rivalry of their groups had disappeared), and Charles Zeleny of the University of Illinois. In addition, Donald Lancefield, who had just finished his doctoral dissertation, and his wife, a cytologist of considerable ability, came out from New York. Lancefield and Sturtevant had some problems to work out together, while Mrs. Lancefield helped Bridges with the cytological work.

Much of the group's time in California was spent working on the Bar-eye phenomenon and triploid flies. Morgan and Sturtevant in particular were interested in trying to find out the nature of the Bar-eye condition, and its strange habit of reverting to normal more frequently than any other known mutation.[14]

Of particular interest during this time was Lilian Morgan's discovery of a strange cytological condition which had considerable implications for the chromosome theory of sex determination. The story of her discovery is told as follows by Sturtevant: "Mrs. Morgan found a very unusual mosaic individual in one of her cultures. As she was examining it, it recovered from anesthetization and flipped off the microscope stage and onto the floor. She searched the floor thoroughly, but was unable to find it. Then she reasoned that flies go towards the light when disturbed, so perhaps the mosaic was on the window; there she found and captured it, and was able to recognize it with certainty because of its unusual appearance."[15] She bred the fly and found that all its female offspring acted as though they had both their X-chromosomes from the female parent. Morgan was quite excited and wrote to Otto Mohr:

[13] Morgan to Otto Mohr, May 4, 1921. Morgan Papers, APS.

[14] For a full discussion of the details of the Bar condition, see chapter 6.

[15] A. H. Sturtevant, *A History of Genetics* (New York: Harper & Row, 1965), p. 55.

"I think someone has probably written you that Mrs. Morgan has a stock that behaves as though the two sex chromosomes are tied together. Rarely a female is found in which they seem to have broken apart again. Whether whole chromosomes are involved, as I think likely, or only a piece of one X tied to another will be revealed by sections [i.e., cytological preparations] that are now ready to study."[16]

When Lilian Morgan made such cytological studies, she found that the two X's were indeed attached together at one end, forming a V instead of the usual two rods.[17] This attached-X strain immediately became very useful as a tool for maintaining sex-linked mutants or combinations in which the females are weak or sterile and for understanding some of the geometrical relationships involved in crossing-over.[18] Bridges's nondisjunction work was moving rapidly because of a new strain obtained from Muller (who was again working at Columbia). Muller's new strain afforded considerable opportunity for pursuing problems regarding both sex determination and the Bar-eye enigma. Morgan wrote enthusiastically to Zeleny trying to persuade him to come to California as early as possible: "Did I tell you that Bridges has a triploid form? This opens up a new world. Muller has a stock in which there is no crossing-over in the X. This, as well as nondisjunction, offers a means of finding out whether the Bar mutation occurs in only one chromosome at a time. I am making up some stocks that will show this. If you like we can work it out together. The idea was Sturtevant's! Doesn't this offer an inducement to join us earlier?"[19]

Especially at Pacific Grove, Morgan took a number of field trips to the tide pools and studied the marine forms of the Pacific coast. Although there is no record that he carried out any specific experiments using these organisms at the time, it was obvious that he was attracted, as always, by the natural habitat and by living marine organisms. When they moved to Stanford in the fall, Morgan frequently made weekend or daylong trips down to the Pacific Grove to collect specimens.

[16] Morgan to Otto Mohr, May 31, 1921; Morgan Papers, APS.

[17] See L. V. Morgan, "Non Criss-Cross Inheritance in *Drosophila melanogaster*," *Biological Bulletin* 42 (1922): 267-274.

[18] See Sturtevant, *A History of Genetics*, p. 55.

[19] Morgan to Zeleny, April 25, 1921; Zeleny Papers, University of Illinois Archives.

While the sabbatical meant mostly work, Morgan did take some time off during the year. The main event was a two-week camping trip for a group of Morgan's scientific associates into the Russian River area north of San Francisco.[20] Of particular importance and excitement was the purchase of the family's first car—a secondhand Overland. This was especially intriguing to Howard who, at fourteen, was itching to learn to drive. Morgan took his first and last driving lessons in California. He never really learned to drive, for to him a car was "just a mechanical thing,"[21] and he never lacked eager drivers. Mrs. Morgan and the children took various side trips in their newly acquired car, especially on weekends. Sometimes Morgan accompanied them on these jaunts, but many times he remained at home to work. The most exciting trip Mrs. Morgan and the girls took, with Howard as driver, consisted of an extensive camping trip with other families of the Stanford faculty to the High Sierras in the Round Top area, as well as to the Yosemite Valley.

The following spring Morgan went abroad to London to deliver the Croonian Lecture before the Royal Society. He left New York May 20, arriving a week later in England. There, he visited Bateson at the John Innes Horticultural Institution, then at Merton College, Cambridge,[22] and the Hope Zoology Department at Oxford. Morgan's visit to Bateson was apparently peaceful, for Bateson by this time had publicly admitted the value of the chromosome interpretation. At Oxford, Morgan visited Huxley and a number of other workers, particularly those studying evolution and protective coloration. He felt that English biology was much too dominated by the old morphological and natural-history school. Nonetheless, he was especially attracted by the work in population genetics and evolution under way there.[23]

In 1925 another honor was bestowed on Morgan: he was invited to give the Silliman Lectures at Yale. The subject was to be the "theory of the gene," and it was published under that title

[20] Howard K. Morgan, "Notes on Thomas Hunt Morgan's Life," p. 5.

[21] Ibid.

[22] See William Coleman, "William Bateson," *Dictionary of Scientific Biography* (New York: Charles Scribner's Sons, 1970), vol. 1, p. 505.

[23] Julian Huxley to Richard Shryock, American Philosophical Society, April 3, 1963; Archives, American Philosophical Society Library, by courtesy Dr. Whitfield Bell.

the next year.[24] In the midst of preparing for the first lecture, however, Morgan received a telegram from his sister Ellen in Lexington that their mother was "very ill, without hope of recovery."[25] Morgan quickly wrote to Harrison explaining that his first lecture might be delayed because of his mother's condition, and left for Lexington on the same day. His mother died a few days later.

Morgan never wrote or said anything that offers a clue as to his relationship with either parent. His mother and sister Ellen came periodically to Woods Hole in the summer, but no mention of any aspect of their visits has been made by Morgan. His mother was described as a beautiful, gracious, southern lady,[26] who was a semi-invalid during much of the last part of her life. After her husband's death she continued to live in Lexington with her daughter in the smaller house, 210 North Broadway, back-to-back with Hopemont.

While in Lexington, Morgan took care of arrangements for his mother's funeral and looked after the closing of her affairs. The estate was small, consisting of a few stocks and the property in Lexington, along with rental fees collected annually for property along railroad right-of-ways in Kentucky and Ohio. The estate was divided between Morgan and his sister; the latter retained the house on North Broadway (Hopemont had been sold some time after Charlton Hunt Morgan's death), where she lived until her death. After a week in Lexington, Morgan returned to New York, and thence a few days later to give the first of the Silliman Lectures. This pattern of activity was typical of him. He did what had to be done at the time of a crisis, executed it efficiently and effectively, getting on with the regular work as soon as possible.

By the late 1920s, several changes had begun to affect Morgan. He was moving further from the day-to-day activity of the *Drosophila* work, even though he continued full force as a writer and publicizer. The details of crosses and the complex problems of some new mutants were beginning to interest him less. The main new ideas had, after all, been worked out, and thus the excitement

[24] T. H. Morgan, *The Theory of the Gene* (New Haven: Yale University Press, 1926).

[25] Quoted from a letter by Morgan to Ross Harrison, January 13, 1925; Harrison Papers.

[26] Mountain, "Notes on T. H. Morgan," pp. 24-25.

was less for Morgan. Besides, the analysis of genetic data was becoming increasingly complex and mathematical, and that did not sit easily with him. As if to give a clue to his thinking, he published in 1927 a monumental treatise, *Experimental Embryology*,[27] which surveyed a number of topics in the modern experimental approach to embryology. This work signaled a revival of interest in the problems of his youth, a revival that was increasingly to occupy his research time as he grew older. He longed to work more closely with marine organisms and to get back into embryological work. He had always kept up some developmental work at Woods Hole, but there were many unsolved problems that seemed fresher to him than the present details of chromosome theory.

Another change was the impending retirement of E. B. Wilson from Columbia. Morgan's career at Columbia had been so tied up with Wilson, both intellectually and personally, that he could not help but look with sadness at his colleague's withdrawal. Things were changing around and within him, and Morgan was never one to let events control his life. He had made a new field vital and opened up its potential. He could now afford to make a change before changes were made for him. It was in this general mood that he received the first tentative inquiries about his willingness to start a new division of biology at the California Institute of Technology.

GENETICS AND EMBRYOLOGY

In the period between 1915 and 1930 Morgan was concerned with bringing the results of the new genetics to bear on two problem areas that had long interested him: embryology and evolution. The former proved difficult and ultimately unsuccessful. The latter involved a struggle, but one that for Morgan produced satisfying results.

The central problem of embryology, posed in genetic terms, was how embryonic cells, which all have the same genes, develop along different lines during embryonic growth, some becoming nerve, others muscle, still others kidney or lung, etc. Morgan defined the relationship between embryology and genetics as that area of physiological processes intervening between the gene and the adult character. As he wrote in 1935: "Between the characters that are

[27] New York: Columbia University Press, 1927.

used by the geneticist and the gene that the theory postulates lies a whole field of embryonic development, where the properties implicit in the genes become explicit in the protoplasm of the cells. Here we appear to approach a physiological problem, but one that is new and strange to the classical physiology of the schools."[28] Morgan and most of his coworkers were convinced that genes acted in some chemical capacity, probably as enzymes, in the physiological processes of the cell. But the experimental evidence as to what form this interaction took was so slight that to Morgan it was a subject of total speculation. It could be that regional cytoplasmic differences in the egg triggered different genes to operate. Or, it could equally well be that regional cytoplasmic differences caused all the genes to operate, but to do so differently in different cells. It could also be that regional cytoplasmic differences in the egg had no bearing on gene action whatsoever. As Morgan noted, however, postulating regional cytoplasmic differences as a triggering mechanism really solves nothing, because we still must explain how the cytoplasmic differences become established. Typical of his experimentalist approach, Morgan wrote, "We must wait until experiments can be devised to help us to discriminate between these several possibilities."[29]

Morgan made some attempt to treat the relationship between embryology and genetics in *Embryology and Genetics* in 1934.[30] In this work, Morgan described a large variety of genetic and embryological experiments, including those of the *Drosophila* school and those of Hans Spemann (1869-1941) and his school of experimental embryology.[31] Each chapter of the book is devoted

[28] T. H. Morgan, "The Relation of Genetics to Physiology and Medicine" (Nobel Lecture) *Scientific Monthly* 41 (1935): 5-18, quotation, p. 8.

[29] Ibid.

[30] T. H. Morgan, *Embryology and Genetics* (New York: Columbia University Press, 1934).

[31] Morgan was impressed with the work of the Spemann school. Spemann and his students had noted that when an early developing tissue comes into physical contact with undifferentiated cells, the latter begin to differentiate. A whole series of such "inductive processes" can be mapped out for the embryo, beginning with a primary inducer, or "organizer," located at the dorsal lip of the blastopore (a region of the gastrula). It was thought that induction was a chemical process, and some workers even claimed to have found the specific type of molecule passed from inducer to induced tissue. Later work showed that all these results were greatly oversimplified, for induction could in many cases be brought about by a very heterogeneous collection of substances (including dust from the floor). The inductive phe-

to a specific topic, either on genetics or embryology. For example, chapter three deals with the egg and sperm, chapter four with the cleavage of the egg, chapter five with gastrulation, chapter ten with multiple chromosome types, chapter eleven with protoplasm and genes, chapter twelve with larval and fetal types, and chapter thirteen with parthenogenesis. Throughout each of these chapters, Morgan sticks rather close to a presentation of the subject in its own terms. A few chapters attempt in some way to discuss both genetics and embryology: for example, chapter two on development and genetics and chapter seventeen on physiological embryology. However, in these chapters only the most general relationships are drawn and are given no more concrete form than hypotheses about successively triggered batteries of genes accounting for differentiation. Even the chapter on physiological embryology is largely devoted to a discussion of the physical changes observed during embryonic development, such as lifting of the fertilization membrane of the egg after entrance of the sperm, or the ameboid movement associated with certain cells during gastrulation. It does not even include a discussion of the chemical embryology of Gavin de Beer and Julian Huxley (i.e., measurements of changes in oxygen consumption or urea production during the growth of embryos).

It is significant that Morgan chose to write a book entitled *Embryology and Genetics* in which so little actual connection could be drawn between the two subjects. A story is told in which one of Morgan's colleagues said that the title of the book had intrigued him, but that the contents had been a disappointment since no relationship was demonstrated about the genetic basis of differentiation. Morgan is reported to have said the man's disappointment was his own fault: "After all, I did exactly what I said I would do in the title: I talked about embryology, and I talked about genetics."[32] This remark illustrates the basic problem with

nomena were real processes; the explanation for them in terms of "organizer substances" has been largely discarded. For further references on the work of Spemann and his school, see Spemann's autobiography, *Forschung und Leben* (Stuttgart: Engelhorn, 1934), and Viktor Hamburger's, "Hans Spemann and the Organizer Concept," *Experientia* 25 (1969): 1,121-1,125. Jane Oppenheimer has written a valuable account of Spemann's work in relation to embryology at the cellular level, "Cells and Organizers," *American Zoologist* 10 (1970): 75-88.

[32] Jack Schultz, personal communication.

which Morgan tried to grapple. That he was unsuccessful is a result born of the complexity of the question itself (it is still unsolved) and the lack of experimental techniques to answer questions about biochemical changes within cells.

Morgan was interested principally in experiments and experimental results. If he failed to offer extensive insight into the relation between embryology and genetics, it was because those insights would of necessity have had to be purely theoretical. Morgan was not opposed to theorizing, and he did indulge in speculation himself on occasion. But he could not go far with speculation alone. As soon as he began to get away from concrete experimental results, his interest began to wane, and his speculations began to lose substance. Morgan's forte was not suggestive theory, but the analysis of concrete results.

The failure to integrate genetic and embryological theory was a disappointment to Morgan. He had always been interested in understanding the mechanism of differentiation, and with his new knowledge about genes it appeared that such a mechanism might be found. But such was not the case with Morgan, Spemann, or even embryologists today. As fascinating as the problem of embryonic differentiation has always been, we are not much closer to knowing its precise mechanism in the 1970s than in the 1930s. Much interesting work has emerged in the areas of tissue induction, nuclear transplantation, biochemistry of genetic control, and other topics that bear on differentiation. But none has demonstrated the exact method of functioning in a full-fledged differentiating system.

Small wonder that Morgan did not achieve his goal; his hopes were undoubtedly too high for his day. Much more gratifying and successful were his attempts to apply the new Mendelian theory to the problem of evolution by natural selection.

GENETICS AND NATURAL SELECTION

Morgan's conversion to Mendelian genetics after 1910 was also a step in his conversion to the Darwinian theory of natural selection. Taking this new road, like that of Mendelism itself, was not easy for the skeptical Morgan. For at least several years after becoming convinced of the chromosomal basis of Mendelism, Morgan remained unwilling to accept a number of points about Darwinian theory. More specifically, he did not see immediately how Mendelism could be applied to the study of evolution in any general way.

It was between 1912 and 1915 that Morgan seems to have made a substantial (but by no means complete) change in his attitude toward natural selection. In 1912 Morgan still held to de Vries's mutation theory, although he was becoming aware of the serious difficulties it had encountered.[33] Morgan still believed, however, that discontinuous variations—that is, mutations of some sort, either large or small—were the raw material for evolution. Nineteen twelve was also the year in which the young Julian Huxley visited Morgan's laboratory at Columbia. Huxley was fresh from studying the evolution of protective coloration and mimicry, which he used as a special case amplifying some of Darwin's basic principles. Huxley was most impressed with the mutant *Drosophila* turning up in Morgan's laboratory and viewed them as throwing considerable light on the origin of variability. To the delight of Morgan's students, he argued openly with Morgan on the issue of natural selection, chastising the older man for his persistent skepticism. Muller recalls that a few days after Huxley's departure, Morgan mumbled something about his English guest as "that young whippersnapper."[34] But by 1916, when he published his Vauxeum Lectures as *A Critique of the Theory of Evolution*, Morgan saw much more clearly that Mendelian phenomena were compatible with, and went a long way toward explaining, certain aspects of Darwinian theory.

A number of factors were responsible for Morgan's change of mind. First, in 1909, just as Morgan's breeding of his *Drosophila* stocks was getting underway, he began to have doubts about the evolutionary role of discontinuous variations in general, and de Vries's mutations in particular. In an article that year he wrote:

[33] See Morgan's article, "Some Books on Evolution," *Nation* 95 (1912): 543-544: Between 1910 and 1912 the cytological studies of Bradley Davis began to suggest that the supposed mutations de Vries observed in *Oenothera* could be explained by complex chromosomal arrangements; see, for example, Bradley M. Davis, "Genetical Studies of *Oenothera*," I. "Notes on the Behavior of Certain Hybrids of *Oenothera* in the First Generation," *American Naturalist* 44 (1910): 108-115; II. "Some Hybrids of *Oenothera biennis* and *O. grandiflora* that Resemble *O. lamarckianiana*," *American Naturalist* 45 (1911): 193-233; and III. "Further Hybrids of *Oenothera biennis* and *O. brandiflora* that Resemble *O. lamarckianiana*," *American Naturalist* 46 (1912): 377-427.

[34] H. J. Muller, "A Biographical Appreciation of Sir Julian Huxley," *The Humanist*, nos. 2 and 3 (1962): 51-55.

> It has often been urged, and I think with much justification, that the selection of individual, or fluctuating variations could never produce anything new, since they never transgress the limits of their species, even the most rigorous selections—at least the best evidence we have at present *seems* to point in this direction. But a new situation has arisen. There are variations within the limits of Linnaean species that are definite and inherited, and there is more than a suspicion that by their presence the possibility is assured of further definite variations in the same direction which may further and further transcend the limits of the first steps. If this point can be established beyond dispute, we shall have met one of the most serious criticisms of the theory of natural selection.[35]

The occurrence of small but definite variations which seem to be inherited in a predictable manner gave Morgan a new insight into natural selection. Contrary to his earlier belief, Morgan now recognized that these small variations were in fact heritable, and could thus be acted upon by selection:

> Darwin knew of cases of sudden mutation, and called them sports or monstrosities. He saw that they could seldom supply materials for evolution because they changed a part so greatly as to throw the organism as a whole out of harmony with its environment. This argument for rejecting extreme or monstrous forms seems to us today as valid as it did to Darwin; but we now recognize that sports are only extreme types of mutation, and that even the smallest measurable degree may also arise by mutation. We identify these smaller mutational changes as the most probable variant that make a theory of evolution possible both because they do transcend the original types, and because they are inherited.[36]

The new type of mutation Morgan observed in *Drosophila* was smaller than de Vries's species-level jumps, but larger than the individual fluctuating variation of the old-style Darwinians. Morgan's observations with *Drosophila* told him that heredity could

[35] T. H. Morgan, "For Darwin," *Popular Science Monthly* 74 (1909): 367-380.

[36] T. H. Morgan, *Evolution and Genetics* (Princeton, N.J.: Princeton University Press, 1925), pp. 129-130. This book is a revised form of *A Critique of the Theory of Evolution*. Differences between the two books are not substantial in terms of Morgan's understanding of natural selection, except in a few particular cases which will be referred to below.

still be discontinuous, and yet be the source of variability on which selection acted. One did not have to choose between species-level variations (de Vries) and infinitely graded variations (the orthodox Darwinians) as the only source of variation. In fact, neither was probably the source of variation in nature at all—it was the small-scale "micromutation" such as scute bristle, vermillion eye, or dumpy wing.[37]

Second, by 1912 the mutation theory had drawn serious criticism, principally because of the work of Bradley M. Davis (see note 33). In that year Morgan wrote that the English edition of de Vries's *Mutation Theory* was already behind the times, largely because Davis had shown that the large-scale "mutations" observed in *Oenothera* appeared to be the result of complex hybrid rearrangements. Muller even showed that the concept of balanced lethals (where lethal genes remain in the population because their lethal nature is only apparent as homozygous recessives), which he had developed to explain certain *Drosophila* results, could also be applied successfully to many cases in *Oenothera.* The concept of balanced lethals was in total agreement with Mendelian principles. It thus appeared that de Vries's variations were not nearly so large as originally thought, and certainly were not species-level differences. The decline in validity of the mutation theory was a strong factor encouraging Morgan to explain natural selection in Mendelian terms.

Third, by 1914 the idea of gene interaction had begun to gain considerable support within the Morgan group. Recall from chapter VI that Muller and Sturtevant had used the idea of modifying factors to explain Castle's results with hooded rats[38] and that according to the concept of position effect genes influence each other's functioning. Muller and Sturtevant seemed to have understood almost immediately that the presence of modifying genes (and the whole concept of gene interaction) could account for

[37] It is perhaps unfortunate that Morgan chose to retain the word "mutation" in referring to those small Mendelian differences he observed in *Drosophila.* Such usage confused the issue of variation by failing to distinguish between the large variations that de Vries considered new species and the small but definite variations that occurred within the limits of an existing species. Under the term "mutation," Morgan included a whole range of variations that bred true. But many of his readers, familiar with the term only as de Vries had used it, were understandably confused about what type of variations Morgan thought were acted upon by natural selection.

[38] Sturtevant, *A History of Genetics,* pp. 60-61.

the apparent occurrence, in any species, of a continuous range of variations which were, nonetheless, inherited discretely. Darwin's assertion that selection acted upon continuous, but still heritable variation, appeared to be more valid than Morgan had once thought.

In *A Critique of the Theory of Evolution* (1916) Morgan devoted considerable space to showing how modifier genes could account for the production of many fine gradations between the definite unit characters of the classical Mendelians. The strength of expression of any character depended on the modifying factors present in the germ cell. Hence, selection could produce almost any intermediate forms by reducing or increasing the number of modifier genes. Selection could stabilize the number of modifiers in a population, it could increase them to a maximum, or it could eliminate them altogether.[39] In this discussion Morgan gave Muller credit for having worked out the modifier-gene conception,[40] and went on to show how many cases of supposed continuous variation could be interpreted in terms of modifier genes (Morgan cited races of Indian corn, pigeon tail size, and cell size in paramecium among others).

The concept of modifying factors allowed Morgan to meet the objection to natural selection arising from Johannsen's pure-line experiments of 1903. Recall that Johannsen's work suggested that selection could only separate out the various pure lines in a heterogeneous population, but that it could not produce any new lines of its own. Because Johannsen's results were experimentally verifiable and quantitative, they had a particularly strong effect on experimentalists such as Morgan. Within the concept of modifier genes, however, limits to the effects of selection were attributable to the fact that modifier factors had been accumulated to their maximum (for that particular population) or reduced to zero. Selection had then to await more mutations or the appearance of more modifiers before it could produce further change. Johannsen's experiments did not really contradict the idea of natural selection in the way that Morgan had been led to believe in the early part of the century (i.e., 1905).[41]

[39] Morgan, *A Critique of the Theory of Evolution*, pp. 165-169.

[40] Ibid., p. 167.

[41] For example, see T. H. Morgan, "The Origin of Species through Selection Contrasted with Their Origin through the Appearance of Definite Variation," *Popular Science Monthly* 67 (1905): 54-65.

Fourth, the work on gene linkage and the chromosome theory led Morgan to another idea that may have been important in allowing him to apply Mendelian principles to natural selection. Violation of the expected linkage ratios, which ultimately led to the concept of crossing-over, had an important bearing on evolutionary principles. Crossing-over would produce new linkage groups in which combinations of characters, perhaps never existing together before in the same organism, could be produced. The idea of crossing-over and recombination showed Morgan and his students that an almost infinite number of possible combinations of characters could (and probably would) be obtained in the course of time in any population. For any given set of characters, crossing-over would give all the possible intermediate combinations between two parental forms. By 1916 Morgan understood the importance of this factor clearly: "By crossing different wild species or by crossing wild with races already domesticated, new combinations have been made. Parts of one individual have been combined with parts of others, creating new combinations."[42] It was the new combinations of old characters, as well as the appearance of additional mutants and their successive recombination, that would provide the raw materials on which natural selection could act.

Fifth, another important factor in Morgan's changed views on natural selection was his slow but increasingly profound awakening to the difference between the *genotype* and *phenotype* of an organism. Genotype refers to the genetic composition of an organism: what genes it carries and can pass on to the next generation. Phenotype refers to how the organism looks, its visible (or measurable) adult character. The genotype and phenotype distinction was made in 1911 by Johannsen, and published in the *American Naturalist* in that year.[43] In the first decade of the twentieth century many workers in both heredity and evolution confused genotype and phenotype (see the previous discussion of Goldschmidt's and Bateson's confusion of vertical and horizontal processes). This confusion manifested itself in failure to understand the difference between the gene passed on from parent to offspring, and the adult character whose development that gene guides. Thus Morgan, for example, opposed Mendelism as "pre-

[42] Morgan, *Evolution and Genetics*, p. 162.

[43] Wilhelm Johannsen, "The Genotype Conception of Heredity," *American Naturalist* 45 (1911): 129-159.

formationist" because he confused the inherited factor with the actual adult character.

In a similar way, failure to distinguish between genotype and phenotype confused Morgan's attempts to understand natural selection. Johannsen had argued that selection acts directly on the phenotype, producing (over a number of generations), changes in the genotype. The phenotype is the result of the combined effect of many genes interacting with one another and environmental influences (e.g., amount of food). Ultimately, then, modifications of the genotype are the basis of the effects of selection. Changes in genotype are permanent in that they are passed on faithfully. Changes in the phenotype merely reflect environmental factors and therefore may not be permanent. Thus for evolutionists the genotype-phenotype distinction helped clear away the old problem of how to determine which variations are inherited and which are not. Breeding tests helped make those distinctions clear. Yet the point is they *could* be made clear, and that was something Morgan had not seen when he wrote about evolution between 1903 and 1909.

There is no direct evidence as to exactly when Morgan first encountered Johannsen's phenotype-genotype distinction, but it was most likely in 1911. Johannsen's paper on this topic was first delivered to the American Society of Naturalists at its Christmas meeting in Ithaca in 1910 (published 1911). Morgan attended that meeting. More important, Johannsen spent part of the following winter term at Columbia.[44] Since we know that Morgan found Johannsen's ideas and personality attractive (he so wrote to Driesch), it would be surprising if the two men did not discuss the issues of evolution, selection, heredity, variation, and Morgan's new work on *Drosophila*. Though it is clear they did not agree on some points (Johannsen was skeptical of the chromosome theory), Morgan may have used the opportunity to explore further Johannsen's genotype-phenotype concept. Ot any rate, it is clear that by

[44] Johannsen had originally been invited to the United States by George H. Shull (see Shull to E. G. Conklin, Summer 1910; Conklin Papers, Princeton University). For the 1911 winter term Johannsen held a lectureship at Columbia for visiting scholars from foreign countries. (L. Kalderup Rosenvinge, "Wilhelm Ludvig Johannsen. I. Liv og Personlighed," and Øjvind Winge, "Wilhelm Ludvig Johannsen. II. Videnskabelig Virksomhed," in *Oversigt over det K. Danske Videnskabernes Selskabs Forhandlinger* (1927-1928): 43-69; and Winge, "Wilhelm Johannsen," *Journal of Heredity* 49 (1958): 82-88.

1915 or so Morgan had made the distinction clearly in his own mind, for he wrote in *The Mechanism of Mendelian Heredity*: "Failure to realize the importance of these two points, namely, that a single factor may have several effects, and that a single character may depend on many factors, has led to much confusion between factors and characters, and at times to the abuse of the term 'unit character.' It cannot, therefore, be too strongly insisted upon that the real unit in heredity is the factor, while the character is the product of a number of genetic factors and environmental conditions."[45]

Sixth, Morgan's students, particularly Muller and Sturtevant, had considerable influence in bringing about a revision in their teacher's thinking on the mechanism of natural selection. As undergraduates, both Muller and Sturtevant had studied R. H. Lock's *Variation, Heredity, and Evolution*, judged at the time by many to be the most up-to-date work relating evolution and heredity.[46] In this book Lock tried to show how Mendelian genetics and Mendelian natural selection fitted together to form a complete mechanism for evolution. From the outset, Muller and Sturtevant were exposed to the idea that there was no fundamental conflict between the theory of particulate inheritance (Mendelism) and the theory of natural selection.

Muller's reports on these discussions with Morgan are most vivid. According to him, Morgan was stubborn; he would not accept the idea that natural selection could create new species, and continually regarded it (as he had done in *Evolution and Adaptation*) as a purely negative force. As Muller put it: "All of us [himself, Sturtevant, and to some extent Bridges] argued with Morgan on that. . . . Morgan would come back and back . . . it seemed to us as if he somehow couldn't understand natural selection. He had a mental block which was so common in those days."[47] He was confused on some very basic issues of Mendelian genetics and natural selection—the effect of dominance on gene

[45] Morgan, et al., *The Mechanism of Mendelian Heredity* (New York: Henry Holt & Co., 1915), p. 210. A few years later, in an article titled "The Theory of the Gene," Morgan himself spelled out the importance of a genotype-phenotype distinction (*American Naturalist*, 51 [1917]: 513-544).

[46] R. H. Lock, *Variation, Heredity, and Evolution* (New York: E. P. Dutton, 1907). Both Muller and Sturtevant have attested to the importance of Lock's book in helping them see the relationship between Mendel's laws and evolution.

[47] Muller interview, 1965.

frequency in a population, for example. He appears to have believed that a dominant gene would take over a population after a while, using this erroneous assumption to try to prove that evolution could occur without natural selection.[48] Muller tells of an instance in which he heard Morgan express this basic confusion in a public lecture:

> The first lecture I ever heard him give was a public lecture in 1909 before I had taken a course with him. Columbia put on a series of lectures by some of its more important faculty members, and this one had to do with evolution. Morgan had just begun to get a few results from *Drosophila* about this time and was, of course, interested in Mendelian inheritance. He had a big, almost terrifying, diagram covering the wall, which began with a dominant mutant here (top) and passing down from one generation to the next, you see. If there was an average of two offspring, he [Morgan] had it inherited by both offspring, and then both of theirs and so on, until finally it swamped out the whole population. This was an illustration of his trying to get around natural selection, having evolution without it. But also [it shows] his not grasping the very simple principle that the mutation only goes down to half the offspring, and mere dominance is not going to change the numbers of genes in the population, which is a very simple mathematical point, almost too simple to be called that. When I had a class under Morgan the following year, and he brought that up again in the class, I had the temerity to point it out to him right there, and he hesitated, and tried to answer me back, and then later, after, not during the class, he admitted it must be right. He asked, "How did you ever think of that?" And I couldn't explain to him how obvious it was.[49]

Morgan also apparently had some difficulty understanding and accepting the modifier-gene hypothesis, even as it originally related to genetics. Muller's friend Edgar Altenburg pointed out that as late as 1910 Morgan attempted to explain multiple-factor cases by assuming a lack of segregation, and considered them as evidence against Mendelian segregation.[50] But understanding the

[48] Ibid. [49] Ibid.

[50] Altenburg to Muller, March 24, 1946; Muller Papers, Lilly Library, Indiana University; cited in E. A. Carlson, "The *Drosophila* Group: The Transition from the Mendelian Unit to the Individual Gene," *Journal of the History of Biology* 7 (1974): 31-48.

multiple factor and modifier hypothesis was crucial to Morgan's understanding of natural selection. According to Muller: "One of the things which must have helped convince Morgan of the validity of natural selection of a Darwinian kind, was the fact that in the selection of the stock for truncate or beaded wing, both of which are due to modifying factors, we gradually got more and more truncate, or more and more beaded wing. They bred truer and truer. And when we analyzed it we found that there was a whole lot of little genes in there besides the major gene. That rubbed it in. . . . That made the small mutations important."[51] It was Muller and Sturtevant who worked on Morgan to convince him of the importance of these results: "We tried to rub those things in, and I think perhaps Sturtevant rubbed it in more than anyone else, because, as I have said, he was Morgan's favorite student."[52] Sturtevant confirmed Muller's account.[53] In the years following 1916, Morgan continued to develop his ideas and understanding of the mechanisms of evolution. He wrote two more books dealing with the subject: *Evolution and Genetics* (1925) and *The Scientific Basis of Evolution* (1932).[54] Taken together, these three works, written at roughly ten-year intervals, help to trace the intellectual refinement of Morgan's views of natural selection. As genetic and other types of data were amassed, Morgan was able to bring together more of the threads of a comprehensive theory.

Most notable in the work following *A Critique of the Theory of Evolution* was Morgan's increasing awareness of the importance of the population (rather than the individual) to the understanding of evolution. In *A Critique*, Morgan did not mention anything directly about populations of organisms and their role in evolution, although earlier in 1909 he had remarked that "each of us is the descendant of a large population . . . a species moves along as a horde rather than as the offspring of a few individuals in each generation."[55] In *A Critique*, Morgan spoke of animal "groups" of the past as being the ancestors of modern groups—but using the term "groups" in a way somewhat different from our present use of "population." Morgan seemed to think of a group as several

[51] Muller interview, 1965. [52] Ibid.

[53] Sturtevant to T. M. Sonneborn, May 5, 1967, pp. 1-2; Sturtevant Papers, California Institute of Technology Archives.

[54] T. H. Morgan, *Evolution and Genetics* and *The Scientific Basis of Evolution* (New York: W. W. Norton, 1932).

[55] Morgan, "For Darwin," p. 376.

species that existed as one part of a complex and continuing wave of organic life. It was a dynamic concept—species were really just groups that were constantly changing. But the term "groups" as Morgan used it in 1916 did not really represent a population approach to evolution.

By 1925, however, Morgan had become aware of the work of J.B.S. Haldane, who was then trying to develop a mathematical procedure for dealing with evolution on the population level.[56] Though Haldane's analytical methods were in their infant stage, Morgan saw the importance of this kind of thinking in clarifying many evolutionary problems. For example, he saw that such an analysis would help to show conclusively that a dominant gene does not necessarily have any better chance of survival in a population than a recessive. He pointed out that questions about the fate of particular genes in a population could only be answered by an analysis of large amounts of numerical data. Morgan had even begun to recognize that what we today would call "selection value" had to be assigned to different genotypes. As he wrote: "These theoretical considerations do no more than suggest certain possibilities concerning the theory of natural selection. Before we can judge as to its actual efficiency we must be able to state how much of a given advantage each change must add to give it a chance to become established in a population of a given number. Since only relatively few of the individuals produced in each generation become the parents of future generations, numbers count heavily against any one individual establishing itself."[57] He goes on to point out that solutions to the problem were only beginning to become apparent: "This is a most difficult problem for which we have practically no data, and as yet only the beginning of theoretical analysis has been made of this side of the selection problem. Haldane has developed a partial analysis of the problem for a few Mendelian situations. He points out that the problem is extremely complex and that there is at present not much quantitative information to furnish material for such a study of natural selection by means of gene mutations."[58] While Morgan was aware of Haldane's work, it is apparent that he never seriously attempted to understand the substance of population genetics as applied to evolution. As pointed out earlier, Morgan was poorly versed in mathematics, and was, in the last analysis, not really interested

[56] Morgan, *Evolution and Genetics*, p. 142.

[57] Ibid., pp. 141-142. [58] Ibid.

in it.[59] He could see the importance of a quantitative approach to evolution, and even the necessity of setting the theory of natural selection on a mathematical base. But he could not follow the complex mathematical arguments of Haldane, Sewell Wright, or R. A. Fisher. Mathematical and statistical analysis was totally foreign to his type of biological training or interests.

Morgan's thinking on evolution and natural selection was refined and expanded along some lines, but there remained certain areas he never fully understood. Among the most important of these was the problem of speciation. As we saw earlier, Morgan believed that species were arbitrary units, created by taxonomists for convenience (in Ernst Mayr's terminology he was a "nominalist," i.e., someone who believes species are not real units in nature). He never really abandoned this belief. Given this bias, it was logical that Morgan continued to minimize the origin of species as a real or significant problem. In his article "The Bearing of Mendelism on the Origin of Species" in 1923, Morgan stated clearly his nominalist position: "How far these new types [Mendelian variations] furnish the variations that make new species may depend on what we call 'species.' If, as some systematists frankly state, species are arbitrary collections of individuals assembled for the purposes of classification; or if, as other systematists admit, there are all kinds of species both in nature and in books, it would be absurd for us to pretend to be able to say how such arbitrary groups have arisen. It is possible that some of them may not have arisen at all—that they may have only been brought together by taxonomists."[60] In his usual manner, Morgan went on to throw down the gauntlet (in a most diplomatic way, of course) by claiming that the systematists might produce one classification system of organisms and the geneticists a completely different one —but both would be valid. The classification system depended, after all, on the *criteria* chosen to draw the boundaries between groups. These criteria depended on one's viewpoint, and thus, in the last analysis, were arbitrary. So certain of the correctness of his viewpoint was Morgan that he refused to use the word "species" in the remainder of his article. As he wrote: "Each of us might, if he wished, erect a species definition of his own, and each would be within his rights in forming such a definition. Whether it would be desirable for the evolutionist to use the word 'species' that

[59] Muller interview, 1965.

[60] *Scientific Monthly* 16 (1923): 237-246, quotation, pp. 237-238.

tradition has assigned to the systematist, is a question for the evolutionist to decide; but, as I have said, it is a perilous adventure for a geneticist to attempt to interpret the historical species in terms of genes. It may also be a work of supererogation."[61]

That Morgan had oversimplified and to a degree even distorted the picture of modern systematics is evident from a response he received to this paper. In a letter from David Starr Jordan, a well-known ichthyologist, proponent of the effects of geographic isolation on the origin of species, and at the time president of Stanford University, Morgan was gently but firmly chastised: "I am much interested in your excellent paper on the bearings of Mendelism. That I may venture on one or two additional suggestions. . . . No modern taxonomist regards species as 'arbitrary collections of individuals assembled for purposes of classification.' Many species are obscured for lack of material or lack of accuracy in published accounts. This is not Nature's fault. Charge it up to the weakness of humanity."[62] Jordan, who had been engaged in a variety of systematic studies over the course of his career, was no nominalist and knew that systematics dealt with something more real than artificial categories.

To Morgan it was adaptation, not speciation, that was the important outcome of evolution by natural selection. What Morgan failed to see was that an interplay existed between the two, making them components of a complex system. The biological advantage of speciation in the long run is adapative: to reduce competition. The divergent specialization that occurs with speciation allows separate populations to adapt specifically to noncompetitive niches. Thus speciation is not only an outcome of adaptation, it is itself an adaptation. Morgan persisted, however, in drawing a distinction between speciation and adaptation, thus ignoring the biological reality. That reality is the existence of distinct biological groups which, judged collectively on a number of criteria, can be distinguished from each other *biologically* (i.e., do not interbreed, have different ecological niches, etc.). By retaining the nominalism of his youth, Morgan persistently downgraded the importance of species as biological units, or of speciation as an integral part of evolution.

Another problem Morgan never fully understood was the role

[61] Ibid., p. 238.

[62] David Starr Jordan to T. H. Morgan, March 28, 1923; David Starr Jordan Papers, Hoover Library Archives, Stanford University.

of isolation in species formation. Although Jordan had made this point clearly in a *Science* article as early as 1906, Morgan never seems to have taken the idea seriously. In his 1923 article, "The Bearing of Mendelism on the Origin of Species," Morgan still argued that the observed infertility between distinct species could be the result of one or two mutations. That he totally failed to understand the problem of barriers and separation of populations, is attested to by a comment Jordan made to him after reading the 1923 paper: "I do not see why geneticists in general ignore the most potent fact in Geographical Distribution, that 'new species' are formed across barriers. Very seldom do closely related forms exist together but *near together* forming 'geminate species.' Of this we have thousands of illustrations. The supposition exists, that perhaps some primal difference or mutant may have existed in the original migrants."[63]

An even more fundamental issue in Darwinian theory was never fully settled in Morgan's mind. Although he did express the view in 1916 that the principle of selection, acting on Mendelian mutations, could produce gradual change in a line of organisms, it is clear that he was never at ease with the idea of selection itself. The concept, perhaps the very term "selection," bothered him; it sounded purposeful, and with his strong dislike of teleological thinking, Morgan reacted against purposefulness in evolutionary theory. Sturtevant recognized Morgan's strong feelings on this matter: "There *was* one respect in which Muller and I had to 'educate' Morgan. He was never fully happy about natural selection, since it seemed to him to open the door to explanations in terms of purpose. We convinced him that there was nothing teleological or contrary to Mendelism in natural selection—but he remained unhappy with it, and arguments had to be repeated again and again—an experience that I think was good for both Muller and me in that it made us very careful about how we stated the case for selection."[64] It was obviously the constant discussion of this issue in the lab that helped Morgan to come around to the idea that selection can work in natural populations without a

[63] Jordan to Morgan, March 28, 1923; Jordan Papers. Geographic isolation was necessary, Jordan went on to say, to preserve the new mutant types that appeared in a particular population. Jordan apparently made little impression on Morgan. Nothing about geographic distribution or the effects of isolation appears in his later book, *The Scientific Basis of Evolution*.

[64] Sturtevant to T. H. Sonneborn, May 5, 1967, pp. 1-2 Sturtevant Papers, quoted with permission.

guiding purpose behind it. In *A Critique*, Morgan finally got the role of selection correct. He carefully sheared away from the term "selection" any reference to mystical or creative forces.[65] Selection is defined by Morgan to mean "both the increase in the number of individuals that results after a beneficial mutation has occurred (owing to the ability of living matter to propagate) and also that this preponderance of certain kinds of individuals in a population makes some further results more probable than others."[66] Yet, in *The Scientific Basis of Evolution* written approximately sixteen years later (1932), Morgan regressed to his earlier, more ambivalent, position on selection. He wrote in that work: "Under the circumstances, it is a debatable question whether still to make use of the term 'natural selection' as a part of the mutation theory [i.e., Mendelian theory], or to drop it because it does not have today the same meaning that Darwin's followers attached to his theory."[67] Elsewhere in the same book, in his endless quest to make certain that no one viewed natural selection as "creating" any new variations, he wrote:

> If all the new mutant types that have ever appeared had survived and left offspring like themselves, we should find living today all the kinds of animals and plants now present, and countless others. This consideration shows that even without natural selection evolution might have taken place. What the theory does account for is the absence of many kinds of living things that could not survive. . . . Natural selection may then be invoked to explain the absence of a vast array of forms that have appeared, but this is saying no more than that most of them have not had a survival value. The argument shows that natural selection does not play the rôle of a creative principle in evolution.[68]

In 1932 no Darwinian claimed that selection itself was a creative principle (though some few did claim so in the early years of the century)—i.e., that it somehow actively created new variations in a particular direction. That Morgan saw fit to come back to this point again and again, and in fact to use it to play down the role of selection in favor of mutation pressure as the main force in

[65] Morgan, *A Critique of the Theory of Evolution*, pp. 184-195.
[66] Ibid., p. 194.
[67] Morgan, *The Scientific Basis of Evolution*, p. 150.
[68] Ibid., pp. 130-131.

evolutionary change, reveals his deep-seated misgivings about the selection principle. Mutations were concrete, familiar events he had encountered every day for over twenty years. Selection still appeared to be a vague, invisible, and slightly teleological-sounding process which Morgan could not fully accept.

The change the modern reader finds between *A Critique of the Theory of Evolution* (and its revised version of 1925) and the much later *Scientific Basis of Evolution* may well indicate as much about Morgan's pattern of working as about a change in his evolutionary thinking. There can be little question that Morgan had always found the idea of selection incompatible with his anti-teleological notions of nature. In preparing his Vauxeum Lectures of 1915 for publication as *A Critique*, Morgan had passed the manuscript around to all the members of the fly group for criticism. Alexander Weinstein reports that he and other members of the group corrected, discussed, and even rewrote sections of the text —all with Morgan's knowledge and approbation. The sections they had particularly criticized were his views on selection. In response to these criticisms, Morgan had apparently revised his thinking on selection, for he let the corrections stand. But it is not clear that he had fundamentally changed his mind. In 1932 Morgan had only Sturtevant (Bridges had never taken as active a part in such theoretical questions as the others) from the old group still working with him. It is not known whether he sought Sturtevant's advice on the manuscript of the *Scientific Basis*, as he had the group's advice on *A Critique*. But if he did, it appears he no longer took advice so readily. Whatever the exact explanation, we know that from the earliest days Morgan's view of selection was always skeptical. His embrace of it in the two books of 1916 and 1925 may represent acceptance of the logic behind the selection principle, as put forward by Muller, Sturtevant, Weinstein, and others. The view expressed in 1932 appears to be more Morgan's true feeling about selection, written without the help of critical advice from the opposing viewpoints.

Although Morgan's own contributions to the progress of evolutionary theory are slight, he did have a profound, if indirect, effect upon the course of evolutionary theory during the twentieth century. By elucidating the nature of Mendelian variations, the process of transmission, and the structure of the germ plasm, he laid the foundation for the quantitative, population approach to evolution that was necessary to understand the dynamics of natural selection. The development of population genetics in the twentieth century was absolutely dependent upon Mendelism. Without quantitative, discrete hereditary units whose transmissional properties were understood and formulated as predictable "laws," no statistical study of the hereditary nature of populations would have been possible.[69]

[69] For details of the development of population genetics in the post-*Drosophila* era, see the following: on the Russian school, Mark Adams, "The Founding of Population Genetics: Contributions of the Chetverikov School, 1924-1934," *Journal of the History of Biology* 1 (1968): 23-39; and "Toward a Synthesis: Population Concepts in Russian Evolutionary Thought, 1925-1935," *Journal of the History of Biology* 3 (1970): 107-129; on the English school, William Provine, *The Origins of Theoretical Population Genetics* (Chicago: University of Chicago Press, 1971).

CHAPTER VIII

Morgan and Scientific Methodology: The New American Biology (1905-1925)

Morgan was not prone to elaborate on philosophical issues. Philosophy was too abstract for him; his interests lay far more in the concrete phenomena around him. However, in a number of places he discussed, in varying degrees, his philosophy of science. From these statements, as well as from an analysis of the kinds of biological ideas and explanations he accepted and the kinds he rejected, it is possible to come to an understanding of the basic philosophical principles upon which Morgan approached his scientific work.

Understanding Morgan's philosophical position is of great importance because of the strong influence he exerted on the "new biology" that was emerging, especially in the United States, during the early decades of the twentieth century. Morgan's philosophy permeated much of the *Drosophila* work and influenced a whole generation of younger workers toward a new way of approaching old biological problems. His philosophy was also the guiding principle behind the founding of the biology division of the California Institute of Technology, which Morgan organized in 1928.

MORGAN'S OPPOSITION TO IDEALISTIC PHILOSOPHY

Morgan's early opposition to the morphological tradition can be seen as a revolt against philosophical idealism in biology.[1] The

[1] Idealism is the philosophical school that tends to place ideas ahead of material phenomena in time and importance. Religion is a form of "idealistic" philosophy in that it sees the material world as merging from the mind of God—i.e., the "idea" of the world preceded historically the material existence of the world. In more modern terms, ascribing war to "human nature" (an idea) instead of material causes such as economic exploitation, is an example of the idealistic mode of thought. Idealism in philosophy is generally contrasted to materialism. See Maurice Cornforth, *Materialism and*

morphology of Haeckel, for example, was often referred to as "idealistic morphology" since much of Haeckel's thinking was dominated by the search for abstract "types" (such as the primeval gastraea, the form from which all higher organisms were supposed to have developed). Haeckel's work, especially in his later years, was based more and more on theory and less and less on the analysis of concrete observations. This was idealism in the sense that it imposed an order and pattern of development on the living world that was largely in the mind of the morphologist, and had little to do with the way things had actually happened. Morgan's familiarity with living organisms and his distrust for idealistic speculation made him prone to accept the materialist point of view in philosophy, even though he would not have called himself a materialist.[2] So distrustful of philosophical categorization was Morgan that he seldom admitted to belonging to any school. However, in his dislike of idealism and metaphysics, in his rejection of the concept of a creator from whose mind the material world emanated, he followed the materialists' approach in most of his thought.

Morgan's opposition to idealists and speculative work carried on long after his own revolt from the morphological tradition of Brooks and Haeckel. Throughout his career, but especially after 1915 when the *Drosophila* work had become established, he continually attacked idealistic and speculative biology in the writings of his contemporaries.

---

the Dialectical Method (New York: International Publishers, 1971), especially chapters 2-5; also V. I. Lenin, *Materialism and Empirio-Criticism* (Peking: Foreign Language Press, 1972), chapters 1-3. Idealism as a philosophical position should not be confused with the term "idealism" used in a colloquial sense—i.e., the undaunted (and often unrealistic) belief in perfectability, or Utopianism.

[2] Materialism in philosophy refers to the belief in the primacy of material factors in the world. Science is materialistic because it is based on the assumption that the real, material world has an existence independent of human beings, and that material reality determines our view of things. Materialist philosophy believes that ideas reflect material reality and can, of course, then, help control that material world. Historical materialism maintains that people's ideas of history, and of the political and social world around them, are derived from the material conditions to which the people are exposed, most notably economic relationships. As with idealism, so with materialism the colloquial usage (someone who is primarily interested in material things) must be distinguished from philosophical usage.

One of Morgan's most open and revealing attacks was directed at the American paleontologist, Henry Fairfield Osborn. While president of the American Museum of Natural History, Osborn had published in 1917 a book, *The Origin and Evolution of Life*, subtitled "On the Theory of Action, Reaction, and Interaction of Energy." This work epitomized for Morgan all the worst features of contemporary idealism in biology. In a detailed letter to Osborn on December 26, 1917, Morgan took to task his colleague's explanation of "the energy concept." This criticism provides such a valuable insight into the criteria by which Morgan judged soundness in scientific thinking that it deserves to be discussed at some length.

One criticism Morgan made was directed at Osborn's careless use of the words "chance," "order," and "law":

> On page 7 in the first chapter you say: 'In other words, in the origin and evolution of living things, does nature make a departure from its previous orderly procedures and substitute chance for law?' This contrast between law and chance occurs in a number of places in the volume and is likely everywhere to mislead the reader who is not informed, as I know you to be, on the whole philosophical conception that lies back of these two words. From the energy point of view there is no such distinction between law and chance conceivable. If, however, you throw over the energy conception and substitute a mysterious and beneficent being who directs all things that are lawful and introduce a devil-of-a-fellow to mess things up generally and take the chance of coming out all right in the end, there might be some chance of such a distinction, but from the energy point of view, as the physicist uses that term, this is nonsense.[3]

Morgan objected to Osborn's implication that a distinction existed between chance and law in the natural world. Morgan himself had struggled with the concept of chance and purpose as applied to evolutionary theory. He had concluded that the term "chance" can have two related meanings. "We mean by chance, in ordinary speech, two main things. 'I chanced to be there' we say, meaning that our being there was not connected with what occurred, not that mysterious forces, instead of two legs, carried us there. The

[3] Morgan to Osborn, December 26, 1917; Loeb Papers, Library of Congress.

other meaning is that of a large number of possible combinations and a particular one happened."[4] In neither use was "chance" opposed to "law." Yet, what Osborn seemed to say was that chance implied a noncausal, while law implied a causal relationship. Moreover, law seemed to imply direction by some purposive or external agent—the impression of ideas on matter (i.e., idealism). To Morgan this distinction between law and chance was philosophically unsound.

Morgan continued his attack by pointing out Osborn's careless use of other terms: "On page 95 you say: 'The chemical or molecular anatomic constitution of the chromatin infinitely exceeds in complexity that of any other form of matter or energy known. As intimated above, it not improbably contains undetected chemical elements.' I understand, of course, that this is a sort of poetic outburst—not an accurate scientific statement, for as you would admit, of course, we know too little about the chemical composition of chromatin or about its method of action to warrant one in saying that it infinitely exceeds in complexity any other forms of matter and energy known."[5] He chastised Osborn for introducing a term from the physical sciences—the energy concept—and then proceeding to treat it in an entirely vague, nonphysical, and idealistic way. Particularly distasteful was Osborn's vitalistic view that living matter had special properties that transcended all other kinds of matter.

Morgan also objected to idealism in the vitalistic theories of his friend Hans Driesch, though for personal reasons was more tolerant of Driesch's attempts in this direction. Out of deference to Driesch as a person and out of respect for his mental ability, Morgan read much of his friend's speculative and philosophical works. From the beginning, however, Morgan could not accept the rampant vitalism:

> I had intended reading again your paper on vitalism in order that I might write to you more specifically about it. I do not dare begin, however, for my letter would never come to an end. I follow you a long way, but cannot truthfully say that I consider you have "proven" the existence of a principle of vitality—except insofar as there is much that we cannot ex-

[4] T. H. Morgan, "Chance or Purpose in the Origin and Evolution of Adaptations," *Science* 31 (1910): 201-210, quotation; p. 203.

[5] Morgan to Osborn, December 26, 1917; Loeb Papers.

> plain. Loeb thinks, and I am almost prepared to follow him, the idea is a sterile one in regard to its value as a working hypothesis; at any rate, it comes dangerously near to metaphysics in our present state of knowledge. That, however, is only a matter of opinion. I do fear, however, that it will be seized upon by those who cannot understand you, and by those who do not want to understand you to show that our work has not led to an advance, but to a retrogression.[6]

To Morgan it was impossible to "prove" the existence of a vital force, for the very methods by which proof is demanded in the ordinary scientific sense are based on a materialist world view; vital principles, entelechies, and the like, by definition, are idealistic conceptions. There is no way that an idealist can prove to a materialist the existence of a "vital principle," or vice versa. Philosophically, the views are mutually exclusive and admit of no reconciliation. As Morgan saw, only experience with concrete phenomena can overcome idealistic philosophy.

Morgan was, of course, concerned with the very problems that eventually drove Driesch to adopt his idealistic position: the problems of growth, regulation, and self-reorganization of the embryo. He could not help but admit that the marvelous powers of differentiation and reorganization that embryos displayed was a source of ceaseless wonder. He wrote to Driesch in 1905: "My results ought, if I have the brains, to have some bearing on the phenomenon of growth, regulation, and formative principle. You have observed that I have kept that possibility open in all my papers. Some days I am a rank vitalist and others I despise vitalism! Do you ever feel so?"[7] But to Morgan any courtship with vitalism was a kind of puppy love. Vitalism was an interesting speculative enterprise and could be indulged in with some profit if it were not taken seriously. But when it *was* taken seriously, it became a detriment to scientific progress. After reading Driesch's lengthy work on vitalism, *The Science and Philosophy of the Organism*[8] (1908), Morgan wrote:

> I can now congratulate you most cordially for having written a masterly presentation of vitalism. It seems to me that you

[6] Morgan to Driesch, September 13, 1899; Morgan-Driesch correspondence, APS.

[7] Ibid., October 23, 1905.

[8] Given as the Gifford Lectures, delivered before the University of Aberdeen in 1907 (London: Adam and Charles Black, 1908).

> make out a far stronger case than ever before and that your book will receive wide attention and no doubt abundant abuse from the "scientists." I myself should abuse it somewhat if I had an opportunity, but nonetheless I admire it very much and regard it as a masterpiece. You have made out the best case for vitalism that in my humble opinion has ever been made, and I like a great deal of what you say. But when it comes to the critical point and the entelechy steps in to alter the physical series of events without itself entering the energy chain, I fail to be convinced. Logically I think you are sound for entelechy must, to my mind, be either that, or else there are other physical properties of matter about which we know nothing as yet that may play the hidden role. My own inclinations are as you know towards the latter assumption. But it is difficult in a letter to put such things briefly and not appear arbitrary. I wish we could meet and have many long talks over these matters.[9]

Morgan admired Driesch's explanatory ability and his critical insight. What he could not accept was the ultimate recourse to idealism, to a nonphysical, nonmaterial force, which was invoked at the point that our ordinary knowledge of matter could no longer account for phenomena.

Despite his fundamental reservations about vitalism, Morgan was not arbitrary or dogmatic on the issue. He not only read Driesch, but in the spring of 1909 organized a symposium based on the work of Driesch and Henri Bergson (exponent of the idea of "creative evolution") at Columbia for the benefit of "the philosophers physicists psychologists and zoologist."[10] The young embryologist A. W. Mayer, and Morgan's friend William Morton Wheeler (then at the American Museum of Natural History) attended, and Morgan wrote to Driesch to "come and defend yourself."[11]

MORGAN'S MATERIALISTIC PHILOSOPHY

Looking back, we can see Morgan's materialist position emphatically stated in his early opposition to Mendelism. When he continually chastised the Mendelians for turning facts into factors and for juggling particles about in the germ plasm for the sake of

[9] Morgan to Driesch, January 30, 1909, pp. 1-3, Morgan-Driesch correspondence.

[10] Ibid., p. 6. [11] Ibid.

saving the theory, he was pointing to the idealistic and metaphysical aspects of the new genetics.[12] To Morgan, a theory was of limited value if one approached it as having no physical reality. The idealistic world view is content with a theory if it makes logical sense; whether or not the theory is physically true is often of less importance. Such a position was unacceptable to Morgan. What was important was to narrow down the range of explanations to those that had a basis in the physically real world.

Morgan's materialist philosophy was born more from his own experience with the world of organisms and natural phenomena than from reading the works of philosophers. From his earliest writings on he displayed a remarkable ability to see processes as a whole and to stay within the bounds of reliable data in devising explanations. Enormously influential, however, in this regard, was Morgan's direct contact with Jacques Loeb, beginning in 1891 at Bryn Mawr. Loeb was the arch exponent of an overt materialist philosophy. Morgan was greatly attracted to Loeb's experimental and scientific works and through them to his philosophical position.

Born in Prussia in 1858, Loeb was a product of the mechanistic materialism of his time.[13] The materialist tradition led him to oppose strongly all vitalistic explanations; his concern with the nature of the human will led him from philosophy to psychology and finally to physiology. At the time he came to America in 1891, Loeb was fully confident that even complex psychological problems could be understood if only they were studied from a materialistic point of view.

[12] Metaphysics is a process of thought in which speculation goes beyond the available data, beyond the observable physical reality (meta = beyond; physics = material world). Both idealists and materialists can think metaphysically. When an idealist points to God as the determiner of all events, it is metaphysics; when a materialist postulates atoms or ultimate particles as components of real matter, it is also metaphysics.

[13] For a more thorough discussion of Loeb's background, work, and influence, see the following: Donald Fleming's introduction to the reprint edition of Loeb's 1912 book, *The Mechanistic Conception of Life* (Cambridge, Mass.: Harvard University Press, 1964); Arnold Gussin, "Jacques Loeb: The Man and his Tropism Concept of Animal Conduct," *Journal of the History of Medicine and Allied Sciences* 18 (1963): 321-336; Nathan Reingold, "Jacques Loeb the Scientist; His Papers and His Era," *Library of Congress Quarterly Journal of Current Acquisitions* 19 (1962): 119-130 (a description, with many excerpts, of the Loeb Papers at the Library of Congress); and G. E. Allen, "Jacques Loeb and the Mechanistic Conception of Life," Sigma Xi National Lecture, 1973, in press.

Loeb was not only a materialist, but also a mechanist.[14] The term "mechanism" meant two things to Loeb: philosophical mechanism and, more concretely, the belief that "the sum of all life phenomena, can be unequivocally explained in physicochemical terms."[15] Philosophical mechanism led Loeb to see complex biological processes such as fertilization or animal behavior in simple one-to-one chemical terms. The physicochemical viewpoint led him to seek the ultimate explanation of all biological and psychological events in terms of molecules in action. Loeb was also a reductionist—one who attempts to understand one set of phenomena by referring it to some other types of phenomena: for example, the attempt to reduce all biological processes to chemical interactions. Biological phenomena are thus "explained" in terms of the laws of chemistry. Part and parcel of Loeb's overall physicochemical viewpoint was his belief that experimentation was the most important method for solving any biological problem. The characteristics of physics and chemistry—use of controlled experiments, collection of quantitative (rather than qualitative) data, repeatability, analysis and predictability—were thus brought to biology by an insistence on rigorous standards of experimentation. Loeb had been strongly influenced by the experimental plant physiologist Julius Sachs (1832-1897) at Würzburg in the later 1880s. Among Europeans, Sachs had been a leader in bringing experimental methods to bear on biological problems.

Loeb's influence on Morgan was direct and unmistakable. Although they worked together for only one year at Bryn Mawr, from 1891 to 1892, they were constant associates at Woods Hole in the summers and in New York after Loeb came to the Rockefeller Institute in 1910. Their families became personal friends,

[14] Mechanism is short for mechanistic materialism. It is one (an older) form of the materialist philosophy which sees all events as the outcome of matter in motion—the random interaction of discrete units (atoms, individual organisms). Mechanists tend to see complex wholes in terms of individual components, believing that the whole is equal to the sum (and no more) of the parts. Loeb was a mechanist in seeing human behavior largely in terms of separate reflex systems or conditioned responses. Mechanists tend to see complex systems as analogous to machines composed of separable, independent parts. Changes within complex systems are the result of random processes (wear and tear) impinging on the system from the outside. Mechanists do not tend to see changes within any system as the result of regular, built-in processes within the system itself.

[15] Jacques Loeb, "The Mechanistic Conception of Life," in Fleming, ed., *The Mechanistic Conception of Life*, p. 5.

as indicated in a letter from Morgan to Loeb on April 17, 1904: "I cannot tell you how glad I shall be to find you in California this summer, for however much I hammer away at you in print (which always turns out to your glory and my detriment, as far as I fear you know), you realize how much I appreciate the value of your work—in fact, I think I realize its true value more than most of your followers and admirers. You see, I am becoming sentimental, and it had better stop."[16] Personally, Morgan was never as close to Loeb as he was to Sturtevant or Otto Mohr. The two men were extremely different in temperament and background. Loeb was more prone to philosophy and to introspection; he was personally shy although intellectually intense and straightforward. He also had a strong interest in social and political matters, and quite often mixed his science and social theory. In all these ways he was quite different from Morgan; the latter was neither shy nor retiring, but had little proclivity for open philosophical discussion or for mixing science and politics. However, the two men had strong intellectual affinities, and Morgan openly expressed his attraction to Loeb's point of view in a letter to Driesch: "Loeb has been here [Woods Hole] . . . all summer and I have learned to know him much better. We agree on so many fundamental views (and differ on these points from most of the people here) that we have become very good friends and strong allies. We have done battle with nearly all the other good morphologists and still survive their united assaults."[17] Later he wrote again to Driesch: "Loeb is here working on ions and getting most interesting results. I see a great deal of him and in fact I find his point of view more like my own than anybody else's here at present. I told him what you said last summer about Johannes Muller, namely, that you thought him greatly underestimated, and he said you are a fine fellow to have found that out. He agrees heartily with you. I hope I was right in that you did tell me that about Muller."[18]

[16] Morgan to Loeb, April 17, 1904; Loeb Papers.

[17] Morgan to Driesch, September 13, 1899; Morgan-Driesch correspondence.

[18] Ibid., July 15, 1901. The next summer found Morgan defending Loeb's approach to Driesch: "It seems to me that you greatly underrate Loeb's work. It is true that he gives to his results the appearance of simplicity when they are in reality often very complicated. But his method gives his results a far-reaching importance and I do not doubt that he has opened up a new field that will help us along." (Morgan to Driesch, March 27, 1902; Morgan-Driesch correspondence.)

Morgan was greatly attracted by Loeb's early work, especially his experiments on artificial parthenogenesis and the colloidal nature of proteins. That Loeb could achieve the cleavage of unfertilized sea-urchin eggs by merely altering the ionic concentration of the water, was a strong step toward understanding the role of the sperm in *chemical* terms. Loeb's claim to have produced whole sea-urchin larvae from unfertilized eggs was to Morgan a very important event in the understanding of what initiated development. It was not something mystical or "vital" about the sperm. The same results could be obtained with certain ions. Morgan saw in Loeb's work an example of the rigorous, physicochemical approach in biology. It was a model for how all future work in biology should be carried out.[19]

Despite his enthusiasm for Loeb's work in biology and for many aspects of the philosophy on which it was based, Morgan could not follow Loeb to the limits of mechanistic philosophy. Although he appreciated the mechanistic viewpoint, Morgan could be better described as more a dialectical than a mechanistic materialist.[20] Morgan was convinced that all biological processes were based on knowable physical and chemical events. But he also believed that in no science was it more important than in biology to study the interaction of parts, as opposed to only the properties of separate and isolated parts. He wrote in 1907: "The most fundamental characteristics of organisms, growth, development, regeneration, seem to involve the organism as a whole in many cases. The interconnection of the parts is one of the chief peculiarities of the organism."[21] Morgan found ample application of this idea in the concepts of gene interaction emerging from the *Drosophila* work between 1910 and 1915. Morgan was not a reductionist. He appreciated the use of physical and chemical methods of analysis,

[19] The relationship between Loeb and Morgan and its effects on the development of mechanistic materialism in twentieth-century American biology have been discussed in detail in G. E. Allen, "T. H. Morgan and the New American Biology," *Quarterly Review of Biology* 44 (1969): 168-188.

[20] Dialectical materialism is another (and historically more recent) form of philosophical materialism. Dialectical materialists are concerned with the complex interactions between parts in a whole. They see the whole as equal to more than the sum of its parts (i.e., to the sum of the individual parts plus their interactions). They reject the simplistic billiard-ball model of physical processes. They see all processes in the world undergoing constant change, motivated by the interaction of various contradictory (dialectical) elements within the systems themselves.

[21] T. H. Morgan, *Experimental Zoology* (New York: Macmillan Co., 1907), p. 366.

and he recognized the importance of studying complex processes initially by breaking them down into their component parts, but he did not believe that every biological problem could find its only satisfactory explanation in purely physical or chemical terms. Physics and chemistry were helpful in understanding biological problems, but an organism was something more than a "bag of molecules." Unlike Loeb, Morgan did not want to make biology over into physics and chemistry. Rather, he wanted to put biology on the same footing (methodologically) with physics and chemistry.

In a study of Morgan's philosophy, Edward Manier has stressed the relative importance of experimentalism (as opposed to simple empiricism) in Morgan's methodology, particularly as it related to his acceptance or rejection of the Mendelian and chromosome theories.[22] Manier points out that the empirical approach is characterized by a concern with a large amount of basically similar kinds of evidence (for example, testing a new mutation against a number of different known stocks by a series of cross-breeding experiments). On the other hand, the experimental approach requires at least two independent types of evidence (for example, determining the existence of a chromosomal deletion both from breeding data and from cytological examination of chromosome preparations).

On these grounds, it is clear that Morgan was motivated more by experimentalist than by purely empiricist considerations. By 1909 considerable empirical evidence was available that Mendel's laws had wide application—i.e., they held for a large number of species. Yet Morgan remained skeptical. What began to change his mind through his own work was not the fact that he could apply the Mendelian theory to yet another organism, such as *Drosophila*, but that he could test the Mendelian theory (investigated through breeding experiments) with evidence from a wholly different area, namely cytology (specifically in the observed behavior of chromosomes during gametogenesis). As soon as he saw that the white-eye mutation acted *as if* it were somehow part of the X-chromosome, he began to view the Mendelian theory in a completely different light.

The fact that Morgan saw Mendel's unit characters as having a possible material basis on the chromosomes does not mean that

[22] Edward Manier "The Experimental Method in Biology: T. H. Morgan and the Theory of the Gene," *Synthese* 20 (1969): 185-205.

he automatically accepted the idea that genes were physical entities whose reality was necessary for the validity of the Mendelian theory. In the preface to *The Mechanism of Mendelian Heredity*, Morgan and his colleagues suggested that Mendel's laws and the superstructure of hereditary theory based upon them *could* be viewed independently of chromosomes, though that was not necessarily the point of view favored by his group:

> We have . . . put our own interpretation on the facts, and while this may not be agreed to on all sides, yet we believe that in what is essential we have not departed from the point of view that is held by many of our co-workers at the present time. Exception may, perhaps, be taken to the emphasis we have laid on the chromosomes as the material basis of inheritance. Whether we are right here, the future—probably a very near future—will decide. But it should not pass unnoticed that even if the chromosome theory be denied, there is no result dealt with in the following pages that may not be treated independently of the chromosomes; for, we have made no assumptions concerning heredity that cannot also be made abstractly without the chromosomes as bearers of the postulated hereditary factors.[23]

On this basis, Morgan and his colleagues could without question be said to have put forth an idealistic theory, one that did not necessarily have any connection with physical reality. Indeed, one gets the sense that many *Drosophila* geneticists in ensuing years took a certain pride in the logical formality and abstraction of Mendelism. In 1915, what Morgan and his colleagues were emphasizing was the fact that the Mendelian analysis of all the *Drosophila* results held sway even if the material reality of genes linearly arranged on chromosomes was denied. And, as late as his Nobel Prize acceptance speech in 1934, Morgan could write:

> What are genes? Now that we locate them in the chromosomes are we justified in regarding them as material units; as chemical bodies of a higher order than molecules? Frankly, these are questions with which the working geneticist has not much concern himself, except now and then to speculate as to the nature of the postulated elements. There is no consensus of

[23] T. H. Morgan et al., *The Mechanism of Mendelian Heredity* (New York: Henry Holt & Co., 1915), p. viii.

> opinion among geneticists as to what the genes are—whether they are real or purely fictitious—because of the level at which the genetic experiments lie, it does not make the slightest difference whether the gene is a material particle. In either case the unit is associated with a specific chromosome, and can be localized there by purely genetic analysis. Hence, if a gene is a material unit, it is a piece of a chromosome; if it is a fictitious unit, it must be referred to a definite location in the chromosome—the same place as in the other hypothesis. Therefore, it makes no difference in the actual work in genetics which point of view is taken.[24]

But it *did* make a difference to Morgan! As was shown earlier, Morgan and his coworkers did feel strongly that hypotheses should be grounded in material reality. The Mendelian theory was not simply a "mathematical formulation"—to be scientifically valid (or, perhaps, satisfying) it was necessary to seek a material foundation for the otherwise hypothetical Mendelian genes.[25]

Although Morgan operated far more as a dialectical materialist than Loeb, he nevertheless joined in the campaign that Loeb led, under the banner of "the mechanistic conception of life," to establish a new approach to biological research in the United States. Along with the physiologist W.J.V. Osterhout, Loeb and Morgan edited a series of biological monographs for Lippincott Company. The Lippincott series was to provide eminent examples of this new approach in biology.

Much of the Loeb-Morgan correspondence now in the Library of Congress deals with the problem of selecting appropriate American authors for the Lippincott books. Repeatedly in these letters the question arises as to whether a particular author is suited for the job, or whether his already prepared manuscript would be appropriate in the series. On one occasion, for example, the inclusion of a work by a German (the series was to be American) cellular physiologist was approved by Loeb because the research was carried out at the Rockefeller Institute: "Since our aim is the furtherance of progress in American science along the lines of quantitative investigation, I think this piece of work which was

[24] T. H. Morgan, "The Relation of Genetics to Physiology and Medicine," *Scientific Monthly* 41 (1935): 7-8.

[25] Morgan et al., *The Mechanism of Mendelian Heredity*, pp. viii-ix.

done here in this country should not be allowed to escape our series."[26]

Morgan and Loeb shared the belief that many of the ills afflicting biology in the nineteenth century had been European—especially German—in origin. At the height of the period of nationalism engendered by the First World War, American biologists were attempting to free themselves from European domination. Neither Morgan nor Loeb rejected European science across the board. Since both had studied abroad (Loeb of course was also German by birth) and had gained much from this experience, they could hardly have done so with sincerity. They did reject, however, the direction so much of European biology—and again they meant primarily German biology—had taken in the latter part of the nineteenth century. The "new" physicochemical biology, free of idealism and metaphysics was to be distinctly American:

> Biology, which not long ago was purely descriptive and speculative has begun to adopt the methods of the exact sciences, recognizing that for permanent progress not only experiments are required, but quantitative experiments. It will be the purpose of this series of monographs to emphasize and further as much as possible its development.
>
> Experimental Biology and General Physiology are one and the same science, in method as well as content, since both aim at explaining life from the physical-chemical constitution of living matter. The series of monographs on experimental biology will therefore include the field of traditional General Physiology.[27]

[26] Loeb to Morgan, May 19, 1920; Loeb Papers.

[27] "Editors' Announcement" for the Lippincott series (the announcement is the same in each volume). A list of the books published in the Lippincott series gives a good indication of the newer type of work Morgan, Loeb, and Osterhout wished to stimulate. Books in the series are alphabetically: E. M. East and Donald Jones, *Inbreeding and Outbreeding* (1919); E. N. Harvey, *The Nature of Animal Light* (1920); Jacques Loeb, *Forced Movements, Tropisms, and Animal Conduct* (1918); Otto Meyerhof, *Chemical Dynamics of Life Phaenomena* (1924); Leonor Michaelis, *Oxidation-Reduction Potentials* (1930); T. H. Morgan, *The Physical Basis of Heredity* (1919); W.J.V. Osterhout, *Injury, Recovery, and Death in Relation to Conductivity and Permeability* (1922); George H. Parker, *The Elementary Nervous System* (1910); George H. Parker, *Smell, Taste and Allied Senses in the Vertebrates* (1922); Raymond Pearl, *The Biology of Death* (1922); Thorburn B.

It is important to note that nowhere in this announcement are the words "mechanism" or "mechanistic conception" used. Undoubtedly Loeb would have liked to have included them. Morgan, however, was more cautious. He was not a mechanist of the same cut as Loeb, and he did not feel that he should allow the series to be associated with an extreme point of view. What was important was the experimental and physicochemical point of view.

Morgan's own volume for the series, *The Physical Basis of Heredity* (1919) indicated clearly that he saw the work of the *Drosophila* group as an example of the success the experimental and materialist methods in biology could offer. There is no doubt that by 1915 or 1916 Morgan and Loeb both considered the Mendelian-chromosome theory as one of the crowning examples of the new biology. Parading his success in a slightly flaunting way to the dissenting biologists, Morgan wrote: "The geneticist says to the paleontologist, since you do not know, and from the nature of your case can never know, whether your differences are due to one change or to a thousand, you cannot with certainty tell us anything about the hereditary units which have made the process of evolution possible. And without this knowledge there can be no understanding of the causes of evolution."[28] Loeb was equally full of praise for the *Drosophila* work. He wrote in 1911 that: "While until 12 years ago the field of heredity was the stamping ground for the theoretician and metaphysician, it is today perhaps the most exact and rationalistic part of biology, where facts cannot only be predicted qualitatively, but also quantitatively."[29] Neither Morgan nor Loeb talked only about a new *theory* of how science should be done; both could point to examples from their own work and that of their younger colleagues as concrete evidence of the success the new methods could bring.

Morgan's commitment to the new biology took another form in addition to the Lippincott series. In 1938 he accepted an invitation to organize the new Division of Biological Sciences at the California Institute of Technology in Pasadena, California. In this undertaking, Morgan sought to organize a biology department that

---

Robertson, *The Chemical Basis of Growth and Senescence* (1923); Donald D. van Slyke, *Factors Affecting the Distribution of Electrolytes, Water and Gases in the Animal Body* (1920).

28 T. H. Morgan, *A Critique of the Theory of Evolution* (Princeton, N.J.: Princeton University Press, 1916), pp. 26-27.

29 Loeb, *The Mechanistic Conception of Life*, p. 25.

would give maximum reign to the most modern lines of work. This meant quantitative, experimental, and physicochemical biology. The guiding philosophy of science in this endeavor was developed through his years of experimental work in embryology and heredity, and given focus through his association with Loeb and the struggle for a new American biology. The Cal Tech undertaking marked the final and in many ways culminating phase of Morgan's scientific life.

CHAPTER IX

The Later Years: Cal Tech, the Nobel Prize, and a Return to Embryology (1928-1945)

By 1928 Morgan was acknowledged as the world's leading geneticist. He was sixty-two years old, a few years from retirement at Columbia. He had behind him thirty-five years of outstanding contributions to biology in general, and eighteen years of trail-blazing research in the field of heredity. He could have remained at Columbia with honor and distinction and continued his work with the Carnegie group, exploring and wrapping up many of the details of the *Drosophila* work.

But that was not Morgan's personality. He seldom thought of himself in terms of retirement—the concept was as foreign to him in spirit at sixty-two as it had been at twenty-two. He was never content with tidying up details of already largely explored territory. It is characteristic of Morgan that at an age when many research workers begin to slow down and rest on their laurels, he set out boldly to chart new courses. In 1928 he left Columbia to organize the biology division at the California Institute of Technology. Concomitant with this move, but in many ways independent of it, Morgan left the active area of *Drosophila* research and returned more and more to the study of experimental embryology.

Morgan did not fall accidentally into either his move to Cal Tech or his return to embryology. There was purpose in both. At the California Institute of Technology, Morgan wanted to give a concrete form to his philosophy of science in general and of biology in particular. He wanted to emphasize the new direction in which he thought biological research ought to move. It was his aim to bring together the best possible people representing the most modern lines of biological research (physiological and physicochemical) and allow them virtually unrestricted possibilities to interact. In returning to embryology, Morgan wanted to get back to those old problems which, from his earliest association with natural history, had always intrigued him the most. These were the problems, once again, on which he could best work with marine forms, those organisms that, as W. K. Brooks had pointed out so long ago, demonstrated most clearly all the great problems of biology. "Life began in the sea and there lie some of the greatest historical problems of biology," wrote Morgan in 1929.[1]

Morgan remained at the California Institute of Technology until 1941, when he officially retired. Retirement meant only that he relinquished administrative duties, however, for he actively continued his research in his laboratory in Pasadena until his death in 1945. Thanks to generally good health throughout his life, his later years remained productive, and he was seldom far in thought or action from his work.

THE CALIFORNIA INSTITUTE OF TECHNOLOGY

Cal Tech had its beginning in 1891 as the Throop Polytechnic Institute, founded by a donation from a California businessman, Amos G. Throop. In 1910 the campus on which the university now resides was acquired under the vigorous direction of a new president of the board of trustees, Arthur H. Fleming.[2] The institution was considerably reorganized, and developed into a university dedicated not only to undergraduate, but also to graduate education and research. It thereby became "a college, graduate school, and research institute of pure and applied science,"[3] and the name was changed to California Institute of Technology. Between 1910 and 1920 the curriculum and research were oriented largely toward engineering and applied sciences. In the aftermath of World War I, however, there was considerable feeling that the basic sciences should be further developed and placed on an equal footing with the applied sciences.

Consequently, between 1920 and 1923 the board of trustees engaged the services of three men to guide the development of

[1] T. H. Morgan, "Outlines of Plans for the Future of Biology at the California Institute of Technology," sent to R. A. Millikan, October 29, 1929; G. E. Hale Papers, California Institute of Technology; microfilm edition, Daniel J. Kevels, editor. A complete holding of the microfilm edition can also be found at the American Philosophical Society Library, Philadelphia.

[2] Fleming (1856-1940) was born in Ontario but became a naturalized U.S. citizen in 1886. After entering the lumber business in California, he also became president of the Johnson Connector Company which made electric safety connectors for railroad cars. In the course of his association with Cal Tech, he donated over $5,000,000 to the institute.

[3] *California Institute of Technology Bulletin* (1928): 50-51.

the basic sciences. These were the chemist Arthur Amos Noyes (1866-1936), the physicist Robert Andrews Millikan (1868-1953), and the astronomer George Ellery Hale.[4] These three men, but

[4] Arthur Amos Noyes was born in Newburyport, Massachusetts, and received his B.S. and M.S. degrees at M.I.T. (in 1886 and 1887). After receiving his Ph.D. at the University of Leipzig in 1890, he returned to M.I.T. as an instructor in analytical chemistry (1890-1894), professor of organic chemistry (1894-1899), and professor of theoretical chemistry (1899-1920). In 1920 he went to Cal Tech as chairman of the chemistry division. As a chemist, Noyes had been particularly interested in thermodynamics and in the development of the ionic theory. He was a member of the National Academy of Sciences and recipient of the Willard Gibbs (1915) and Theodore William Richards (1932) awards of the American Chemical Society, and the Davy Medal of the Royal Society (1927). During World War I he was acting chairman of the National Research Council, and in 1927 president of the American Association for the Advancement of Science.

R. A. Millikan was born in Morrison, Illinois, the son of a preacher and grandson of one of the early settlers of the Mississippi River country of Illinois. He received his undergraduate training at Oberlin College (B.A., 1891) and obtained a Ph.D. at Columbia in 1895. After a year's residency at the universities of Berlin and Göttingen, he was appointed in 1896 to an assistantship at the then recently established Ryerson Laboratory at the University of Chicago. Here he was associated with A. A. Michaelson, who had pioneered in testing certain aspects of the theory of relativity. Millikan rose through the academic ranks at Chicago and was appointed professor of physics in 1910. It was at about this time that Millikan showed, through the ingenious "liquid drop experiment," that the charge on a single electron was a constant value. Moreover, the charges were exact and discrete. In 1914 he carried out another set of exhaustive experiments to determine the exact value of Planck's constant. The value he obtained agreed exactly with that computed by Planck from other types of data, namely heat radiation. Millikan was elected to the National Academy of Sciences in 1915 and was awarded the Nobel Prize in physics in 1923, "for his work on the elementary charge of electricity." (H. Schuck et al., *Nobel, The Man and His Prizes*, Norman, Okla.: University of Oklahoma Press, 1951, p. 436.) During World War I Millikan served as vice-chairman of the National Research Council, was an advisory member of the antisubmarine board, and chairman of the science and research division of the U.S. Signal Corps. Millikan was chairman of the executive council of Cal Tech, a post he held from 1921 until his retirement in the summer of 1945.

George Ellery Hale was a statesman and entrepreneur of science of whom it was once said that he could, like his father who ran a successful elevator company in Chicago, "sell an elevator to anybody." (Quotation from F. R. Lillie, see T. H. Morgan to G. E. Hale, July 3, 1929; Hale Papers.) Born in Chicago, Hale received his B.A. from M.I.T. (1890) while also working at the Harvard College Observatory (1889-1890). At about this time he helped organize the Kenwood Observatory in Chicago (1891), where he invented the spectroheliograph (a device for studying the spectral emissions

especially Millikan, were eminent scientists as well as sound administrators with many connections in the government and private foundations. As members of the executive council, they worked with the board of trustees to expand all areas of basic research. It was they who, in the late 1920s, decided that a division of biology must be organized.

While all three men contributed significantly to the reorganization of Cal Tech, it was Millikan who brought the institution to its prominent position in basic science. Some saw Millikan's greatest contribution to American education as the building of the California Institute of Technology. What Millikan saw at Cal Tech was the opportunity to carry out an educational experiment that would have the greatest ramifications for the future of basic research. Millikan had grown impatient with the complex and cumbersome academic administration characteristic of so many universities. He wanted to abolish departments and all other artificial barriers to scientific cooperation.

The United States, and southern California in particular, were ripe for such an experiment in the early 1920s. The economy was growing astonishingly as a result of the war, industrial profits were increasing, and the expansion of empire abroad had begun. Southern California was growing agriculturally and industrially, and Millikan was able to tap the vast potential there. He saw the need for more electric power and began at Cal Tech an important proj-

---

of the sun). In 1892 he became an associate professor of astrophysics at the University of Chicago and began immediately to organize the Yerkes Observatory of which he remained director from 1895 until 1905. In 1904 he organized the Mount Wilson Observatory outside Pasadena, under the sponsorship of the Carnegie Institution of Washington. He remained its director until 1923. He served Cal Tech from the early 1920s as a member of the board of trustees, and as chairman of the observatory council, the executive group linking the physics department with the Mount Wilson Observatory. In 1895 he established the *Astrophysical Journal* and became a member of the National Academy of Sciences, later serving as foreign secretary for many years. In 1916 he was chairman of the Academy Committee to organize the National Research Council, and was active in the postwar years in the reorganization of international science. In his astrophysical researches, Hale showed that not only does the sun as a whole have a magnetic field, but that sun spots themselves have magnetic fields and are the centers of enormous electrical activity. Although widely known as an astronomer, Hale's forte was management and administration, and as a fund raiser he served as an important link between Cal Tech and both government and private industry.

ect in high voltage engineering research (leading to the production of electric power from the Hoover Dam on the Colorado River for faraway Los Angeles). He foresaw the growth of the aviation industry and initiated a program in aerodynamic engineering. In addition, and perhaps most important, Millikan saw the need to attract to Pasadena basic scientists of the first rank, for that was the chief ingredient of a first-rate research institution. He did just that. At the time Morgan was invited to join the Cal Tech staff in 1927, the faculty in physics included Millikan, R. C. Tolman, and a young professor, J. Robert Oppenheimer; in chemistry it included Noyes and Linus Pauling; and in aerodynamics, Theodore von Karman (1881-1963).

But in southern California there was also a growing need for the scientific development of agriculture. A state agricultural school already existed, and Millikan had no desire to become a competitor. He did see, however, that basic biological research was important, albeit indirectly, to the further growth of scientific agriculture. He also recognized the Cal Tech curriculum was one-sided in its emphasis on the physical sciences. Thus there were two strong reasons for building a program in the life sciences: its ultimate value for agricultural and its role in rounding out and balancing the scientific curriculum.

There were several reasons Millikan, Hale, and Noyes settled on Morgan as the logical person to organize the division of biology. First, Morgan's writings on the new directions of biological research had emphasized the important relationships biology bore to physics and chemistry. His view that the future of biological research rested on the utilization of physical and chemical methodology, "and even mathematics" (a revolutionary idea in those days), corresponded perfectly with Millikan's belief in the integration of the sciences. Moreover, Morgan always emphasized the importance of research and, like Millikan, thought American universities by and large catered too much to teaching and too little to basic investigation. By the mid-1920s Cal Tech had become the seat of the new physics and chemistry in the United States. It was obvious that in the development of biology they should seek a man who saw biology proceeding toward an integration with the physical sciences.

Second, by 1928 Morgan was well known to each of the members of the executive council through his activities with the American Association for the Advancement of Science and the National

Academy of Sciences. Millikan, Noyes, and Hale were all members of the National Academy, and in 1927 had, in fact, participated in the election of Morgan to the presidency of that body. Morgan was also vice-chairman of the executive board of the National Academy, with Millikan the other vice-chairman. Millikan, perhaps better than the others, knew that Morgan was a capable administrator and a creative scientific worker.

Third, of course, Morgan was acknowledged by the mid-1920s to be one of the leading biologists in the country, if not the world. As his involvement with the Lippincott series shows, Morgan was deeply committed to the integration of biology with physics and chemistry. Millikan and others could see that Morgan's chief function at Cal Tech might lie less in the research he himself continued to carry out than in the stimulation he provided to the investigators he assembled to work together in this new enterprise.

The first formal feelers were put out to Morgan by Hale in the winter and spring of 1927. Morgan was a little unsure about the proposal at first. He felt he might be getting a little old for such an ambitious project, but even more important, as he wrote to Hale, he had always considered himself "a laboratory animal who has tried hard most of his life to keep away from such entanglements."[5] On May 11 Morgan had lunch with Hale and Noyes in New York. Noyes presented him with a preliminary plan for developing a biology department that would be integrated with physics and chemistry. As a result of the meeting with Noyes, Morgan felt much more inclined to accept the proposal.

Several factors influenced Morgan positively toward the Cal Tech adventure. First, it was a chance to put into practice the kind of biological philosophy he had been developing over the previous quarter century. Here, in close and intimate association with physics, chemistry, and mathematics, biology could be pursued along quantitative and rigorous physicochemical lines. Founding a division of biology at Cal Tech would provide a concrete way in which the new biology could influence the younger generation of biological research workers. Second, building a department from the ground up would afford Morgan the opportunity to bring together a group of biologists, biochemists, biophysicists, and others whose works could be pursued in a cooperative, rather than a competitive, atmosphere. Morgan saw in Cal Tech the opportunity to build an organization that would encourage the kind of

[5] Morgan to Hale, May 9, 1927; Hale Papers; quoted with permission.

social interaction that characterized the fly room at Columbia. As he wrote to Hale in the spring of 1927: "The idea of the participation of a group of scientific men . . . in common venture for advancement of research fires my imagination to the kindling point."[6] Third, Morgan had been attracted to California ever since his first visit there in the summer of 1904. He not only found California a beautiful and exciting climate, but was much attracted by the high quality and variety of marine forms that could be obtained throughout the winter on the California coast. In addition, Morgan found as he grew older that he increasingly disliked the northeastern winters (especially the snow)[7] and that the mild and luxuriant climate of Pasadena was an attractive place to spend the last years of his life.

In late May, Morgan received a formal letter from Millikan, as chairman of the executive council at Cal Tech, offering him the post of director of the Division of Biology. By the end of the spring term, in early June, Morgan was still uncommitted; most of the details had still to be worked out, as Morgan wrote to his daughter Isabel: "I had a letter from Dr. Millikan offering me the position. I am to meet him in New York July 4 or 5. This looks like business, but details must be settled first."[8] Still unsettled, the Morgans packed up and went to Woods Hole in June. By early July, Morgan made his decision: he would accept the Cal Tech offer, but he wanted to wait another year before making the move. This was partly because he wanted time to work out the many details of the new organization, but also out of consideration to Columbia, for, as he wrote to Hale, "I have always had the most friendly relations with Columbia and they have put up with my idiosyncrasies."[9] When Morgan broke the news of his new venture to his old friends Conklin and Lillie at Woods Hole, he said gleefully, "They didn't tell me that I am an old fool, but actually congratulated me."[10]

In late July, Morgan went to the Dry Tortugas Laboratory (an island about 100 miles off the coast of Key West, Florida) and returned through New York in order to inform President Butler

[6] Ibid.

[7] T. H. Morgan to Isabel Morgan, September 21, 1932, p. 2; from the private papers of Isabel Morgan Mountain.

[8] Ibid., June 1927.

[9] Morgan to Hale, July 8, 1927; Hale Papers; quoted with permission.

[10] Ibid.

of Columbia about his decision. Since Butler was not in, he penned the following letter of resignation:

> Dear President Butler,
>
> The President and Trustees of the California Institute of Technology at Pasadena have asked me to organize a Department of Biology on the same footing as the Departments of Physics and Chemistry that are at present in existence there. It is their wish to have the new department devoted largely to research and to that end they have pledged ample means for the undertaking. Much as I regret to leave Columbia University, I feel that I cannot decline this opportunity that will enable me to build from the bottom up a department whose purpose is to carry on, and to train men devoted entirely to modern research in the biological fields. Whether I am already too old for such an effort, or lack the capacity of putting through such an ambitious program, later events can only show, but since the Trustees have been courageous enough to entrust the responsibility to me I should be just a little ashamed to refuse to undertake it.
>
> As you know, perhaps, I have consistently avoided administrative undertakings and, now, I find myself tempted into an adventure that will make serious inroads to my personal interests in research. Possibly I am deceiving myself into thinking that this is an exceptional opportunity to organize the kind of research that seems to me to be in the lines of future advance in biology, but as I have said, I have the temerity to try to do it, and to hope it leads to moderate success.
>
> The Trustees of the University would, I understand, have been willing to have me begin at once to get the organization of the new department underway, but I have felt that it would be unfair to withdraw so abruptly from Columbia University where I have received every consideration and kindness. It also seemed to me that it would be little short of desertion to retire at once and leave the responsibility of the reorganization of the department to Dr. Wilson. I have, therefore, stated that it would not be possible for me to leave Columbia before the end of the current academic year. If this meets with your approval, I beg to place my resignation in your hands to take place at that time [June 1928].

> My association with Columbia for 23 years has been in all respects delightful and profitable, and I feel under deep obligation for the very exceptional opportunities that I have enjoyed in the complete freedom to carry out my work as the best interests of the Department seemed to require. I appreciate also the cordial relations with my colleagues in the Department, and with the administrative offices of the University who have done everything within their power to promote the steady growth of our work. I shall leave with many misgivings, and hope that it may not seem ungrateful on my part to leave my post after so much time. . . .[11]

Butler replied that he was sorry to lose Morgan, but realized that this was an opportunity of great magnitude for the furtherance of biological research. In an interesting aside, however, Butler remarked that he thought that the current trend was toward *concentration* of work in the sciences, rather than dispersal. Apparently Butler recognized neither the attractions of the West Coast for academic, specifically scientific, work, nor the fact that Morgan was forming a new kind of concentration of workers which, because of rigid departmental and institutional lines, could never have been realized at an institution like Columbia.

There was much work to be done between the fall of 1927 and the summer of 1928, when Morgan was to take up his official residence in Pasadena. Morgan outlined four areas of planning and development that would require advance work: designing a biological program, hiring personnel, designing and preparing the buildings and facilities, and behind all of this, fund raising. Of these four areas, only the last was distasteful to Morgan. He enjoyed aspects of each of the others. For instance, he liked planning the buildings, overseeing each detail—even to the extent of insisting that the doors open into the hallway so as not to take away valuable space inside the laboratories.

Morgan's program for biology at Cal Tech developed over a period of several years and underwent various modifications and expansions. The basic principles of the new biology division, however, remained the same from the early conversations with Noyes, Hale, and Millikan in the spring of 1927, through the years of development between 1930 and 1940. Only the details were modified to meet new circumstances or unexpected problems.

[11] Morgan to Nicholas Murray Butler, August 4, 1927; Morgan Papers, Cal Tech; quoted with permission.

In preparing to go to Cal Tech, Morgan developed his ideas around five major points. One was that research, rather than simply teaching, should remain the major focus of the institute.[12] As Millikan had emphasized earlier, research was not to exclude teaching, but rather was to go hand in hand with it. Since Cal Tech emphasized graduate education, teaching was to be more a matter of collaboration than formal lecturing.

Second, research at Cal Tech would focus on pure (basic), rather than applied, problems. It was for this reason, for example, that Morgan rejected the idea of building a hospital in conjunction with Cal Tech; he did not want the Division of Biology to become the training grounds for people who would solve practical problems in medicine or agriculture.[13] Recognizing the problems that would develop if the institute were to become primarily a technological research institute tied to agricultural, medical, or industrial enterprises, Morgan argued that its principal job was to push back the frontiers of basic knowledge, for the applications of such knowledge could be carried out by others: "The California Institute is particularly fortunate in the relatively great concentration here of physicists, chemists and biologists. The close association of all departments of the Institute provides an opportunity unsurpassed for their cooperation. The *application* of discoveries in pure science is much less difficult, and such knowledge becomes rapidly diffused on account of the appreciation of the medical profession [Morgan was writing particularly in relation to the establishment of a hospital] of their importance."[14]

Morgan argued that Cal Tech should not worry about developing research that was immediately useful, for all basic research can be beneficially used for human needs; without basic knowledge, however, the solution of practical problems becomes impossible. As he wrote in his program of 1930, "The study of the metabolism of animals and plants is a typical example of the enormous unexpected benefits which have accrued from the development of pure science."[15] But the matter was more serious than simply one of personal preference or the setting of priorities. To Morgan, pro-

[12] See A. A. Noyes, "Draft of Paper Explaining the Purposes of Cal Tech," Hale Papers.

[13] Morgan to Hale, December 4, 1930; Hale Papers.

[14] Morgan's "Program for Biology," sent to Hale with a letter, December 4, 1930; Hale Papers.

[15] Ibid.

grams for applied research had serious long-range implications for the future of the institute. For instance, applied research programs could set up a pattern of funding for specific problems or projects that would be highly restrictive. Such programs could also create a distinction or division between two types of research workers; if this distinction were intensified by a differential availability of research funds, the institute as a whole would suffer. Most important, Morgan felt it was necessary to convince benefactors and business leaders that their own interests would best be served in the long run by generous financial support of basic rather than applied research.[16]

A third aspect of Morgan's program involved setting priorities for the fields of biological research to be supported. Since it was impossible to develop all areas of biology equally, Morgan felt it was necessary to concentrate. Concentration should be on those areas, he felt, where physics, chemistry, mathematics, and biology could cooperate—and this meant at the most fundamental level, problems of a physiological nature. What Morgan meant by physiology was not simply human or vertebrate physiology as it was currently practiced in biology departments or medical schools, but rather the study of functional problems, primarily at the cellular level.[17] It was such functional problems, approached from the physical, chemical (biophysical and biochemical), genetic, embryological, and evolutionary points of view whose study should be encouraged in modern research. Morgan had no compunction himself about eliminating areas of descriptive biology and morphology, including systematics, from the curriculum. However, he found a convenient public rationalization for what might appear to be a one-sided approach to biology. As he wrote to Hale, "We can counter the argument that we are neglecting the old aspects of biology by saying that we are not trying to duplicate what is being done well elsewhere, but rather furnish the opportunity for new directions of research."[18]

A fourth element of Morgan's program was that biology at Cal Tech should be rigorous and analytical in character. With personnel trained in biophysics, biochemistry, and mathematics, biology

[16] Morgan to Noyes, June 4, 1928; Hale Papers.

[17] See T. H. Morgan, "The Relation of Biology to Physics," *Science* 65 (1927): 213-220.

[18] Morgan to Hale, May 15, 1927; Hale Papers; quoted with permission.

should be able to exist on the same footing as physics and chemistry.

The fifth and final feature of Morgan's program, upon which the realization of all the others in many ways depended, involved the selection of personnel. It was his aim to gather together the best workers in every field and encourage their continual cooperation and interaction. Morgan informed Hale and Millikan that he had no interest in bossing the new division of biology, but only in coordinating the work of men who knew what they wanted to do.[19]

In setting forth his program, Morgan was adamant that the new organization at Cal Tech should be called a division or department of *biology* rather than the traditional departments of botany and zoology. As he wrote, this insistence "calls for a word of explanation":

> It is with a desire to lay emphasis on the fundamental principles underlying the life processes in animals and plants that an effort will be made to bring together, in a single group, men whose common interests are in the discovery of the unity of the phenomena of living organisms rather than in the investigation of their manifold diversity. That there are many properties common to the two great branches of the living world is becoming almost daily more manifest, as shown for example in the discoveries that the same principles of heredity that obtain among flowering plants apply also to human traits, and that, in their response to light, animals and plants conform to a common law of physics.
>
> It is true that, at what may be called the biological level, an immense diversity of form and function may manifest itself, but enough insight has already been gained to make evident that this diversity is in large part due to the permutations and combinations of relatively few fundamental and common properties. It is in the search for these properties that the zoologist and botanist may profitably pool their interests. The animal physiologist today who wishes to have a broad outlook over his field can as little neglect the physiology of bacteria, yeast, and higher plants as the bacteriologist and plant physiologist can ignore the modern discoveries in animal physiology. The geneticist who works with animals will

[19] Ibid.

> know only half his subject if he ignores the work on plants, and both plant and animal geneticists will fail to make the most of their opportunities if they overlook the advances in cytology and embryology. It is, then, to bring together in sympathetic union a group of investigators and teachers whose interests lie in the fundamental aspects of their subjects, that a department of Biology will be organized.[20]

Along the same lines, Morgan suggested that perhaps the name of the division should be simply "The Biological Laboratory," rather than "Biological Department." Morgan was anxious that the program be set up with the greatest amount of flexibility and with the greatest scope and ambition. At the same time, however, it should not promise what could not ultimately be delivered. For this reason he was willing to proceed only along general lines, with no strings attached and no concrete promises as to the exact form and structure of the division and its program. The academic year 1927-1928 was to be a time of planning and sifting through various proposals.

Morgan saw the biological research of the new division proceeding along five basic lines. The first was to be genetics and evolution. This was to include not only a continuation of the work on the genetics of *Drosophila*, but also to embrace the new fields of the genetics of viruses and bacteria. Furthermore, genetics and evolution were considered to be two sides of the same coin, interacting in Morgan's mind in a dialectical fashion: the one preserving stability, the other promoting innovation. The second line was to be experimental embryology, which, in Morgan's view, was to focus on the cellular and subcellular events during embryonic differentiation. By conscious or unconscious choice, it did not include either in the initial planning stages or in later developments, the experimental work at the tissue and organ level being carried out at the time by the Spemann school. A third line was general physiology, which included particularly the study of cell physiology, but also ranged through general plant and animal physiology. A fourth line was biophysics. This meant to Morgan

[20] T. H. Morgan, "Development of Biology at the California Institute of Technology," draft of a paper sent to George Ellery Hale; Hale Papers. See also a printed version of this same quotation in T. H. Morgan, "A New Division of Biology," *California Institute of Technology Bulletin* (1928): 16-17.

the study of such problems as the quantum absorption of plant pigments in photosynthesis and the response of plants and animals to light, an extension of Loeb's older work on phototropism. The fifth area was to be biochemistry. To Morgan at the time this meant not only the study of anabolic and catabolic reactions within cells but also, more specifically, the study of enzymes and their effects on biochemical reactions. To Morgan especially, the study of enzymes and proteins in general was an important aspect of the new physicochemical approach to biology.

A last line of work, to be added at a somewhat later date, was experimental psychology. Influenced by J. McKeen Cattell and some of the work that had emerged from Clark University's famed psychology department in the preceding several decades, Morgan felt that psychology and physiology were two sides of the same coin, and should have a closer relationship in the future. However, Morgan felt that it was wise to leave psychology out of the initial organizational plans, lest it appear that the Cal Tech program was betraying its initial high aim of pursuing quantitative science. Later, when the general biological disciplines were established in a rigorous fashion, truly good, experimental psychology could be brought in without jeopardizing the overall dedication of the institute to rigorous and analytical science.

The five divisions Morgan envisioned were only guidelines for the types of work he wanted to see carried out at Cal Tech. The areas of research that were ultimately pursued would depend in large part on the people Morgan was able to attract to Pasadena. For example, he wrote to his friend Gary Calkins at Columbia in 1930 about the developing plans for a physiology group at Cal Tech: "Just how our physiological groups are going to develop, for example, is also a question, dependent again to some extent on what men are available."[21] And, in other cases, the kind of work developed in one area, for example physiology, would have definite implications for the kind of work carried out in another area, such as biochemistry and cytology.[22] Since the program was fluid in Morgan's mind, depending in large measure on personnel, faculty recruitment was the chief administrative job that lay ahead in the development of biology at Cal Tech.

[21] Morgan to Gary Calkins, February 26, 1930; L. C. Dunn Papers, American Philosophical Society.

[22] Ibid.

RECRUITMENT OF FACULTY FOR CAL TECH

One of the principles Millikan and the executive council of Cal Tech had established right from the beginning was the necessity of obtaining "the best men possible." On October 19, 1927, Millikan wrote a long letter to Morgan explaining the policy of the institute regarding faculty personnel and recruitment. He made a number of points which to him were important in determining the character of the faculty. The first point was stated simply: "We expect to have as good men at the Institute as are found anywhere in the United States, and to pay as good salaries."[23] The maximum salary paid anywhere in the United States at that time, Millikan stated, was approximately $10,000. Cal Tech could go up to or beyond that figure if necessary. However, "We wish to be very careful not to pay such salaries to men who would not be in demand at Harvard, the University of Chicago, and institutions of like grade, on similar terms."[24] Salary was not to be a limiting factor in building a good faculty. In fact, Millikan saw Cal Tech as an able competitor to the older and prestigious institutions of the East Coast. To insure, however, that faculty recruited from the East Coast should not lose contact with their former colleagues and centers of work, the salary figure included enough for a trip east at least once a year.

A second principle Millikan set forth was that the most important phase in recruiting faculty was the initial selection process. It was much more important, Millikan felt, to be highly selective in bringing people to the institute in the first place, than in "weeding out" candidates for permanent positions after four or five years. As he wrote: "The time to do the selecting, especially of the more highly paid men, is, in general, not after a man is on the grounds, but *before he comes here.* [F]or in any institution the pressure of one sort or another to hold men who are already on the grounds is usually pretty nearly irresistible."[25] A third principle was that recruitment should be aimed much more at bright, promising, and young research workers than at already established luminaries. The institute would grow most successfully, Millikan felt, if the people who were hired were younger workers of promise rather than those who were, in his words, "over the hill." A fourth and

[23] R. A. Millikan to T. H. Morgan, October 19, 1927; Morgan Papers, Cal Tech; quoted with permission.

[24] Ibid.

[25] Ibid.

last principle was that Cal Tech had no fixed scale of salaries. As a result, "Men who hold the rank of full professor are being paid all the way from $3,500 to $10,000."[26] The actual salary was determined by how much the individual was in demand, by the competitive nature of the field in which he or she was working, and by a variety of more indefinable criteria. Millikan assured Morgan that he would have a free hand in offering salaries (within budget limitations, of course) and that the institute would impose no restriction on how much he offered individuals whose participation he wished to secure.

These guidelines, which were not enunciated by Millikan alone but had been developed over several years by the executive council, had the advantage of considerable flexibility. The aim of attracting people of high quality was admirable. On the other hand, the lack of a fixed salary scale, and the sometimes indefinite criteria used to determine actual salary, lent themselves to paternalism and to what were considered by some to be unfair practices. For example, a large number of young workers came through the institute for two- to five-year periods as research assistants, fellows, and graduate students. As Millikan pointed out, they usually subsisted on very low stipends and did not normally expect to stay. However, by using a fluid salary scale, the very best of these people would be kept on at faculty positions, either with a relatively low salary if their work was just beginning to be recognized, or at a higher salary if their work made them more in demand. "In the course of years we can select with certainty from this group [the fellows, national research fellows, etc.] the outstanding men of the next generation, and our policy has been to watch these younger men like hawks and not to push up their compensation beyond, or even to, its natural economic level until we are very certain that we have found the man of altogether outstanding calibre. As soon as we are convinced of that, the Institute tries to carry out the policy of anticipating the pressure for increased compensation."[27] In at least one case this policy caused Morgan to lose a prospective member of his genetics group. In the spring of 1928 he offered Alexander Weinstein, who was then spending a year at Columbia studying the effect of X-rays on mutation, a position as assistant professor at Cal Tech.[28] Weinstein refused on the grounds that Morgan was offering him a lower

[26] Ibid. [27] Ibid.

[28] Morgan to Millikan, May 4, 1928; Hale Papers.

salary ($3,500) than was being offered to other men of his age and experience.[29] This was the kind of attitude that was likely to annoy Morgan anyway, even if it had not gone against the general hiring principles of the institute. Morgan stood firm with his offer, and Weinstein did not join the Cal Tech staff. Morgan was not a person to haggle over financial matters, yet it is apparent that he felt a sliding scale, adjusted to fit competitive and other standards at any particular time, was perhaps the best way to proceed. Nonetheless, the introduction of discrepancies between individuals at an institution, especially where salary is concerned, are always likely to produce resentments which can, under certain circumstances, influence the cooperative nature of the work.

The kind of people Morgan was searching for were, in his words, "like Jacques Loeb or A. V. Hill."[30] Both Loeb and Hill had carried out highly quantitative, physicochemical studies on complex organismic problems. Both were esteemed in their respective fields, and their work was regarded throughout the biological world as prime examples of the new experimental, physicochemical approach to biology. Morgan recognized the difficulty of obtaining men of this caliber, even among younger workers, in the five general areas he had outlined in his program. For genetics and embryology he felt that he could take responsibility himself for recruitment. In the other areas, however, he sought help from colleagues both at home and abroad.

Recognizing that a considerable amount of time would be required to select appropriate staff members in the various disciplines, Morgan, Hale, and Millikan decided that the department of genetics would be the only group set up for the first year (1928-1929). Morgan began the task of collecting this group. Sturtevant was offered a professorship in genetics, and Bridges was induced

[29] Ibid., May 28, 1928.

[30] Archibald Vivien Hill (b. 1886) is an English physiologist whose research interests included quantitative studies on oxygen consumption during muscular action. Hill had demonstrated that oxygen is required only for the recovery phase of contraction, not for the contraction phase itself. He had been a professor of physiology at the University of Manchester (1920-1923) and at University College, London (1923-1925), before becoming Foulerton Research Professor of the Royal Society in 1926 (a post he held until 1951). Although Morgan had vaguely considered Hill for a position at Cal Tech, he never pursued the matter actively. Hill was, however, a prototype of the kind of general organismic physiologist Morgan would have liked to have attracted to Cal Tech.

to come as a Carnegie Research Fellow. That first year Morgan also brought Ernest Gustav Anderson (b. 1891), who had received a Ph.D. in plant genetics under R. A. Emerson at Cornell in 1920.[31] Morgan also hired Sterling H. Emerson (R. A. Emerson's son) who had received a Ph.D. in genetics from Michigan in 1928. Both Anderson and Emerson were to carry out work in the botanical side of genetics. In addition, Morgan also brought Theodosius Dobzhansky from Columbia to Cal Tech, the first year as an international research fellow, and the second year as an assistant professor of genetics. Morgan also tried to coax the Yugoslav-born Milislav Demerec (1895-1966) from the Carnegie Institution at Cold Spring Harbor, and the German-born Curt Stern (b. 1902) from the Kaiser-Wilhelm Institute in Berlin-Dahlem.[32] Demerec preferred to remain at Cold Spring Harbor, and Stern was interested at that time (1928) in forging a career in Germany.[33] It is interesting to note that Morgan chose prospective members for the genetics work from two groups: those who had worked with him and his group at Columbia or those who had worked in R. A. Emerson's group at Cornell. These were the two focuses for genetics research in the United States in the

[31] After finishing his degree, Anderson had worked on *Drosophila* genetics in Morgan's lab at Columbia from 1920 to 1922 as a research associate of the Carnegie Institution of Washington. From 1922 to 1923 he had been an instructor of biology at the City College of New York, and from 1923 until 1928 had been a fellow of the National Research Council at the University of Michigan. He remained at Cal Tech until his retirement in 1961.

[32] Demerec had received his Ph.D. in plant genetics under Emerson at Cornell in 1923; he had then gone on to become a research associate at the Carnegie Institution at Cold Spring Harbor, where he had begun to work on the so-called "unstable genes" in *Drosophila*. (See Bentley Glass, "Milislav Demerec [1895-1966]," *Biographical Memoirs, National Academy of Sciences* 42 [1971]: 1-27.) Stern had received his degree in cytology and general biology from the University of Berlin in 1923. Although he had no formal training in genetics at that time, he had become impressed with Morgan's *Physical Basis of Heredity*, which had just been translated into a German edition in 1921. He had received a postdoctoral fellowship to study with Morgan at Columbia, where he began the work on crossing-over that culminated in direct evidence (1931) that crossing-over between homologous chromosomes does take place. (See Curt Stern, "From Crossing-Over to Developmental Genetics," *Stadler Symposia* 1 and 2 [1971]: 21-28, especially pp. 21-22.)

[33] T. H. Morgan to Stern, May 18, 1928, and July 3, 1928; both from the personal papers of Curt Stern, University of California, Berkeley.

mid-1920s—the one (Morgan's) dealing with the genetics of animals, the other (Emerson's) with the genetics of plants.

Although genetics was to be the only area set up in a working fashion for the first year, Morgan began to seek out possible faculty for the other areas in 1927. He used his location in New York as a means of meeting and interviewing likely candidates. The most important of these areas to Morgan was physiology, followed by biochemistry and biophysics. Morgan expected that physiology was going to be difficult, partly because it was farther from his own experience.

In late October of 1927 he sought the aid of Sir Charles Sherrington (1857-1952) who was at the time in New York. Morgan wanted to get some suggestions for good general physiologists, both plant and animal. Sherrington was apparently of little specific help.[34] Among Morgan's own choices were William John Crozier (1892-1955) of Harvard and Selig Hecht (1892-1947) of Columbia.[35] Morgan made specific offers to both Crozier and Hecht in the early spring of 1928, and both men wavered. President Lowell of Harvard, however, offered Crozier a physiological empire to stay, to which he acceded.[36] Columbia, too, sat up on its haunches and "made unheard of efforts" to retain Hecht, which may have slightly annoyed Morgan since no such attempts were made to keep him.[37] Despite this failure to achieve his two top choices, Morgan felt that his offer at least had salubrious effects on the institutions involved and thus could not help but improve the state of physiology in the United States. As he wrote to Hale:

> At this end, Hecht is also marking time. To our surprise the University has suddenly exerted itself and is trying to make him an offer that he will find difficult to refuse. I am sitting tight, and will make no further move in Hecht's case should the University succeed in offering him counter attractions. Crozier's decision will probably influence Hecht's. Probably in a day or two I shall know what is going to happen. Little Pasa-

[34] Morgan to Hale, October 31, 1927; Hale Papers.

[35] Crozier had received his Ph.D. from Harvard in 1915, and was well known for his studies on animal behavior and the central nervous processes involved in temperature regulation. Hecht had received his Ph.D. from Harvard in 1917 and was a pioneer in studying biochemical aspects of the photosensory process.

[36] Morgan to Millikan, May 28, 1928; Hale Papers.

[37] Morgan to Hale, March 10, 1928; Hale Papers; quoted with permission.

> dena is stirring up two of the big establishments in the East so that they may be forced to do unheard of things. All this is to the good of general physiology, so that, while we may not immediately profit ourselves, we can at least have the satisfaction of knowing that we have helped things along, in that we have picked out two men whom these Eastern moguls are willing to take extraordinary measures to retain. I shall, of course, be disappointed, but by no means downcast. There are other people in the world still to be considered.[38]

Morgan then went on to conclude with his own touch of humor: "As I told Crozier and Hecht, I should not only have to do something different, but something even better."[39] Morgan also considered the possibility of making an offer to Leonor Michaelis (1875-1949), then of the Johns Hopkins University.[40] Morgan knew Michaelis from Woods Hole, where the latter had come for several summers, and where he "has the esteem of the best men."[41] Michaelis's appointment at Hopkins was a temporary one (he left in 1929 for the Rockefeller Institute), and Morgan was certain that he would come to Cal Tech if invited. However, he held off largely because Michaelis was not a young man, and Morgan felt the importance of trying to get as many younger workers as possible.[42]

In both biochemistry and biophysics, Morgan saw no immediate prospects in the United States. It seemed to him wisest at this time (1927-1928) to defer any decisions on both areas until he had taken up actual work in Pasadena. It appeared to Morgan that he would do best to make a trip to Europe within the next several years to seek both advice and interviews with appropriate physiologists and biochemists. Two countries in particular would perhaps provide fruitful leads: England for physiology and Holland for plant physiology and biochemistry. Morgan felt strongly

[38] Ibid., March 7, 1928; quoted with permission.

[39] Ibid.

[40] Michaelis received an M.D. degree from Berlin in 1896, coming to the United States in 1926 to Johns Hopkins. Michaelis's special interest was the application of physical chemistry to medical and biological problems, and he is perhaps best remembered today for deriving the equations for enzyme kinetics that bear his name.

[41] Morgan to Millikan, May 28, 1928; Hale Papers; quoted with permission.

[42] Ibid.

the importance of proceeding slowly and getting the right people rather than making hasty and expedient decisions simply to fill a post for the next year. He was following very carefully not only Millikan's guidelines, but his own sense of procedure.

Embryology was low in Morgan's recruitment priorities, reflecting, perhaps, his belief that the field was not ready for a breakthrough of the sort *Drosophila* had represented in genetics. Aside from himself, the only other worker with strong embryological interests was Morgan's own graduate student Albert Titlebaum (later changed to Tyler), who had begun his graduate studies at Columbia and moved with Morgan to Cal Tech. Tyler's interest was largely in the type of questions Morgan had studied some thirty years previously: the factors influencing the development of cleavage planes in early embryos, the effects of centrifugation on egg development, and the factors influencing fertilization. Morgan simply put off the difficult decision of whom to recruit for embryology. He appears to have felt that neither the Spemann school nor the biochemical embryologists (e.g., de Beer, Needham) offered enough real potential to seek out any of their representatives for Cal Tech.

STAFFING THE LABORATORY

By the time he took up residence in Pasadena in the fall of 1928, Morgan had a staff of only four faculty members (himself, Sturtevant, Anderson, and Sterling Emerson), two international research fellows (Yoshitaka Imai and Dobzhansky), and one research fellow of the Carnegie Institution of Washington (Bridges). There were also five or six graduate students, including Albert Tyler and Carl C. Lindegren. The basic operation was small, but it was to grow as Morgan got his feet on the ground and acquired a sense of where the division should be headed as an integral part of the Cal Tech community.

In the ensuing years, Morgan built the areas of biochemistry and plant physiology with a stellar array of research workers. Biochemistry and plant physiology came first, and most easily. In 1929, Henry Borsook (Ph.D., University of Toronto, 1924) was appointed assistant professor of biochemistry. In 1930 three important appointments were made: Herman E. Dolk (Ph.D., University of Utrecht, 1927) as assistant professor of plant physiology, Robert Emerson (Ph.D., University of Berlin, 1927) as assistant professor of biophysics, and Kenneth V. Thimann (Ph.D., University of London, 1928) as assistant professor of biochemistry. In 1932 Frits Went (Ph.D., University of Utrecht, 1927) was appointed assistant professor of plant physiology, and in 1934 Cornelius A. G. Wiersma (Ph.D., University of Utrecht, 1929) was appointed assistant professor of physiology.[43] All these men had strong backgrounds in chemistry and physics and were interested in applying the physical sciences to biological problems. Went in particular expressed from the outset a strong interest in the physiological aspects of genetics and embryology. "Over the years," he wrote to Morgan just before accepting the invitation to Cal Tech, "I have come to certain conceptions about the physiological base of heredity, which could only be chequed [*sic*] by the facts. And where to find them if not with you?"[44] To Morgan, the study of the physiology of growth, as in Went's or Thimann's work with plant hormones, was an important approach to questions that ultimately related to gene action in development.

In addition to these appointments, which were meant to be the permanent faculty, provisions were set up in the Division of Biology (as in the other divisions) for a large number of temporary appointments as research fellows, associates, assistants, and visiting professors.[45] Under these programs many, mostly young, workers came to Cal Tech for a period of one to five years. Between 1930 and 1936 this list included: George Beadle, Hans Gaffron,

[43] Borsook was interested in the applications of thermodynamics to physiology and biochemistry, particularly aspects of energy transfer in biological systems. Dolk worked on biochemical aspects of plant physiology. Robert Emerson (no relation to Sterling) investigated the physical aspects of photosynthesis, especially the efficiency of energy transfer. Thimann was interested in the general biochemistry of plants, including that of proteins and amino acids, growth hormones, and the physiology of bacteria and protozoa. Frits Went, who had served as director of the Foreigners' Laboratory at Buitenzorg, Dutch Java, before coming to Cal Tech, was interested especially in the effects of plant hormones on growth; in 1937 he and Thimann wrote a book, *Phytohormones* (New York: Macmillan Co.). Wiersma was especially interested in neuromuscular transmission mechanisms in invertebrates (crustacea).

[44] F. W. Went to Morgan, July 28, 1932; Morgan Papers, Cal Tech; quoted with permission.

[45] Six different categories were established: research fellows, national research fellows, international research fellows of the Rockefeller Foundation, associates of the Rockefeller Institute of New York, research assistants, and visiting professors.

Barbara McClintock, K. J. Lindestrom-Lang, Boris Ephrussi, P. D. Darlington, and Johannes van Overbeek (research assistant, 1934-1937). One of these, George Beadle, stayed on as an instructor (1935-1936) and returned in 1946 after teaching for a year at Harvard and for nine years at Stanford to head the biology division. Graduate students in the department during these years included Lindegren, Sidney W. Fox, and James F. Bonner.[46] While this is an impressive list of names, the importance of their presence at Cal Tech lies less in their individual contributions than in the atmosphere their presence helped to create and maintain. They were mostly younger men and women just beginning their careers. They were bright, innovative, and dedicated to the task of applying physics, chemistry, and mathematics to biological problems. They mixed freely with one another, and not a few later collaborations resulted from contacts made initially at Cal Tech.[47] The laboratory at Cal Tech had an international atmosphere, with workers from other countries constantly passing through. It was, in some sense, the same atmosphere as Naples in 1895 or Woods Hole more recently, which always excited Morgan's imagination.

THE PRACTICAL SIDE: FACILITIES AND FUNDS

Long before Morgan took up residence in Pasadena, the problem of buildings and physical facilities had occupied a considerable amount of his attention. The general plan of the biological laboratories was to be that of a quadrangle, built in Spanish Colonial style. The trustees had initially secured B. G. Goodhue Associates of New York as architects. Although the senior architect, B. G.

[46] Lindegren was a graduate student at Cal Tech from 1928 to 1931, when he received his Ph.D. He later became research associate at Washington University, St. Louis, and professor of biology at the University of Illinois, Carbondale. Fox received his Ph.D. in 1940 and, after several industrial positions, became assistant and later professor of chemistry at Iowa State University (1943-1955), professor of chemistry at Florida State University (1955-1964), and professor of biochemistry and director of the Institute of Molecular Evolution at the University of Miami after 1964. Bonner received his Ph.D. in 1934 from Cal Tech; he then served as a research fellow at the universities of Utrecht and Zurich (1935-1937) and returned as assistant professor at Cal Tech from 1938 to 1942; he became associate professor in 1942 and professor in 1946.

[47] For example, Beadle later went to Paris to work with Ephrussi on biochemical and hormonal aspects of genetics; the association between Thimann and Went on plant hormones has already been mentioned.

Goodhue, had died in 1924, the trustees, with Hale's urging, retained the firm and the original plans. Once Morgan decided to accept the Cal Tech offer, he began immediate consultations with the architects in New York on a myriad of details on each of the several sides of the quadrangle. Morgan thought that the initial plans were quite exciting: "We are going to have the most beautiful group of buildings conceivable, and the most efficient laboratories that can be constructed."[48]

As with the program and personnel, Morgan felt that the physical facilities were integral to the entire development of biology at Cal Tech. The physical relationship of the various laboratories was intended to reflect the interrelationships between the branches of biology that Morgan wished to promote. For example, the biology wing was to be contiguous on one side with the organic chemistry laboratories, rather than with geology as Hale had originally proposed. As Morgan wrote to Millikan in 1930, it was with organic chemistry that the biologists hoped "to have close contacts."[49] A second matter on which Morgan insisted as the plans were being drawn up was that the teaching areas and library should be kept as far from the research laboratories as possible. The reason for this was to reduce the possibility of intrusion on the research work. Thus, the library was designed to be a separate building attached to one end of the biology quadrangle, and the undergraduate lecture rooms were to be placed on one wing "as near the exit as possible, in order that the undergraduates should not interfere with the research work by using the hallways."[50] But most important, Morgan wanted to make sure that the building plans were not rushed through before it was clear what staff was going to be hired to use them. Morgan felt very strongly that the details of the various laboratories should be designed in conjunction with those who were going to work in them. He found himself opposed initially to Arthur Fleming, chairman of the Cal Tech Board of Trustees, who wanted to build all three sides of the quadrangle at once. Morgan staunchly opposed this plan from the outset.[51] Morgan ultimately had his way, though not without first

[48] Morgan to Arthur Fleming, September 10, 1927; Hale Papers; quoted with permission.

[49] Morgan to Millikan, July 27, 1930; Hale Papers; quoted with permission.

[50] Morgan to Hale, November 2, 1927; Hale Papers; quoted with permission.

[51] Ibid., September 10, 1927; see also Morgan Papers, Cal Tech.

proposing that Fleming's intervention could be circumvented by requiring all building plans to be approved by the executive committee before any action was taken on them.

In addition to the basic laboratory buildings, there were several other facilities Morgan insisted Cal Tech should provide. The first was a fresh- and salt-water aquarium in which many California and Pacific aquatic species could be kept for display and research purposes. After some negotiation, the architects made what Morgan called a "beautiful modification" of the plans by adding a small aquarium building onto one end of the three main biology buildings.[52] The second was a garden in which the botanists could carry out breeding experiments. Although there was some talk of using the existing gardens of the Huntington estate, a large tract of land in Pasadena with formal gardens and a house converted into an art museum, the legal technicalities were too restrictive to make this possible. Thus Morgan insisted that some additional land be purchased for later development as an experimental garden. The third was a marine station on the Pacific Coast not far from Pasadena, to be operated by the institute. Morgan did not feel that such a station should be developed immediately, as he himself wanted to oversee the selection of a site and construction of the facilities. All of this was, however, written into the plans for the biology division before Morgan ever set foot onto the grounds of Cal Tech itself.

An additional practical problem Morgan had to face during the organizing year 1927-1928 was building a sound library for the biology division. In the fall of 1927 he raised the issue with Fleming, especially concerning details of the purchase of books, journals, and other items necessary for a working research library. Fleming advised Morgan to draw up the list of books and periodicals he felt were essential and have it checked by the library staff at Cal Tech for possible duplication. The institute would then take charge of ordering the items through their customary dealers and distributors both in the United States and abroad.[53] Morgan's ideal of a good working library in biology was the one at Woods Hole. Consequently, in the summer of 1928 (which the Morgans spent as usual at Woods Hole), Morgan engaged Mrs. Selig Hecht (who had some training and experience in library work) to draw

[52] Morgan to Hale, September 10, 1927; Hale Papers; quoted with permission.

[53] Fleming to Morgan, November 18, 1927; Morgan Papers, Cal Tech.

up a list of the essential biological books and periodicals that every research library should have. This list was sent on to Pasadena. However, several years were required to build an effective working library in biology at Cal Tech.

In Morgan's mind, none of these matters—the physical facilities, the program plans, the library, or personnel—could be separated from each other in planning the new division; all were interrelated. His interest was obviously in the people and in the intellectual interactions he hoped he could promote. The program was dependent on the people; the physical facilities influenced the way people met and interacted, and the people who were there ought to have a say in how the facilities were organized.

The aspect with which Morgan had the least to do directly was fund raising. Funding for the establishment of the division of biology was planned along similar lines as that for the institution as a whole. As a private university, Cal Tech received donations from individuals, foundations, and industry.[54] Fleming had been one of the most regular private donors. Among the foundations that had previously supported the development of physics and chemistry were the General Education Board and the Rockefeller Foundation, and Southern California Edison and the Daniel Guggenheim Fund for the Promotion of Aeronautics, Inc., had made substantial contributions toward the development of specific research programs.

Several private donors had pledged substantial contributions for the construction of the division of biology. These included A. C. Balch and William G. Kerckhoff.[55] It was Kerckhoff's donations,

[54] See *California Institute of Technology Bulletin* (1928): 51-52.

[55] Balch (1864-1943) was an electrical engineer born in New York and educated at Cornell (M.E., 1889). Practicing first in Seattle, he moved to Los Angeles in 1896 and became president of the Greenwich Investment Corporation, the Loredo Land Company, the Summit Lake Investment Company, and the Meridian Limited and a director of the Southern California Gas Company. He was also a member and president of the board of trustees of Cal Tech and a member of the observatory council. William George Kerckhoff (1846-1929) was an industrialist born in Terre Haute, Indiana who went to California at the age of nineteen, settling there permanently after 1879. After a start in the lumber business, he organized the San Gabriel Power Company with Balch and ultimately in association with Henry E. Huntington became president of the Pacific Light and Power Company. Kerckhoff, Balch, and others began the tradition of damming up rivers in the southwestern United States to provide the Los Angeles vicinity with electric power.

later continued by his wife after his death in 1929, that made possible the completion of the biological laboratories at Cal Tech that bear his name. Several foundations supported the development of biology. The Carnegie Institution of Washington continued its $12,000 per year grant for genetics research; the General Education Board made a total donation of $2,100,000 for enlarging the facilities, with special reference to the biology division; and the Rockefeller Foundation, through the National Research Council, provided $20,000 per year for the national research fellows program. Somewhat later, largely through the efforts of Millikan, several steel companies made contributions to Cal Tech, and at Morgan's suggestion insurance companies were asked to contribute to studies in biology (which Morgan craftily argued should have bearing on the data of mortality).[56]

Negotiations for the funding necessary to establish the biology division were carried out largely by Millikan, Hale, Noyes, and Fleming, though Morgan frequently joined in on specific aspects of this project when called upon. For example, he had several meetings with Max Mason, director of the natural sciences division of the Rockefeller Foundation from 1928 to 1929, and president of the Rockefeller Foundation from 1929 to 1936.[57] Morgan also spent some time discussing the Cal Tech program with Wycliff Rose of the General Education Board, talks that ultimately led to rather large donations from that foundation.

In their fund-raising efforts, Hale, Millikan, Noyes, and Morgan emphasized the practical (industrial, agricultural, medical) advantages that can arise from the existence of a strong program in basic science. Morgan, for example, drew up a long list of

[56] Morgan to Raymond Pearl, March 2, 1928; Hale Papers. Morgan had written to Pearl asking him to provide any "impartial evidence" about biological research on problems such as the physiology of aging which would ultimately generate conclusions useful to the insurance companies.

[57] Mason (1877-1961) was an American mathematician who received his Ph.D. from Göttingen in 1903. After twenty-three years as a member of the faculties of M.I.T., Yale, and Wisconsin, he became president of the University of Chicago in 1925. In 1928 he joined the Rockefeller Foundation. He maintained a strong interest in the developing programs at Cal Tech, becoming a member of the executive council in 1926 (a post he held until 1949). He was also a member of the National Academy of Sciences and the National Research Council, and served with the Naval Experimental Station in New London from 1917 to 1919. Mason's work in mathematics involved developing the relationship between algebra of matrices and integral equations.

practical medical benefits that had been made possible only by the prior existence of basic research in the fields of physiology and chemistry.[58] Hale made the point more explicitly in discussing the increased profit that could accrue to those industries that supported basic science: "The 'practical man,' with his haste for immediate returns, often fails to see the vital need for pure science. Not so with the great leaders who have developed such corporations as the General Electric Company, the Du Pont Company, the Eastman Kodak Company, the Westinghouse Company, the American Telephone Company, and others who contribute so much to national prosperity and progress. Elihu Thompson, in whose research laboratory the General Electric Company had its origin, ranks as one of the most enlightened pioneers in the utilization of science by industry."[59] Morgan did not object to science having practical utility—he was, of course, glad that it did. It was basic science, however, that was crucial to him. If he had to make occasional arguments to justify the work at Cal Tech to wealthy entrepreneurs, he would do what was necessary. In contrast to Hale, who appeared to thoroughly enjoy his role as fund raiser and apologist for pure research, Morgan found the whole process tedious. He knew himself well. He was, in fact, a "laboratory animal" who never enjoyed any other environment quite so much as the laboratory bench covered with *Drosophila* culture bottles, Chaetopterus or sea urchins.

PASADENA, 1928-1935: BUILDING THE DIVISION OF BIOLOGY AND ESTABLISHING A MARINE STATION

So exciting was the prospect of working with Noyes, Hale, and Millikan in the common adventure that Cal Tech offered, that Morgan had accepted the invitation without ever having seen the physical facilities. He first visited Pasadena officially in February of 1928 to discuss the myriad details involved with organizing the division of biology. Budgets, building plans, library plans, personnel, as well as general procedures within the university community, were all discussed there with Hale, Millikan, Noyes, and Fleming. Morgan returned after a ten-day stay refreshed and con-

[58] T. H. Morgan, "Program for Biology," enclosed in a letter to Hale, December 4, 1930; Hale Papers.

[59] G. E. Hale, "Science and the Wealth of Nations," *Harper's Magazine* (January 1928): 223-251; quotation, p. 243.

vinced that the Cal Tech program was going to be one of the most exciting innovations in biology in the twentieth century.

After spending the summer of 1928 at Woods Hole, the Morgan family left for the West Coast on September 6, via the Santa Fe Railroad. They arrived in Pasadena on September 16, and moved not immediately, but soon after, into the house that Hale and Millikan had helped to secure for them (at 1149 San Pasqual Street, directly across from the laboratory). The Morgans shipped most of their belongings by boat through the Panama Canal, from New York to Pasadena, including the old player piano and its hundreds of rolls. When they arrived, a note welcoming the Morgans to Pasadena and to Cal Tech awaited them from Hale. Morgan was touched by the kindness shown him by his new colleague and overwhelmed by the progress that had been made in finishing the first of the biology buildings. In thanking Hale for his welcoming letter, Morgan wrote in a brief memo: "The new laboratory is splendid—I am very happy that it has all turned out so well. Two rooms are ready today and three more tomorrow; we'll settle down in them and go to work while the rest of the building is being finished."[60] With the genetics group ready to move in and get to work, Morgan was optimistic that the biochemistry and physiology buildings could be planned during the ensuing year with full reference to their specific occupants and the programs in physiology to be pursued in them. The first year was to be one of planning, expanding, and developing the program on the spot in a number of concrete ways.

The Morgans found life in Pasadena warm and congenial compared to New York City. Furthermore, for the first time, all their children were away from home, both the youngest daughters having entered colleges in California. Living so close to the laboratory, Morgan could easily come home for lunch, a custom both he and Lilian appreciated. They entertained more frequently than had been their pattern in New York, their friends in Pasadena, as in New York, tending to center around the university. Aside from Sturtevant, Morgan's friends included Millikan, Hale, and especially Noyes, whose flair for the unusual and sense of humor greatly attracted Morgan. Morgan and the physicist Richard C. Tolman (b. 1881) also became unusually close friends, as did their wives

[60] Morgan to Hale, September 18, 1928; Hale Papers; quoted with permission.

Lilian and Ruth, who shared a common interest in music and played violin and piano duets together at least once a week. The two couples were neighbors and spent many evenings together.[61]

In the way of recreation, Morgan substituted croquet for handball, which he had played regularly in New York. In addition, the Morgans had a full-sized billiard table in the Pasadena house, a game that they all—including their daughters when home on vacation from college—enjoyed thoroughly. Morgan also continued breeding and raising pigeons, an activity he carried out until the end of his life. Living in California, it was impossible not to make trips into the mountains, something that particularly thrilled Lilian but that, as Tom grew older, he did less frequently. One trip, in the fall of 1933, involved a horseback trip in the mountains behind San Diego with a group of friends. Although unaccustomed to riding, Morgan found this an exhilarating experience, but, as he wrote, it was "quite a stunt for the ladies."[62]

All in all, Morgan found California the "paradise" he had felt it to be in his previous trips to the West. It was an ideal climate and intellectual environment. The more relaxed pace and the pleasant natural environment did not mean that he curtailed his academic work in any serious way. Work remained his primary occupation and administration his second. However, planning the development of an institute that could grow and develop on its own from a solid foundation required continual effort during the first several years in Pasadena. Morgan did not let up in his drive to achieve the excellence he thought was potentially inherent in the institute.

After one year in Pasadena, Morgan summarized the future needs of the Division of Biology. Funds for the buildings had been obtained. However, on the basis of figures from the MBL, Morgan estimated that an additional $50,000 would be required for buying books, and as well as an additional $20,000 for an endowment to cover annual subscriptions to various journals. This was a must, Morgan emphasized, if the institute was to take its place among the foremost research establishments of the country. Second, Morgan felt it was time to begin thinking about a marine station. Pasadena was in an extremely favorable location to use

[61] Isabel Morgan Mountain, personal communication, August 1974.

[62] T. H. Morgan to Isabel Morgan, November 12, 1933, from the private papers of Isabel Morgan Mountain.

marine organisms for the study of biological problems. Morgan estimated that approximately $150,000 to $160,000 would be required to get such a station set up.

Morgan saw as his main and immediate task during the coming year the securing of appropriate people to lead the physiology group. He also wanted ten additional foreign fellowships to bring to Cal Tech the brightest young workers from Europe who had just received their Ph.D.s: "The presence of such a group of younger men would bring into the laboratory a spirit of international understanding that would broaden the somewhat narrow outlook of our own investigators."[63] Of these many needs, only those relating to the library were to materialize during the next two years. The marine station did not begin operating until 1934, and while Morgan did make several appointments in biochemistry and plant physiology before 1932, the main group in general physiology had to wait until 1934.

Of all these future projects, the one Morgan relished the most was the marine laboratory. During his first year at Pasadena, Morgan began looking around for suitable seashore sites. He found in the area of Newport Beach the small townlets of Corona del Mar and Balboa Beach, which "surpassed even my fondest hopes."[64] Corona del Mar was only about one hour's drive from Pasadena, and the fauna obtainable there was "the best I have seen anywhere, including Naples."[65] However, the land was very expensive, and for the laboratory to be successful, it was necessary to have a director. It was not until 1932, in the aftermath of the first shock wave of the depression, that enough money was available either to hire a director or to purchase land and build facilities. In that year Morgan appointed as director George MacGinitie (b. 1889), a young ecologist and marine naturalist, then at Stanford.[66] MacGinitie had had considerable experience in directing

[63] Morgan to Hale, August 29, 1929; Hale Papers; quoted with permission.

[64] T. H. Morgan, outline enclosed in letter to Millikan, August 29, 1929; Hale Papers.

[65] Ibid.

[66] George MacGinitie received his A.B. degree from California State Teacher's College in 1925, and his master's from Stanford in 1928. He had served as an instructor in biology at California State Teacher's College, Fresno (1925-1938), of zoology at the Hopkins Marine Station (of Stanford, 1928-1929), and subsequently, as assistant professor (1929-1932). He had also served as acting director of the Hopkins Marine Station in 1930 and as assistant to the director from 1930 to 1932.

a marine laboratory at Stanford, and had in Morgan's eyes a considerable breadth of knowledge about the marine fauna of the Pacific coast. MacGinitie did not have a Ph.D., but that was perfectly acceptable to Morgan, who maintained that degrees meant little in judging the quality of people.

MacGinitie's first task was to find an exact location and develop plans for constructing a marine laboratory. In his searchings through Newport Beach, he found an old boat club at Corona del Mar which had closed because of the depression. The main building had cost $100,000 to build, but in 1932 it was for sale at less than $50,000 (see figure 19). Morgan was convinced at once that this building could be renovated to serve as a laboratory, and MacGinitie's job was to see that the remodeling was carried out along lines that would be most useful to the Cal Tech staff. Once the building was in use, MacGinitie's work involved making arrange-

Figure 19. The Marine Station of California Institute of Technology, located at Corona del Mar in the town of Newport Beach, about an hour's drive south of Pasadena. The main building, originally a yacht club, was purchased in 1932, renovated, and converted into a laboratory; it was opened in 1934. (From the A. H. Sturtevant Papers, Courtesy of the Archives, California Institute of Technology.)

ments with local collectors to obtain specimens, meeting the needs of investigators who wished to use the laboratory, and organizing summer field courses taught at Corona del Mar. In addition, a general caretaker was hired for maintaining the facilities on a day-to-day basis.

From the first, Morgan envisioned the lab at Corona del Mar becoming a West Coast Woods Hole. In fact, several Cal Tech people worked there regularly in the summers: for example, Tyler (embryology of marine invertebrates), Wiersma (neurophysiology of crustaceans), Noyes (chemical problems using sea water), and Morgan (self-fertilization in the tunicate *Ciona*). Because it was so expensive to live in the towns of Newport Beach or Corona del Mar, Morgan had cots installed on the top floor of the laboratory building, so as to make it possible for workers to spend more than one day at a time there. MacGinitie was the only one, however, who worked regularly at the station throughout the year. Morgan encouraged his colleagues to make as much use of the station as possible for themselves and their graduate students, and to this end insisted that no charges whatsoever be made for the use of the station by any Cal Tech staff member.

Morgan himself used the marine laboratory perhaps more than anyone. He would often drive down on weekends during the winter months with Tyler, sometimes staying over Saturday night and working through the day on Sunday. Particularly gratifying to Morgan during his embryological studies at Cal Tech, especially at Corona del Mar, was his association and friendship with Tyler. Born in 1906 in Brooklyn, Tyler had obtained his B.A. from Columbia in 1927 and his M.A. in 1928. After moving with the Morgan group to Cal Tech, he obtained his Ph.D. there in 1929. Remaining at Cal Tech throughout his career, Tyler worked in various areas of developmental biology, especially on fertilization, immunology, and experimental embryology. Tyler, more than anyone else in the later years, was Morgan's protégé, working on many of the same problems (relation between sperm entry into the egg and planes of cleavage; the effects of centrifuging eggs before and after fertilization) that had intrigued Morgan forty years earlier. Tyler spent a year abroad as a National Research Council Fellow (1932-1933) at the Kaiser-Wilhelm Institute and at the Naples Zoological Station investigating some of these problems. Like Morgan, he was convinced that problems of development, such as embryonic differentiation, were amenable to chemical and physical analysis if only the right organism was discovered on which experiments could be carried out. This belief formed part of Tyler's later interest in immunobiology. In many ways far ahead of his time, he saw the immune response (where cells of the immune system must manufacture antibodies specifically in response to invading foreign substances) as a possible model for investigating how cells differentiate.[67] Morgan felt that in many ways Tyler's approach to modern embryology was a more fruitful one than those of either the biochemical embryologists (Needham, Huxley, and de Beer) or the "organized" school of Spemann.

Despite the special place the Corona del Mar Station occupied in Morgan's heart, it never met his expectations of becoming a California Woods Hole. Several factors may have contributed to this. First, Stanford's Pacific Grove Station was already in existence and well known and may have reduced Corona del Mar's drawing power. Second, the association of the Corona del Mar Station with a particular university reduced its general appeal. Third, the kind of work that could be carried out at Corona del Mar was often very far from the mainstream of biological research being carried out at Cal Tech. Morgan himself had insisted on bringing to Cal Tech investigators who worked in highly rigorous and quantitative areas of physiology and biochemistry. What these workers often needed were well-equipped laboratories with the most modern equipment. The Corona del Mar Station was not set up to provide such facilities. Morgan's interest in marine organisms and their particular biological prospects stemmed partly from his love of natural systems and organisms in nature. This was a feeling shared by many members of the older generation of biologists at the time, but it was becoming increasingly rare in the younger workers (as it is in many cases almost totally absent today). Tidepool biology was perhaps an interesting diversion, but it was not the kind of science the men Morgan was assembling in Pasadena wished to pursue professionally. Fourth, even in his later life, Morgan's own allegiance still lay with Woods Hole; until 1942 he continued to take his family there for the summers. In Morgan's eyes Corona del Mar had not replaced Woods Hole; Woods Hole was still the Naples of North America.

With the marine laboratory established by 1933, Morgan could turn his full attention to the final major institutional problem, the

[67] In recent years the immune system has become increasingly important for studies in the molecular aspects of cell differentiation.

rounding out of the group in general physiology. This involved hiring one or more physiologists concerned with problems at the cellular or subcellular levels. To Morgan this meant someone concerned with neurophysiology, and in particular, problems such as neuronal transmission. He was unsettled on the exact field of neurophysiology that would be most appropriate, however, and prepared to make a trip abroad in 1933 or 1934. Like Gilman searching for his first Hopkins faculty, Morgan felt that he should look as far and wide as possible, especially in Europe, to secure the very best investigators he could find, whether in Edinburgh, London, Belgium, the Netherlands, or Sweden. After several disappointments, he finally secured the services of Cornelius Wiersma from Utrecht whose work on interneuronal connections in crustaceans was as close as Morgan felt he could come to the kind of physiology he had originally envisioned. Thus, by 1934, the physiology group had been built up to its expected level, and the academic and personnel issues seemed to be well in hand.

Morgan had always approached his administrative duties at Cal Tech as something of a necessary evil in order to achieve his goal of integrated physicochemical research. His conception of the job of administrator was that he should serve as a catalyst, a facilitator who brought people together and provided the basic opportunities and wherewithal for their continued interaction. He wanted neither to supervise nor to control, but only to put the right ingredients together and let them react on their own. This he appears to have accomplished effectively. If the interactions did not always prove to be as extensive as he had hoped, or as successful as those of his own *Drosophila* group at Columbia had been, Morgan nonetheless felt that he could do no more than to try and let people do their best in congenial circumstances.

Morgan's original agreement with the executive council and board of trustees of Cal Tech had been that he would see through the organization of the Division of Biology for the first five years (1928-1933), after which time he could voluntarily retire, or the institution could retire him, without prejudice. In 1933, however, the program was not developed to Morgan's full satisfaction, although the institute was more than pleased with the progress that had been made. He was thus asked to remain for an additional five-year period (1933-1938) which was ultimately extended until 1942. Morgan retired officially in June of 1942 at the age of seventy-six, being appointed at that time to the position of professor and director emeritus. From 1942 until 1946, the division was run by a biological council, of which Sturtevant was chairman. In 1947, George Beadle came back to Cal Tech from Stanford and successfully guided the department's activities until he left in 1961 to become chancellor of the University of Chicago.

THE FRUITS OF SUCCESS: THE INTERNATIONAL GENETICS CONGRESS (1932) AND THE NOBEL PRIZE (1933)

Morgan's position as the dean of American genetics brought him in the early 1930s two outstanding tokens of recognition: presidency of the Sixth International Congress of Genetics (1932) and the Nobel Prize in Physiology and Medicine (1933). The genetics congress was held in Ithaca at Cornell University from August 24th to the 31st, 1932. Despite the financial restrictions imposed on it by the depression, the meeting was a gala affair, as most international congresses have always tried to be. The two reigning figures at the congress were Morgan, as president, and R. A. Emerson, of Cornell, as chairman of the local committee (see figure 20). By the early 1930s the work of Emerson's group had achieved for botanical genetics what that of Morgan's group had achieved for animal genetics. Emerson's work with maize had established firmly the validity of the chromosome theory for plants. In many ways, it was the work of these two men and their respective colleagues that determined the overall theoretical framework within which the many different topics discussed at the congress were placed. Even the numerous papers dealing with agricultural plant or animal breeding were now cast in terms of the Mendelian-chromosome theory.

In his presidential address to the congress, Morgan tried to emphasize the relationships between the new genetics and embryology, cytology, physiology, and evolution. In particular, Morgan emphasized the important light the Mendelian-chromosome theory had thrown on the theory of evolution. In many ways his own thinking had progressed on this subject since 1926 when he had published the second edition of his book *Evolution and Genetics*.[68] What Morgan emphasized particularly was the fact that by mutation of discrete Mendelian genes, small individual variations of the type

[68] Originally published as *Critique of the Theory of Evolution* (Princeton, N.J.: Princeton University Press, 1916).

Figure 20. T. H. Morgan (left) and R. A. Emerson (right), conversing at the Sixth International Congress of Genetics, Cornell University, 1932. Morgan was president, and Emerson chairman of the local committee of the congress. Emerson had pioneered in genetic analysis of chromosome structure, mapping, and studies of hybridization in *Zea mays* (corn). His work provided the kind of detailed genetic information for a single plant species that Morgan's work had provided for a single animal species. (Courtesy of Isabel Morgan Mountain.)

Darwin had emphasized could be achieved. Partly through such mechanisms as modifier genes, but also through pleiotropy (in which one gene is recognized to have effects on a number of different characters), Morgan came to realize that the kind of inherited traits that were envisioned by Darwin as the raw material of evolution were in fact produced by the intricate operation of Mendelian inheritance. While a gene could be identified by its principal effect, such as eye color, it could also have numerous less visible and dramatic effects still subject to the pressures of selection. At this stage in his career Morgan's emphasis on evolution and genetics came almost as a confession, an insistence on making clear a point he himself had struggled with for so many years.

At the end of his presidential address, Morgan remarked that he had been challenged recently to state what seemed to be the most important problems of genetics for the immediate future. Although recognizing that such predictions were of little real value, he nonetheless proceeded to list the following: (1) the physiological processes involved in the replication of genes; (2) the cytology and genetics of the physical events during synapsis and crossing-over; (3) the relations of genes to characters: the way the information of genes is translated physiologically into adult characters; (4) the physical and chemical changes involved in the mutation process; and (5) the application of genetics to horticulture and animal breeding. Morgan's predictions were amazingly accurate. Several of the problems such as the physical events in crossing-over were already being intensively studied at the time (by Stern who had earlier been associated with Morgan's group, and by Creighton and McClintock from Emerson's lab). Others, such as the physiological relations of genes to characters, the nature of mutation, and the process of gene duplication were to see breakthroughs by the 1940s or 1950s (some of them, indeed, coming out of Morgan's own group at Cal Tech). But how were these new discoveries to be made, Morgan asked? "Should you ask me this question, I should become vague and resort to generalities. I should then say: by industry, trusting to luck for new openings, by the intelligent use of working hypotheses (by this I mean a readiness to reject any such hypotheses unless critical evidence can be found for their support). By a search for favorable material, which is often more important than plodding along the well-trodden path, hoping that something a little different may be found."[69] And then Morgan added as his final statement, summing up his view of how to do scientific work: "And lastly, by not holding genetic congresses too often."[70]

Particularly delightful to Morgan and his wife was the fact that Otto and Tove Mohr came from Norway expressly to attend the congress. One day during the course of the congress itself, the idea struck Otto Mohr of getting the old *Drosophila* group from 1918 back together again for lunch. According to Tove Mohr the plan met with great enthusiasm: "Mohr is going to gather the old-timers for lunch."[71] Muller, Morgan, Mohr, Bridges, Sturtevant, Alexander Weinstein, Donald Lancefield, and H. H. Plough, among others, attended. Morgan was seated between Muller and Mohr, and "the old spirit between the 'boss' and the 'boys' quite naturally reappeared."[72] It was a spirited luncheon with many jokes and much good humor. Yet, like all reunions, it was a passing episode. The original fly-room group had changed—the members had gone their separate ways.

The ultimate honor for Morgan was the award of the Nobel Prize in October of 1933. Morgan had been considered for the Nobel Prize as early as 1919, when he had been nominated by Ross Harrison, and in 1922, when he was nominated by Otto Mohr.[73] But the committee had been hesitant on the grounds that the prize was supposed to be for contributions relating to medicine; all previous prizes had, in fact, been in areas of physiology or physiological chemistry. The Mendelian-chromosome theory seemed far enough removed from physiology and medicine at the time that, in the eyes of many committee members, it did not qualify.[74] Moreover, recipients of the prize before Morgan had been, with one exception, medical men or physiologists who were members of medical faculties.[75] Thus on two counts—the apparent

[69] T. H. Morgan, "The Rise of Genetics," *Proceedings, Sixth International Congress of Genetics* 1 (1932): 87-103; quotation, p. 103; reprinted in *Science* 76 (1932): 261-267; 285-288; quotation, p. 288.

[70] Ibid.

[71] See Tove Mohr, "Hermann J. Muller, 1890-1967; An Appreciation by a Friend," *Journal of Heredity* 63 (1972): 132-134; quotation, p. 133.

[72] Ibid.

[73] See letter from R. G. Harrison to the Nobel Foundation, December 18, 1919; Harrison Papers, Yale University. Also, Otto Mohr to the Nobel Foundation, 1922; Morgan Papers, APS.

[74] H. Schuck et al., *Nobel, The Man and His Prizes*, p. 135.

[75] Ibid., p. 142. The sole exception prior to 1932 was August Crogh (1874-1949) a Danish physiologist in Copenhagen who received the 1920

distance of Morgan's discoveries from physiological and therefore traditional medical work, and his own background as a nonmedical biologist—Morgan's previous nominations had been rejected.

Two factors may have contributed to the change of view on the part of the committee. One was that work in several countries, but most notably that of Otto Mohr in Norway and Gert Bonnier in Sweden (both ardent geneticists), had focused on applications of the Mendelian-chromosome doctrine to the study of human heredity. Mohr's work, and especially his Institute for the Study of Human Genetics at the University of Oslo, had shown the importance of the Mendelian theory for the diagnosis of human hereditary disease.[76] Second, in June of 1933, Morgan had received an honorary M.D. degree from the University of Zurich in direct recognition of his contributions to genetics and human medicine. While it is unlikely that the committee would be swayed by such a trivial matter in itself, the award from Zurich was an overt indication of the growing understanding in the medical community about the relationship between genetics and human health.

On the morning of October 20, 1933, Morgan was at work in his laboratory when he received a long-distance telephone call from New York. Thinking it was one of his children, Morgan went to the phone and was taken aback by the relay of a telegram to him from Stockholm. The telegram read: "On behalf of the Caroline Institute I beg to inform you that the Nobel Prize for physiology and medicine for 1933 has been awarded to you for your discoveries concerning the hereditary function of the chromosomes. [signed] Gunnar Holmgren."[77]

The news, of course, generated considerable excitement, but Morgan seemed to take it all with a certain bemused calmness. Lilian Morgan described the day in a letter to her children:

> It happened this morning at about 11. Father came into my room [at the Kerckhoff laboratories] with the news. He had been called to the telephone for long-distance from New York and thought, of course, that it was one of you, but it

---

prize for his discoveries relating to the motor-neuronal control of capillary opening and closing.

[76] See, for example, Mohr's book *Heredity and Disease* published in 1933. It should be pointed out that Mohr's impact on the Nobel Committee was enhanced by the fact that he was an M.D., as well as a Ph.D., and thus spoke from a "professional" point of view.

[77] Quoted in L.V. Morgan to Isabel Morgan, October 20, 1933; from the personal papers of Isabel Morgan Mountain.

was a relay of a cable from Stockholm! . . . Dr. Mohr told me last summer that many people had wanted this, and their message today is one of the nicest. . . . The Tolmans came over; in the office it has been a hectic day between the reporters and the messages, not to mention the regular business. . . .

I forgot to say that we shall have to go to Stockholm. I do not know when. I have made all sort of castles-in-the-air, but Father says it is too soon to think about that yet. . . . Father is quietly sitting in his chair reading Anthony Adverse just as though nothing had happened. I am sure if he could pause a moment he would send his love to you all with that from

Mother[78]

Both Morgan and his wife appreciated the significance of the award as an indication that the importance of genetics to medicine and physiology was at last being recognized. In a P.S. to the letter quoted above, Lilian Morgan wrote: "The really nicest thing is that the value of Biology to medicine is recognized to such an extent that they are going out of their way to give the prize for medicine to a biologist."[79] Morgan, of course, was pleased, though it was not in his nature to indulge that pleasure openly. It is also probably true that ultimately the greatest satisfaction to Morgan was not the recognition of his own personal work so much as a recognition that general biology, through one of its central disciplines, genetics, was accorded its rightful place as a part of physiology and medicine. As a result of the honor, Morgan even agreed to a filmed interview which was incorporated into the Pathé newsreels projected weekly into the local theaters. Never one to seek flashy publicity, Morgan nonetheless found this incident amusing, as he wrote to his daughter Isabel: "I wonder if the Pathé news in New York carried me? We saw it here. I wouldn't have had the courage, but Mother insisted. Well, it wasn't in the comic but it was certainly funny to *see* yourself *talk*."[80] More in line with Morgan's character, he agreed to have his picture taken by newsmen on the day of the announcement only if the picture included a group of neighborhood children who were hovering about, attracted by the excitement (figure 21).

[78] Ibid. [79] Ibid.

[80] T. H. Morgan to Isabel Morgan, November 12, 1933; from the private papers of Isabel Morgan Mountain.

Figure 21. T. H. Morgan surrounded by a group of neighborbood boys the day it was announced he had received the Nobel Prize (October 20, 1933). When a local newspaper reporter asked to take Morgan's picture, he agreed on condition that the young people who were around be included. (Associated Press photo, courtesy of Tove Mohr.)

Although the Nobel Prize is generally conferred in December of the year in which the award is announced, Morgan was unwilling to journey to Stockholm at that time. On the afternoon of October 20, a few hours after receiving notification of the award, Morgan wrote to the Caroline Institute saying that he would be unable to attend the ceremonies scheduled for December 10. Morgan's chief reason was that pressing duties in regard to the organization of the physiology group made it imperative for him to remain in Pasadena through the winter term. A secondary reason was that Morgan wanted to combine any trip to Stockholm with a trip to various laboratories in Europe where he could interview potential physiologists for Cal Tech. A trip in December would have left little time to properly plan such a trip. And besides all this, Morgan wanted to take his family abroad when he went and to use the occasion to visit some European friends as well. Morgan's message to the members of the Nobel Committee was read at the banquet on December 11, 1933, in Stockholm, by Laurence Adolph Steinhardt, the U.S. ambassador to Sweden:

> It is with great regret that I am unable to be present at the award of the Nobel Prizes. Circumstances here in connection with the establishment of a new group in physiology and with the immediate future of biochemistry in genetics make it imperative for me to remain, otherwise I should be present despite the distance; please express my appreciation of the honor conferred. I expect to go to Stockholm in May or June to meet my friends and colleagues there. The many letters which have come to me are unanimous in expressing satisfaction in the far-sighted recognition by the Caroline Institute of genetics as contributory to medicine and physiology. Personally I realize, of course that the work in genetics has not been accomplished by any one individual or group of individuals, but has been world-wide and the outcome of many hands and minds. It should, I think, give great satisfaction that the contributions to genetics have been international and that Swedish geneticists have done more than their share in bringing the results to a successful issue.[81]

Morgan scheduled his trip to Stockholm for June of 1934 and with his wife and daughter Isabel sailed from New York for London on April 25 on the steamship *Manhattan*.[82] While in England, Morgan visited William Bateson's old coworker, R. C. Punnett, discussed English physiologists with Sherrington and A. V. Hill, and went to Edinburgh to interview a potential physiology candidate.[83] From Edinburgh, the Morgans went to Newcastle and caught a boat for Bergen, Norway on May 22. From Bergen they journeyed by rail across the central mountain chain to Oslo to spend a few days with the Mohrs. While Otto Mohr had asked Morgan to speak at the university, Morgan gracefully and humorously, but firmly, declined: "Many thanks for your invitation to speak, but I am not going to spoil my visit to you, for myself, by anticipating an address, or, for you, by having you listen to one. So far I have kept out of public speaking and with the exception of the necessary one in Stockholm I hope to reach home with a

[81] From "Address Delivered by Mr. Steinhardt at the Nobel Banquet;" Morgan Papers, Cal Tech.

[82] Morgan to Otto Mohr, March 26, 1934; Morgan Papers, APS.

[83] Morgan to Mohr, May 1934; Morgan Papers, APS.

clean slate. Nevertheless, I appreciate your kindness in giving me a chance."[84] The Morgans had looked forward to such a visit with the Mohrs for many years. While the Mohrs had visited them on several occasions in the United States, the Morgans had never been to Norway. Morgan and Otto Mohr had much in common, which accounts, perhaps, for the easy communication between them. Both were gentle and refined people who loved their work. Both had a dedication to rigorous thinking but always approached even biological problems with a kind of wisdom that saw to the heart of an issue. Both had good senses of humor and were eminently at ease with themselves. Both derived from the upper middle class; they could be at ease with the world because life had not been an uphill struggle.[85] Furthermore, the Morgans were equally close to Otto Mohr's wife, Tove, an M.D. with a long and busy obstetrical practice which ended with complete retirement only in 1970.[86]

The Morgans spent a full day in Oslo visiting the Mohrs and then Tove took Lilian and Isabel to her father's country estate, about fifty miles from Oslo near the small town of Fredrickstad. There they spent a second day walking through the beautiful Norwegian countryside while Morgan himself went ahead to Stockholm to work on his Nobel Lecture. In a few days they reunited in

[84] Ibid.

[85] Mohr was, by the early 1930s, a distinguished biologist, geneticist, and scholar in his own right. He had been appointed professor of anatomy at the University of Oslo in 1919, immediately upon his return from the year with Morgan at Columbia. He had been elected to the Norwegian Academy of Sciences in 1920 and became its president in 1940. He was a member of the Danish and Finnish Academies of Science, and had been chairman of the international committee for the International Genetics Congress of 1932. After prison and later house arrest during the Nazi occupation of Norway, he was elected rector of the University of Oslo in 1946. He was a member of the American Philosophical Society. In 1933 he gave the Edward K. Dunham lectures at Harvard on heredity and disease. Mohr received honorary doctor's degrees from the Caroline Institute in Stockholm and from Edinburgh University. See A. M. Dalcq, "Notice sur la vie et l'oeuvre de M.O.L. Mohr, correspondent étranger," *Bulletin de L'Académie Royale de Médicine de Belgique* 7 (1967): 691-698.

[86] Tove Mohr was a prototype of the active and emancipated European woman of the twentieth century. In addition to her medical practice she raised a family of three children and, for a decade, managed an extensive farm, Thors, which had been in her family for four generations. She was active in a number of social and political reforms in Norway from the early 1920s on.

Stockholm for the special Nobel ceremony. Morgan wrote back hastily to Otto Mohr from Sweden: "You gave us a wonderful time in Oslo and we shall always remember it."[87] Lilian Morgan in her letter to Tove Mohr, however, captures a little more of the sense of personal warmth the visit to Norway engendered in them all: "How can we tell you how much we all enjoyed seeing you and Otto and your splendid children in your own home in the lovely land of Norway? Every day brought something new and ever to be remembered. The day at your old house made me know Venke [the Mohr's first daughter, born in New Bedford, Massachusetts] especially a little better and all the days are going to make me understand even better than before all that you will write me in the future about your friends and the children and the hospital."[88]

In Stockholm the Morgans stayed in the Hotel Reisen, with rooms looking out onto the waterfront. Morgan's Nobel Lecture (something that is expected of all Nobel Laureates) was scheduled for Monday, June 4, 1934 at 2:00 p.m. There was no special banquet or ceremony other than the presentation of the award, the lecture, and a small reception.

Morgan's Nobel Lecture was titled "The Relation of Genetics to Physiology and Medicine."[89] In his lecture, some parts of which have already been referred to in chapter 7, Morgan took the opportunity to hammer home a crucial point: that genetics is closely related to physiology and thus to medicine. After describing the Mendelian-chromosome theory as it related the facts of heredity to those of cytology, Morgan gave several examples of how genes could be viewed as influencing an organism's physiology: embryonic differentiation and biochemical processes in development (such as Beadle and Ephrussi's very recent studies on the chemistry of eye-pigment development in mutant strains of *Drosophila*).

Morgan then went on to relate the general work of his group on *Drosophila* to human genetics and the importance of this for medicine. He pointed out that the main contribution so far of genetics to medicine had been "intellectual."[90] By this he meant

[87] Morgan to Otto Mohr, May 29, 1934; from the personal papers of Tove Mohr.

[88] Lilian Morgan to Tove Mohr, May 29, 1934; from the personal papers of Tove Mohr.

[89] Published as T. H. Morgan, "The Relation of Genetics to Physiology and Medicine," *Scientific Monthly* 41 (1935): 5-18.

[90] Ibid., p. 16.

that the whole area of human heredity had been up to this time so vague and "tainted by myth and superstition" that a scientific understanding of the subject was an achievement of the first order. These myths to Morgan included the concepts of maternal inheritance (the idea that the egg contributes more to the offspring than the sperm) and the inheritance of acquired characters. Morgan emphasized that the study of genetics with *Drosophila* had taught a great deal about the important interactions that occur between heredity and environment. Genes are not fixed entities uninfluenced by what goes on around them. The expression of a phenotypic character is a result not only of the complex interaction of all the elements of the genome, but also of the environmental influences that can modify in one way or another the physiology of gene action. Morgan emphasized that what makes people the way they are—physically, mentally, and emotionally—is very complex, and cannot be explained by jumping on either side of the nature-nurture argument. The problems of human hereditary defects are not to be eliminated (as some enthusiasts supposed) by controlled breeding, but rather by understanding the physiological nature of certain hereditary defects, and providing means, where possible, of offering a cure (as in diabetes). Genetic counseling could also be a valuable tool, although Morgan pointed out that it must always be based on probabilities and therefore had to be used with caution. He concluded his address with an urgent plea: medical science must take the lead in improving human health by eliminating the causes of human disease (psychological as well as physical); genetics is not a panacea. However, he was firmly convinced that genetics could "at times offer a helping hand."[91]

On the evening after he delivered his address, the Morgan family left Stockholm for Svalöf in southern Sweden, the home of an important plant-breeding station Morgan wanted to visit. From Svalöf the Morgans went to Belgium, Holland, and England to interview more physiologists. The trip was successful, for it was at Utrecht that Morgan invited Wiersma to come to Cal Tech the following September. Wiersma accepted, and thus Morgan could return home satisfied that his physiology group was launched. From England the Morgans returned to the United States in mid-June, going directly to Woods Hole for the summer of 1934.[92]

[91] Ibid., p. 18.

[92] Lilian V. Morgan to Tove Mohr, June 13, 1934; from the personal papers of Tove Mohr.

MORGAN'S RESEARCH WORK AT CAL TECH

At Cal Tech the *Drosophila* work continued, funded by an annual grant (approximately $12,000) from the Carnegie Institution of Washington. Increasingly, the *Drosophila* work focused on mapping all groups of chromosomes in a highly detailed and systematic way. In the course of such work, as Morgan had emphasized, new genetic principles were still being discovered (such as "repeats," where whole sections of a chromosome consist of duplicated parts).[93] Most of this work was carried out by Bridges and Sturtevant, although as usual Morgan kept in touch with the overall plans. It was thus with a particular sense of personal and professional loss that Morgan reacted to Bridges's death in 1938. Bridges had a serious heart attack (with endocarditis) on December 6, 1938, and died three weeks later on the 27th.[94] Morgan wrote of Bridges's illness to Otto Mohr:

> Calvin had a heart attack last summer at Cold Spring Harbor, but recovered partially. After being here a few days, he had a much more severe attack and was taken to a hospital in Los Angeles. He was very depressed but in no great pain except now and then which was relieved by sedatives. He died quietly, unconscious, on the 27th. The post-mortem showed that the valves of his heart had become infected and were gradually breaking up. . . . We shall miss him terribly and his death is a serious set-back to the work that he was carrying on. . . .
>
> He left, of course, an immense body of unpublished material but in such shape that nobody but Calvin himself could possibly interpret the data. Fortunately, under pressure he was obliged each year to give a summary of his work in the Carnegie reports and these in a way must tell the story of what he accomplished. He had on hand at the time of his death quite a large amount of material concerned with the revision of the salivary maps, a much more complete identification of loci than had ever before been attempted, but

[93] T. H. Morgan, "Calvin Blackman Bridges, 1889-1938," *Biographical Memoirs, National Academy of Sciences* 22 (1941): 31-48; especially p. 38.

[94] T. H. Morgan to Isabel Morgan, December 6, 1938; from the personal papers of Isabel Morgan Mountain.

> whether anything can be done with his drawings I very much doubt.[95]

Inasmuch as Morgan himself was quite distant from the details of the mapping work at this point, and Sturtevant was involved in additional projects as well, Bridges's loss considerably slowed down the completion of the genetic work with *Drosophila* which had begun so many years before. But Bridges's cytogenetic work could not be replaced. Bridges's death hastened even more Morgan's gradual withdrawal from active involvement with the *Drosophila* work.

Additional problems relating to the *Drosophila* work were from this time on carried out by Sturtevant, Jack Schultz, a research associate of the Carnegie Institution, and Lilian Morgan. Lilian's work in particular turned up a new curiosity: the so-called "closed-X" or "ring chromosome."[96] As with most of the *Drosophila* work, discovery of new phenomena began with observation of an unusual ratio in a mating for known factors. In this case, it was the appearance of an unexpected class of males and females where recombination between two sex-linked traits (forked bristles and Bar eye) was much less than expected. Cytology was immediately employed to determine what chromosomal abnormality could account for these new findings. Working with Dobzhansky (who did the cytological work), Lilian Morgan found that many of the X-chromosomes in this strain were closed—i.e., bent around onto themselves to form a ring. Again, in accordance with the chromosome theory of heredity, the alteration in breeding results was correlated with a change in physical structure of the chromosomes. The finding of an unusual circular chromosome in *Drosophila* is interesting, in light of modern genetics, where the chromosomes of many bacteria such as *Escherichia coli* (today's *Drosophila* for experimental genetics) are normally circular.

Morgan's own active research interests returned increasingly to experimental embryology during his years at Cal Tech. These included studies on the factors determining the first cleavage plane in eggs, the factors affecting polar spindle formation and location during the first cleavage, and the studies on cross- and self-fertili-

[95] Morgan to Otto Mohr, December 31, 1938; Morgan Papers, APS.

[96] L. V. Morgan, "A Closed X-chromosome in *Drosophila melanogaster*," *Genetics* 18 (1933): 250-283.

zation in the simple Ascidian *Ciona* (which Morgan referred to on one occasion as "my blessed *Cionas*.")[97] The work on fertilization was probably the most important to Morgan himself during these later years, and he spent countless weekend hours pursuing it at Corona del Mar.

It had long been noted that *Ciona*, which is a hermaphroditic organism, is usually self-sterile. This meant that the sperm of one *Ciona* will not fertilize eggs of the same individual. Morgan made from this the following deduction: "If the statement were strictly true that the spermatozoa of the hermaphroditic Ascidian *Ciona intestinalis* rarely or never fertilized the eggs of the same individual, but will fertilize all the eggs of all other individuals, a highly paradoxical situation would seem to present itself. Such a statement would imply that no two individuals are ever alike [in terms of the mechanism for fertility-sterility]. If two like individuals were to occur, they would be expected to be cross-sterile."[98] Morgan then went on to point out the following problems: If two like individuals occur, they would be expected to be cross-sterile. There can be no doubt that if they do occur they are extremely rare. There are two problems here; one, the nature of the physiological reaction between eggs and sperm in relation to self- and cross-fertilization; the other, the kind of genetic situation that will account for the individual differences."[99]

In a series of six papers published in the *Journal of Experimental Zoology* between 1938 and 1944, Morgan explored both the physiological and the genetic sides of the question. One of his first findings was that the concentration of sperm fluid had an effect on self-fertilization: the more concentrated the sperm solution, the greater the frequency of self-fertilization even in individuals that were normally self-sterile.[100] Morgan's second finding was that egg water (water in which the eggs of one individual had been allowed to stand for some period of time) does not interfere with cross-fertilization, but actually enhances self-sterility (with sperm of the same individual): e.g., makes self-fertilization more difficult to

[97] T. H. Morgan to Isabel Morgan, November 3, 1944; from the personal papers of Isabel Morgan Mountain.

[98] T. H. Morgan, "The Genetic and Physiological Problems of Self-Sterility in *Ciona*. I. Data on Self- and Cross-Fertilization," *Journal of Experimental Zoology* 78 (1938): 271-318; quotation, p. 271.

[99] Ibid.

[100] Ibid.

achieve even with increased sperm concentrations.[101] Morgan suggested that some substance produced by the egg, or at the surface of the egg, which was later released into the surrounding water, could be responsible for the phenomenon of self-sterility; it could be regarded as a "self-sterility factor." On the other hand, sperm water (liquid in which sperm had been allowed to swim for a period of time and later strained off) had no effect on either self-sterility or self-fertilization.

Morgan's third finding was that certain acids can greatly enhance the prospects of self-fertilization. According to one writer, Morgan hit on this unusual relationship quite by accident while working one day at Corona del Mar in 1938. Morgan had a hunch that acidified sea water might overcome the block to self-fertilization, and Tyler suggested that weak acids might be better than strong ones. Since no bottles of weak acids were immediately available at the station, Morgan took a lemon from his lunch basket, squeezed out the juice, and measured portions of it into dishes of sea water containing *Ciona* eggs. In a large number of samples, self-fertilization did, in fact, occur.[102] In addition, certain proteolytic enzymes such as trypsin (and also the stomach juice of crabs) could induce self-fertilization when applied to a suspension of eggs. Even when the solution of enzymes such as trypsin was neutralized (trypsin normally occurs in an alkaline environment in the lower vertebrate gut), it could overcome the block to self-fertility.[103] Other agents, however, such as ether or alcohol, produced erratic and unpredictable results.[104] Therefore, Morgan concluded that weak acids and protein-digesting enzymes had a similar effect on the eggs and were in some way blocking the self-sterility factor.

A fourth finding was basically negative. He had no success in extracting from either eggs or sperm any materials that would facilitate self-fertilization or prevent cross-fertility. A fifth was that

[101] T. H. Morgan, "The Genetic and Physiological Problems of Self-Sterility in *Ciona*. II. The Influence of Substances in the Egg Water and Sperm Suspensions in Self- and Cross-Fertilization in *Ciona*," *Journal of Experimental Zoology* 78 (1938): 319-334.

[102] Howard J. Teas, ed., "Introduction," *Genetic and Developmental Biology* (Lexington, Ky.: University of Kentucky Press, 1969), p. 5.

[103] T. H. Morgan, "The Genetic and Physiological Problems of Self-Sterility in *Ciona*. III. Induced Self-Fertilization," *Journal of Experimental Zoology* 80 (1939): 19-54; especially pp. 52-54.

[104] Ibid.

when sperm from one individual were placed in egg water from that same individual, the chances of self-fertilization (all other conditions being equal) were lowered (even when sperm concentration was high).[105] A last item of data was obtained for Morgan by some of his biochemical associates at Cal Tech. The surface of *Ciona* eggs was found to be highly positive when tested for protein, and less so when tested for other major organic compounds. Morgan concluded that chemically the surface of the egg cell—its membrane—was composed largely of protein material.

Although these data did not provide an exact mechanism for understanding the nature of self-sterility and cross-fertilization, Morgan drew from them some general conclusions. It appeared that the egg, rather than the sperm, was the gametic agent determining whether fertilization would or would not occur. More specifically, Morgan focused on the egg membrane as perhaps the key site where the block to self-fertilization might lie. It seemed likely that the egg produced some substance—very likely a protein—which was localized in or on the membrane and which specifically prevented sperm of the same individual from passing the barrier. It was unclear, of course, exactly how this could happen, for Morgan's experiments threw no light on the actual interaction between sperm and the egg membrane. The fact, however, that a specific protein or proteins could actually determine whether fertilization with particular sperm did or did not take place, raised interesting physiological, biochemical, and genetic questions.

In the final two papers of his series in the *Journal of Experimental Zoology*, Morgan turned to the genetic aspects of the *Ciona* problem. Here he made some highly interesting observations:

1. Considerable variation existed between individual *Ciona* as to whether they would self-fertilize at all, and if they would, how easily.
2. The offspring from an individual capable of self-fertilization were themselves more prone to self-fertilize than the average *Ciona*.
3. The range of variability among the offspring of a single self-fertilized adult was very large: some offspring showed 100 percent self-fertilization capability.
4. Self-fertile individuals showed no reduction in cross-fertility

[105] T. H. Morgan, "The Genetic and Physiological Problems of Self-Sterility in *Ciona*. IV. Some Biological Aspects of Fertilization," *Journal of Experimental Zoology* 80 (1939): 55-80.

with other individuals, whether those individuals were self-fertile or self-sterile.[106]

From these observations Morgan drew several conclusions. First, the genetic systems governing self-fertility and self-sterility appeared to be independent of one another. Second, self-sterility and self-fertility were not either/or conditions, but like most inherited traits showed a wide range of phenotypic expression—a continuum in which absolute self-sterility or self-fertility existed only at the extremes. Third, Morgan concluded that more than one pair of genetic factors must be involved in determining the characteristics of self-sterility or self-fertility. If only one or even two pairs were involved, there would be many more cases of identical offspring from any one cross and much less range of variability.[107] Morgan assumed that there may be as many as six pairs of factors involved, but he had no quantitative data to suggest the exact number. The great range of variability in the percentage of self-fertility possible among different self-fertilizing adults could be explained by assuming that mutation in one or more of the factors could occur early or late in the development of the gonads of the original adult. If a mutation allowing for self-fertility occurred early in development, a large number of eggs of that individual would therefore be self-fertile. If the mutation occurred late, a small number would show this characteristic. This was not far different from a hypothesis invoked by Goldschmidt, and later by Bridges, to explain the different degrees of gynandromorphism and sexual mosaics observed in *Drosophila*.

Despite the fact that this later work of Morgan's appears to go back to problems reminiscent of his early career, it has many elements that reflect the philosophy of science he hoped to put into practice at Cal Tech. Morgan had wanted the work at Cal Tech to be cooperative; he had wanted Cal Tech to be a place where biologists, physiologists, biochemists, embryologists, and geneticists could work from their respective viewpoints on common problems. Morgan tried to carry this out in his research on *Ciona* and enlisted the help of a number of colleagues from the Pasadena

[106] T. H. Morgan, "The Genetic and Physiological Problems of Self-Sterility in *Ciona*. V. The Genetic Problem," *Journal of Experimental Zoology* 90 (1942): 199-228; especially p. 224.

[107] T. H. Morgan, "The Genetic and Physiological Problems of Self-Sterility in *Ciona*. VI. Theoretical Discussion of Genetic Data," *Journal of Experimental Zoology* 95 (1944): 37-59; especially pp. 58-59.

laboratory. For example, Boorsook helped with the physiology, especially with the determination of the degree of acidity (pH) necessary to induce self-fertilization; James F. Bonner and Sidney Fox helped with the biochemical analyses of the composition of the *Ciona* egg surface; Tyler helped with the embryology, especially in raising the cultures of *Ciona* at the Corona del Mar laboratory; and Sterling Emerson helped both with the genetics and with the interpretation of the theoretical questions. The *Ciona* work was an example, crude in some ways but far reaching in others, of starting with a biological problem and trying to use physics, chemistry, genetics, and embryology as paths toward a common solution. Morgan's work did not really solve the problem. In fact, one of the most interesting implications of Morgan's conclusions—namely that the block to self-fertility was genetic and appeared to be determined at the egg surface by proteins—was almost completely overlooked at the time.[108] Despite the incomplete nature of the *Ciona* work, Morgan's studies made clear some of the facts, and at least started to show how a problem that began with a simple observation could be attacked on all sides by using the integrated disciplines of physics, chemistry, and biology.

On the personal level, Morgan's work with *Ciona* represents his last major research effort. His last paper, published posthumously in 1945, concerned the normal and abnormal development of the eggs of *Ciona.*[109] This paper in particular, and the *Ciona* work in general, was a significant close to Morgan's long and productive career. In the first place it was a testimonial to his abiding love for organisms—to the fundamental biological problems of living things in their natural environments. For all of his writings about the necessity of quantitative and laboratory experimentation in biology, it was the field, the organism in the tide pool, that most attracted his interest. Yet the problems observed in the tide pool could only be answered, in most cases, by experiments done in the laboratory. Morgan himself was most inclined to work on the less quantitative and more generally biological aspects of the problem. Although he believed that the use of physics and chemistry was essential, it is significant that Morgan's final work was pub-

[108] This implication was especially interesting in that it was becoming apparent by the early 1940s that genes function biochemically by producing certain proteins.

[109] "Normal and Abnormal Development of the Eggs of *Ciona*," *Journal of Experimental Zoology* 100 (1945): 407-416.

lished in the hundredth volume of the *Journal of Experimental Zoology*, a periodical devoted to studies on whole organisms, and one he helped found in 1904.

Interestingly enough, the types of problems he wrote about in his first paper for that journal[110] dealt with the same subject as his last. Biology as a general field of research had undergone enormous change during that forty-year period, and Morgan had been intimately involved with that change in direction. Yet he himself had remained true to his own interests and abilities. He was still seeking solutions to those problems that loomed large in embryology and development, from the day of Roux and Driesch to that of molecular genetics, already developing at Cal Tech.

DROSOPHILA AND THE DEVELOPMENT OF BIOCHEMICAL GENETICS AT THE CALIFORNIA INSTITUTE OF TECHNOLOGY

It is often said that the *Drosophila* work represents the end of the line for classical genetics—that the newer biochemical and molecular genetics which came to fruition in the 1950s and 1960s evolved from a wholly different tradition and represented a separate and independent line of development. While there is some truth to this statement, it neglects the very real roots biochemical and molecular genetics had in the classical *Drosophila* school.

It was Morgan's explicit aim at Cal Tech to forge an intellectual and working relationship between the studies on genetic transmission and those of genetic translation. As he wrote to Max Mason on May 15, 1933: "Genetics is broadening its scope. The gross structural features of inheritance are today fairly well known, and the workers are turning to the physiological aspects of heredity. Here we are endeavoring to bring our genetic group into closer contact with the physiological group. The success of this effort will depend in large part on the presence of progressive and thoughtful men familiar with the most recent advances in physiology. The geneticists stand ready to cooperate."[111] Morgan knew throughout his period at Cal Tech that there were no simple and

[110] T. H. Morgan, "Self-Fertilization Induced by Artificial Means," *Journal of Experimental Zoology* 1 (1904): 135-178.

[111] Morgan to Mason, May 15, 1933; Morgan Papers, Cal Tech. Mason was at this time president of the Rockefeller Foundation, from which Morgan was seeking financial support for the work at Cal Tech.

obvious experimental tools by which to investigate the physiology (by which he meant the biochemistry and molecular basis) of genetics. But he did realize that the most fruitful way to attack the problem was to bring together under one roof people who could approach heredity from several points of view. To this end he had brought to Cal Tech in the early and middle 1930s several workers whose interests lay in this direction. Among these were Beadle, Ephrussi, Hans Gaffron, and Max Delbrück. These individuals gave the study of biochemical and molecular genetics its first experimental push.

Beadle, a product of the Emerson school at Cornell, was a botanist who learned *Drosophila* genetics as a postdoctoral fellow in Morgan's laboratory at Cal Tech. Morgan was highly pleased with Beadle's abilities and his quick comprehension of the basic problems in the *Drosophila* field.[112] Beadle originally began working on the cytogenetics of *Drosophila*; while at Cal Tech he met Ephrussi, an international fellow of the Rockefeller Foundation (1932-1933), and later visiting professor (1935-1936). Ephrussi came from Paris with an interest in studying the developmental aspects of *Drosophila* genetics. In particular, he focused on the biochemistry of pigment formation as a developmental phenomenon governed by specific genetic factors. Ephrussi developed the technique whereby the eye discs (developing eye buds) of *Drosophila* larvae of one color could be transplanted into the abdomen of adults of the same or a different eye-color strain. He then observed the phenotype of the transplant. By using many different mutant strains as transplant donors and hosts, Ephrussi, at first working alone, and later with Beadle back in Paris, demonstrated the specific biochemical steps governed by each eye-color allele for a whole series, from white through orange, vermillion, and red.[113] They showed that eye color is produced by a stepwise series

[112] Morgan to J. B. Johnston, College of Science, Literature and Arts, University of Minnesota, October 3, 1934; Morgan Papers, Cal Tech. Morgan wrote of Beadle in this letter of recommendation: "Dr. Beadle has been with us four years, during two of which he held a National Research Fellowship. We have formed a very high opinion of his ability, both as a research man and as a teacher. He was trained as a botanist and a geneticist, and he acquired a thorough knowledge of the genetics of *Drosophila* and is prepared to teach genetics from both the animal and plant point of view. He is an excellent cytologist in these fields. . . . Personally he is cooperative and has good executive ability." Quoted with permission.

[113] Boris Ephrussi, "Chemistry of 'Eye Color Hormones' of *Drosophila*," *Quarterly Review of Biology* 17 (1942): 327-338.

of chemical reactions, each apparently controlled by a certain gene. This was both a developmental and a genetic problem, which served to draw together elegantly the implicit relationship between genes and the control of steps in biochemical pathways.

Hans Gaffron (Cal Tech research fellow from 1930 to 1932) was a biochemist who had received his Ph.D. at the University of Berlin in 1925 and, after spending a year at Cal Tech, returned to his home university with ideas about the relationship between genetics and biochemistry. In Berlin, Gaffron met Max Delbrück, a physicist who had pioneered in the development of quantum mechanics in the late 1920s and early 1930s. Delbrück had become interested in biology through the influence of another physicist with far-reaching and imaginative ideas. Niels Bohr.[114] Bohr had emphasized (among other things) the vast number of important problems biology offered. Bohr had suggested to Delbrück that the simple reduction of biological problems to physics and chemistry (especially the atomistic type of physics and chemistry adhered to by most biologists) was inadequate to the solutions of critical problems such as those of heredity and development. Those imbued with the new ideas of quantum physics recognized that the act of breaking an organism down into its individual component parts destroyed the very organizational basis that ought itself to be the object of future study. Thus Delbrück, after obtaining his Ph.D. from Göttingen in 1930 and serving as a Rockefeller Foundation Fellow in Physics with Bohr at Copenhagen, had gone to the Kaiser-Wilhelm Institute for Chemistry (1932-1937) with an interest in approaching biological problems. Through his association with Gaffron and others, his attention became specifically oriented toward genetics:

> My interest in biology was first aroused in Copenhagen by Bohr, . . . the move to Berlin in 1932 was largely determined by the hope that the proximity of the various Kaiser Wilhelm Institutes to each other would facilitate a beginning of an acquaintance with the problems of biology. This good intention eventually materialized. A small group of physicists and biologists began to meet privately (mostly in my mother's house) beginning in about 1934. To this group belonged G. Moliere, Werner Bloch, Lamla, Werner Kofink (from the

[114] See Gunther Stent, "That Was the Molecular Biology that Was," *Science* 160 (1968): 390-395; also Robert Olby, "Schrodinger's Problem: What Is Life?" *Journal of the History of Biology* 4 (1971): 119-148.

physics side), Curt Wohl (physical chemistry), Hans Gaffron (biochemistry), K. G. Zimmer (biophysics), and most vital of all, N. W. Timofeeff-Ressovsky (genetics).[115]

Gaffron's report on the genetic work at Cal Tech, along with a visit from representatives of the Rockefeller Foundation in 1936, stimulated Delbrück's desire to go to work at Cal Tech:

> The move to the U.S. in 1937 was catalyzed by a visit of a representative of the Rockefeller Foundation, from which I had six years earlier received a fellowship in *physics*. It was suggested to me during this visit that I again apply for a fellowship to permit me to pursue with greater freedom and effectiveness my interest in *biology*. At that time it was the declared policy of the Division of Natural Sciences of the Rockefeller Foundation to encourage the entry of physicists and chemists into biology. I chose Caltech as the place to go to because of its strength in *Drosophila* genetics, its sympathetic attitude towards my scientific interests in general, and to some extent because of its great distance from the impending perils at home.[116]

Delbrück spent two years at Cal Tech, but his work centered less on *Drosophila* genetics than he had anticipated. As he wrote: "Even though I had come to Caltech planning to learn more about *Drosophila* genetics, I soon teamed up with E. L. Ellis doing phage research."[117] It was through this association that Delbrück began his important work on phage genetics which led to the development of a molecular approach to heredity.

These men contributed to the two main areas of genetic work that became important in the 1940s, 1950s, and 1960s: biochemical genetics on the one hand, and molecular genetics on the other. Biochemical genetics is that aspect of genetics dealing with the relationships between genes and biochemical reactions in the cell. Essentially it deals with the question of how the gene produces its

[115] Introductory remarks by Max Delbrück to the Delbrück Papers, California Institute of Technology Archives, "Guide to the Papers," p. 4.

[116] Ibid.

[117] Ibid. Ellis (b. 1906) received his bachelor's degree from Cal Tech in 1930 and his Ph.D. in 1934. After working for the Food and Drug Administration (1934-1936), he returned as a research assistant in 1936. His interests were in the biochemistry of bacteria and bacteriophage ("phage" for short, are viruses that infect bacteria rather than other types of cells).

effect in physiological terms (sometimes called the "translation" of gene information into actual phenotypic characters). Molecular genetics is that aspect of genetics dealing with the actual molecular structure of the gene and its direct products, as well as with the process of gene replication.

Beadle and Ephrussi's work showed that genes appeared in some way to control the individual steps in multistep biochemical pathways. However, Beadle and Ephrussi recognized that *Drosophila* was not the most favorable organism for trying to determine the nature of this relationship. After he returned to the United States, Beadle eventually took a position at Stanford University, where he met the microbiologist E. L. Tatum (b. 1909). While still a graduate student at Cornell, Beadle had heard a lecture on genetic segregation in the mold *Neurospora* (which grows on bread) by Bernard O. Dodge of the New York Botanical Garden.[118] Dodge had apparently tried to persuade Morgan, before 1925, of the advantages of using *Neurospora* for genetic studies. Eventually Dodge was persistent enough that Morgan took some *Neurospora* cultures with him to Columbia and brought them out to Cal Tech in 1928.[119] When Lindegren arrived at Cal Tech as a graduate student, his background in bacteriology made him an excellent prospect for "cleaning up" the *Neurospora* stock; that is, freeing it of bacterial contamination. Lindegren's work brought Beadle's attention once again to *Neurospora*; the seed doubly planted finally took root when Beadle and Tatum began looking for a favorable organism with which to pursue studies in biochemical genetics.

Using *Neurospora*, Beadle and Tatum were able to show conclusively that genes functioned by producing enzymes (proteins). The enzymes in turn were envisioned as controlling the specific steps of multistep chemical pathways. Thus, for example, the production of eye-pigment molecules might involve six or eight chemical reactions, each controlled by a specific enzyme, in turn produced by a specific gene. Although the one gene–one enzyme hypothesis has been modified in recent years, it represented a startling breakthrough in physiological genetics in the early 1940s:

[118] As related by Beadle in "Biochemical Genetics: Some Recollections," in *Phage and the Origins of Molecular Biology*, ed. John Cairns, Gunther Stent, and J. D. Watson (Cold Spring Harbor, U.Y.: Cold Spring Harbor Laboratory of Quantitative Biology, 1966), p. 24.

[119] Ibid., p. 25.

it at once accounted for the relationship between genes and metabolic reactions, gave a physiological explanation to multigene systems, and provided a basis for understanding the whole range of phenomena embraced by the idea of gene interaction. For this important work Beadle and Tatum received the Nobel Prize in 1958.[120]

Delbrück, too, realized soon after arriving at Cal Tech that *Drosophila* was not a suitable organism for genetic work at the molecular level. His work with Ellis, however, suggested that bacteriophage, which had an even shorter life cycle than *Drosophila* (often a matter of a few minutes), might be the organism he was seeking. Delbrück was most interested in determining the physical nature of the gene: what molecules actually composed a gene and how the properties of those molecules could account for the unique characteristics genes displayed—the ability to replicate themselves faithfully but also to introduce errors or mutations occasionally. Soon after leaving Cal Tech for Vanderbilt University in 1940, Delbrück met Salvador Luria and Alfred Hershey. According to Gunther Stent, "With this meeting the American phage group came into being."[121] The members of the group were united by a single common interest: the desire to understand how the bacteriophage can infect a bacterial cell, and within a half-hour or so reproduces several hundred progeny.

Development of the original phage group was slow, but after Delbrück organized the first annual summer bacteriophage course at the Cold Spring Harbor Laboratory in 1945, the "new gospel" began to spread among physicists and chemists.[122] The story of the various lines of work culminating in the Watson and Crick model of DNA, has been told a number of times from a number of points of view.[123]

[120] See Beadle and Tatum, "The Genetic Control of Biochemical Reactions in *Neurospora*," *Proceedings of the National Academy of Sciences* 27 (1941): 494-506.

[121] Stent. "That Was the Molecular Biology that Was," p. 393.

[122] Ibid.

[123] Ibid. In addition, the following articles are of some value: Eugene L. Hess, "Origins of Molecular Biology," *Science* 168 (1970): 664-669; Linus Pauling, "Fifty Years of Progress in Structural Chemistry and Molecular Biology," *Daedalus* 99 (1970): 998-1,014; John T. Edsall, "Protein as Macromolecules: An Essay on the Development of the Macromolecule Concept and Some of its Vicissitudes," *Archives of Biochemistry and Biophysics,* supplement 1 (1962): 12-20; J. D. Watson, *The Double Helix* (New York: Atheneum, 1968); and Gunther Stent's "DNA," *Daedalus* 99 (1970):

It is most important here to understand how the Cal Tech influence spread through two lines of work, initially quite separate: biochemical genetics and molecular genetics. It was the convergence of these two lines that produced what J. D. Watson has so effectively labeled the "molecular biology of the gene."

The molecular biologists influenced by Delbrück and others at Cold Spring Harbor in 1945 were not much concerned with the biochemical genetics of Beadle, Ephrussi, and Tatum. What was of great interest to them was a recently published article (1944) by O. T. Avery (1877-1955), Colin MacLeod (b. 1909) and Maclyn McCarty (b. 1911) on the "transforming principle."[124] Using bacteriophage, Avery, MacLeod, and McCarty showed that a nonvirulent strain of bacteria could be "transformed" into a virulent strain by infection with phage that had previously infected the virulent strain. The phage appeared to carry something with it from the virulent strain that genetically converted nonvirulent into virulent bacterial cells. That something appeared to be either deoxyribonucleic (DNA) or protein (the only two components of phage), but it was not totally clear which. In 1952, however, the phage workers A. D. Hershey (b. 1908) and Martha Chase (b. 1927) made a simple and profound discovery.[125] Using radio-

909-937. Robert Olby has dealt from the historian's point of view with the development of the DNA idea in "Francis Crick, DNA, and the Central Dogma," *Daedalus* 99 (1970): 938-987; Olby has also traced the influence of Erwin Schrodinger on the development of molecular biology in "Schrodinger's Problem: What is Life?" Stent has also provided some interesting information in his introductory remarks, "Waiting for the Paradox," in *Phage and the Origins of Molecular Biology*, pp. 3-8. Delbrück has given his side of the story in "A Physicist Looks at Biology," in the same volume, pp. 9-22, as has Beadle in "Biochemical Genetics: Some Recollections," pp. 23-32. Olby has collected much information on the development of the DNA story in his book, *The Path to the Double Helix* (London: Macmillan & Co., 1974), a much more objective account than Watson's. A critical review of Olby's book has been written by Seymour Cohen, "The Origins of Molecular Biology," *Science* 187 (1975): 827-830. Anne Sayre's *Rosalind Franklin and DNA* (New York: W. W. Norton, 1975) is a sober account which goes a long way toward rectifying the image of Franklin created by Watson in *The Double Helix*.

[124] O. T. Avery, C. M. MacLeod, and M. McCarty, "Studies on the Chemical Nature of the Substance Inducing Transformation of Pneumococcal Types," *Journal of Experimental Biology and Medicine* 79 (1944): 137-158.

[125] A. D. Hershey and Martha Chase, "Independent Functions of Viral Proteins and Nucleic Acids in Growth of Bacteria Phage," *Journal of General Physiology* 36 (1952): 39-54.

active tracers, they showed that it was the DNA, not the protein, of phage that supplied the genetic material. DNA must be the molecule of heredity. A year later Watson and Crick proposed their now-famous model for the molecular structure of DNA.[126] That model was able, in one sweep, to account for all the major aspects of heredity that needed explanation in molecular terms: replication, transmission, information transfer, translation of genetic information into metabolic processes (and ultimately the observable phenotype), variation (mutation), and even genetic controls (turning on and off of genes). The work of a host of phage investigators and biochemical geneticists from the 1930s through the 1960s accomplished a revolution in molecular biology equal in import to that accomplished by the *Drosophila* workers a generation earlier. Indeed, when many of the recently accumulated principles of molecular genetics were gathered together in J. D. Watson's *Molecular Biology of the Gene*,[127] that book became to modern genetics what *The Mechanism of Mendelian Heredity* had been to classical genetics.

The revolution in molecular genetics of the past twenty-five years came, as had the revolution of the Morgan group, through the convergence of two separate lines of work. The biochemical genetics of Beadle and Tatum did not stand as a full satisfying explanation without a molecular understanding of how genes actually controlled the production of enzymes. Similarly, the molecular genetics of Watson and Crick did not stand complete without an understanding of how gene products are involved in the cell's metabolic processes (i.e., as enzymes and structural proteins). What is interesting from our point of view is that both lines of work—biochemical and molecular genetics—were nurtured directly by Morgan and the genetics group at Cal Tech (through the work of Ephrussi and Beadle on the one hand and Delbrück on the other). Although many other influences were at work, Morgan provided the atmosphere in which these approaches to heredity could begin to develop. It would be erroneous to suggest that Morgan himself started these new lines of work. But it is true that Ephrussi, Beadle, and Delbrück, among others, all came to Cal Tech specifically to learn about genetics and its relationship to

[126] J. D. Watson and F.H.C. Crick, "Molecular Structure of Nucleic Acids," *Nature* 171 (1953): 964.

[127] New York: Benjamin, 1966; 2d ed. 1970.

physiology. The various workers involved in these new fields developed their own organisms and their own techniques; in this respect perhaps the only debt they owed to their work with the Morgan group was the recognition that *Drosophila* was not the most suitable organism for the next phases of genetic research and that something like *Neurospora* might be. Nonetheless, that the relationship of biochemical and molecular genetics to the *Drosophila* genetics of the Morgan school was both direct and conscious seems clear and unmistakable.

THE FINAL YEARS

Morgan's health was generally good throughout his life. Hence, he was able to keep up his work at a vigorous and consistent pace until shortly before his death. However, in his later years, two accidents made serious inroads into his working schedule and for some time undermined his general health. The first occurred in Los Angeles in October of 1931. On the way to a concert after a dinner party attended by him and his wife, the limousine in which Morgan was riding (Lilian was in another car) was hit broadside by a vehicle coming from a side street. The limousine was turned over, and shattered glass was everywhere. While everyone else was helped out, Morgan remained inside in considerable pain and partially stunned. He found he could not move because of a severe pain in his back. A bystander, a young medical student named Leon Baker, came running over to help. Climbing into the car, he found that a long sliver of glass had penetrated quite far into Morgan's back. Morgan was sufficiently conscious that he discussed with Baker what should be done, and together they concluded that the young man should remove the glass sliver. Baker slowly pulled the glass stiletto out, with a consequent heavy hemorrhage. However, by skillfully placing his thumb over the wound, and exerting constant pressure, Baker was able to prevent a serious loss of blood during the complicated process of removing Morgan from the damaged vehicle and throughout the trip to the hospital.[128] He was admitted to Huntington Memorial Hospital on October 7, remaining until the 11th. Two weeks later, serious hemorrhaging necessitated a return, this time for two

[128] Events summarized by H. K. Morgan, "Notes on Thomas Hunt Morgan's Life," p. 6.

months (until December 22, 1931).[129] On two different occasions during the long recovery period, Morgan encountered complications that almost brought death. But the damage was finally repaired, and he could begin to return to his work after the first of the year. He remained in a considerably weakened condition, however, and was only able to resume a full work schedule after a few months.

Characteristically, Morgan did not forget the critical aid the young medical student had given to him. Morgan learned that Baker was having a difficult time meeting his expenses in medical school. The next fall Baker found, quite to his surprise, that he was the recipient of a new scholarship. Because Morgan left no records on such matters, it cannot be stated with certainty that he was the donor. However, Howard Morgan reports that he once asked his father about the unusual coincidence of this scholarship and got "only a knowing grin."[130] Morgan's gratitude and personal feeling for Baker were considerable. Two years later (1934), when Baker unexpectedly died, Morgan wrote to his parents in Ethel, Missouri:

> I heard a few days ago with great sorrow of the death of your son Leon. He did me a very great service two years ago, at the time of my automobile accident: he acted promptly and intelligently, and in consequence of his help, they got me to the hospital in time to make a blood transfusion. I saw your son at the hospital when I was recovering, and later Mrs. Morgan and I had the pleasure of having him take a meal with us.
>
> He was a very fine boy, and it is very sad that he was not destined to what, I am sure, would have been a fine career.[131]

The other accident occurred in the early 1940s (probably 1941 or 1942) at Woods Hole. Because of gasoline rationing, Morgan, in a mood of independence, said one morning that he would ride his bicycle to the laboratory instead of being driven the three-quarters of a mile. He was over seventy years old, and as a result of the automobile accident of ten years earlier, his sense

[129] From Records of Huntington Memorial Hospital, Pasadena, California, courtesy Josephine Cox, Medical Secretary.

[130] H. K. Morgan, "Notes on Thomas Hunt Morgan's Life," p. 6.

[131] T. H. Morgan to Mr. and Mrs. W. L. Baker, April 14, 1934; Morgan Papers, Cal Tech.

of balance was imperfect. Besides that, he had not ridden a bicycle for many years. Mrs. Morgan and his daughters set up a howl of protest, but as was often characteristic with Morgan, the greater the insistence against doing something, the greater the likelihood that he would do it. So he got on his bicycle and began the ride downhill toward the MBL. Less than a block away, at a sharp turn, Morgan's bike overturned, and he was thrown to the ground unable to get up. He was bruised, but with no apparent injuries. Morgan's daughter, Isabel, who had run behind her father fearing just such a spill, insisted that he not move until Dr. Robert Loeb (son of Morgan's old friend Jacques Loeb) could examine him. Although Loeb normally made no medical calls at all while at Woods Hole (his vacation spot), as a favor he acted as the Morgan's family doctor there, and he came quickly to examine Morgan. He found no overt signs of serious problems, but ordered the "patient" to remain in bed for several weeks. Nothing could have been worse news to Morgan, since it greatly altered his work plans for the summer, but he reluctantly complied. Although again Morgan seemed to recover fully, such shocks at an advanced age weakened his physical condition. He was fortunate, in fact, that during his long and productive career nothing more serious or permanently debilitating than these two accidents took him away from his work.

As work was his life, Morgan never really ceased scientific investigation until virtually a few days before his death. Though occasionally in letters of his later years, especially in the early 1940s, we find references to his "getting old," the spirit and humor that characterized him from his youth on always remained part and parcel of his personality.[132]

By the mid-1940s Morgan could well feel that the old school of which he had been such a champion was passing. Noyes had died in 1936 and Hale in 1938. In addition to Bridges's death in 1938, probably the most saddening loss of that year was his friend E. B. Wilson. We are left with no personal statement from

[132] For example, on February 11, 1939, Morgan wrote to his daughter Isabel about the prospects of selling the barn from the Woods Hole house. There was an advantage, he said, to doing it *now* so as "to have it out of the way when we are gone to heaven." Morgan to Isabel Morgan, February 11, 1939; from the personal papers of Isabel Morgan Mountain. In a letter to Otto Mohr on June 21, 1945, Morgan wrote, "Mrs. Morgan and I are well, but getting older." From the Morgan Papers, APS.

Morgan about his feelings at the end of this long and productive friendship, but in a way it was implicit in their relationship that nothing needed to be said.[133] Though so different in many ways in personality, these two men had a remarkably similar career in biology. From Hopkins to Bryn Mawr to Columbia, and through nearly fifty summers together at Woods Hole, they had developed a bond that superseded their common scientific interests. They were products of the same generation and era. And though Morgan had actively created a new generation of biologists at Cal Tech, they were not in his image. It was with men like Wilson that he shared the most fundamental emotional and conceptual understanding of biological phenomena.

Because of the worsening political situation in Germany and Central Europe in the late 1930s, and especially because of the war after 1940, Morgan lost touch with two of his most cherished European friends, Hans Driesch and Otto Mohr. Contact with Mohr was reestablished after the war for a brief time before Morgan's death. However, he never heard from Driesch again, although the latter survived him by a year.

The summer of 1942 was the final summer the Morgans spent in Woods Hole. Because of gasoline rationing and other wartime transportation problems, travel across the continent became extremely difficult in the next three years. It had been sixty-three years since Morgan had first gone to Woods Hole as a young student from Johns Hopkins. The end of his association with that institution marks symbolically, perhaps, more than anything else the end of the era in which Morgan, Loeb, Wilson, and their associates at MBL had championed the rise and development of experimental and physicochemical biology. Morgan and his wife missed the life at MBL in the summers of 1943, '44, and '45, but contented themselves with visits from those of their children who could make it to the West Coast.

Morgan continued active through the summer and fall of 1945, until a recurrent stomach ulcer in November confined him to the hospital. He died on December 4, 1945 of a ruptured artery at the site of an old ulcer. There was no public ceremony, only the

[133] Morgan wrote Wilson's biography for the National Academy, but it is largely factual and relatively brief, expressing Morgan's appreciation of Wilson's scientific and administrative abilities. See Morgan, "Edmund Beecher Wilson, 1856-1938," *Biographical Memoirs, National Academy of Sciences* 21 (1940): 315-342.

gathering of a few friends in Pasadena. Morgan had lived a full and productive life which had personally and intellectually left its mark in so many ways on all who knew him.

In a brief note to Lilian Morgan, E. G. Conklin expressed perhaps most clearly the sense of loss among Morgan's scientific friends:

> I was overcome with sorrow when I received your message this morning saying that Tom had died yesterday. . . . What a wonderful life he had! What happy memories to you and the children and all his friends and associates that are left! My memories of him go back to our first meeting at Johns Hopkins in the fall of 1888. During all the years since then I have regarded him as my ideal of a scientist and as a sincere friend. Without him Woods Hole and all the places and meetings where we used to see each other will seem lonesome. Indeed, I missed both of you very much last summer for your room at the Laboratory was occupied by strangers.[134]

Lilian Morgan best expressed the sense of personal loss. In response to a letter from L. C. Dunn, who had organized a gathering of Morgan's old friends at Columbia in late December of 1945, she wrote:

> I wish that I might have been present with Mr. Morgan's old and true friends at the informal gathering at Columbia which I am sure your guiding hand made a most fitting memorial. . . .
>
> There is also the appreciation by Mr. Morgan's friends of his wit and humor, without which there could be no real understanding. He kept all those qualities to the very end—as you say they live on in their influence on us. . . .
>
> If there are people in the world (as there are) who can give what we feel we have received (and can keep) from Mr. Morgan, is there not hope that the better and constructive side of human nature may in the end prevail?[135]

It is in the qualities of the interaction of human and intellectual spheres that Morgan has contributed much to our understanding not only of genetics, but also of the process of science. Morgan

[134] Conklin to Lilian V. Morgan, December 5, 1945; Morgan Papers, Cal Tech; quoted with permission.

[135] L. V. Morgan to Dunn, January 1, 1946; Dunn Papers; American Philosophical Society.

was a product of his times, perhaps more than he would have been willing to admit. But he left his imprint on modern biology; he influenced the subsequent course of work in the field in a way that made it quite different from what had gone before. A lucid experimentalist himself, he also possessed that special personal quality that enabled him to work cooperatively. It is in this way, more perhaps than in his technical finds, that he represented the most profound wave of the future.

APPENDICES

Family Genealogy of the Hunt-Morgan Lines

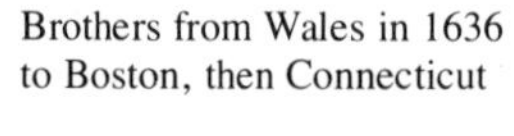

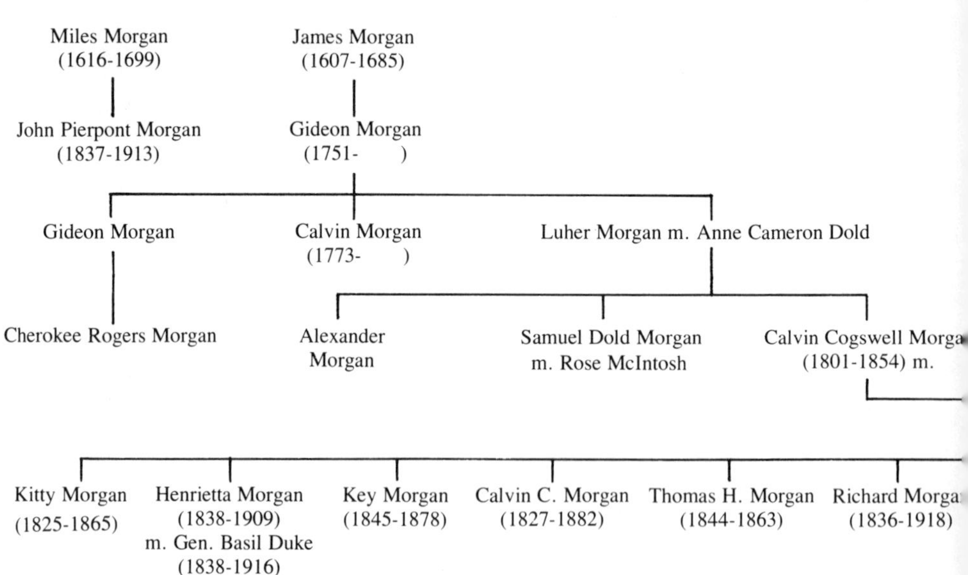

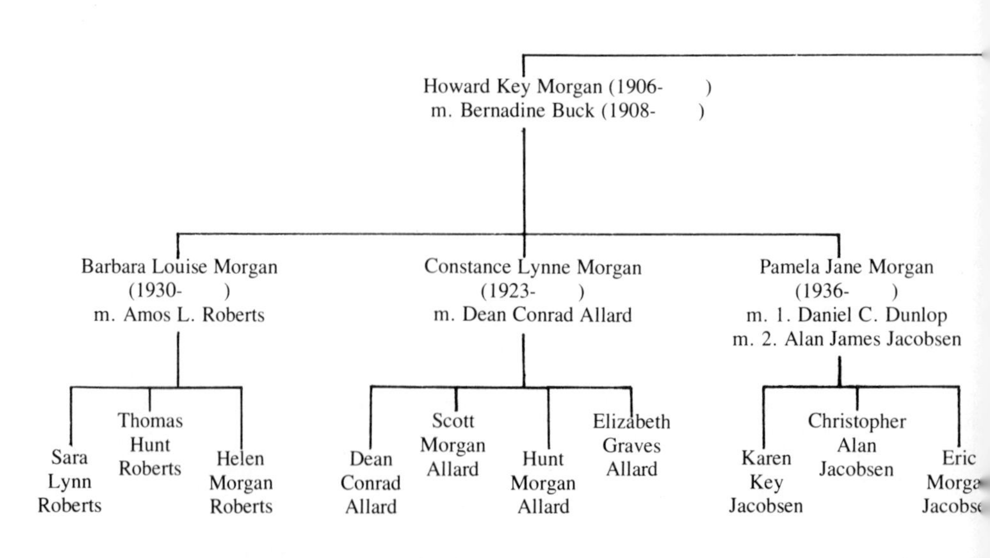

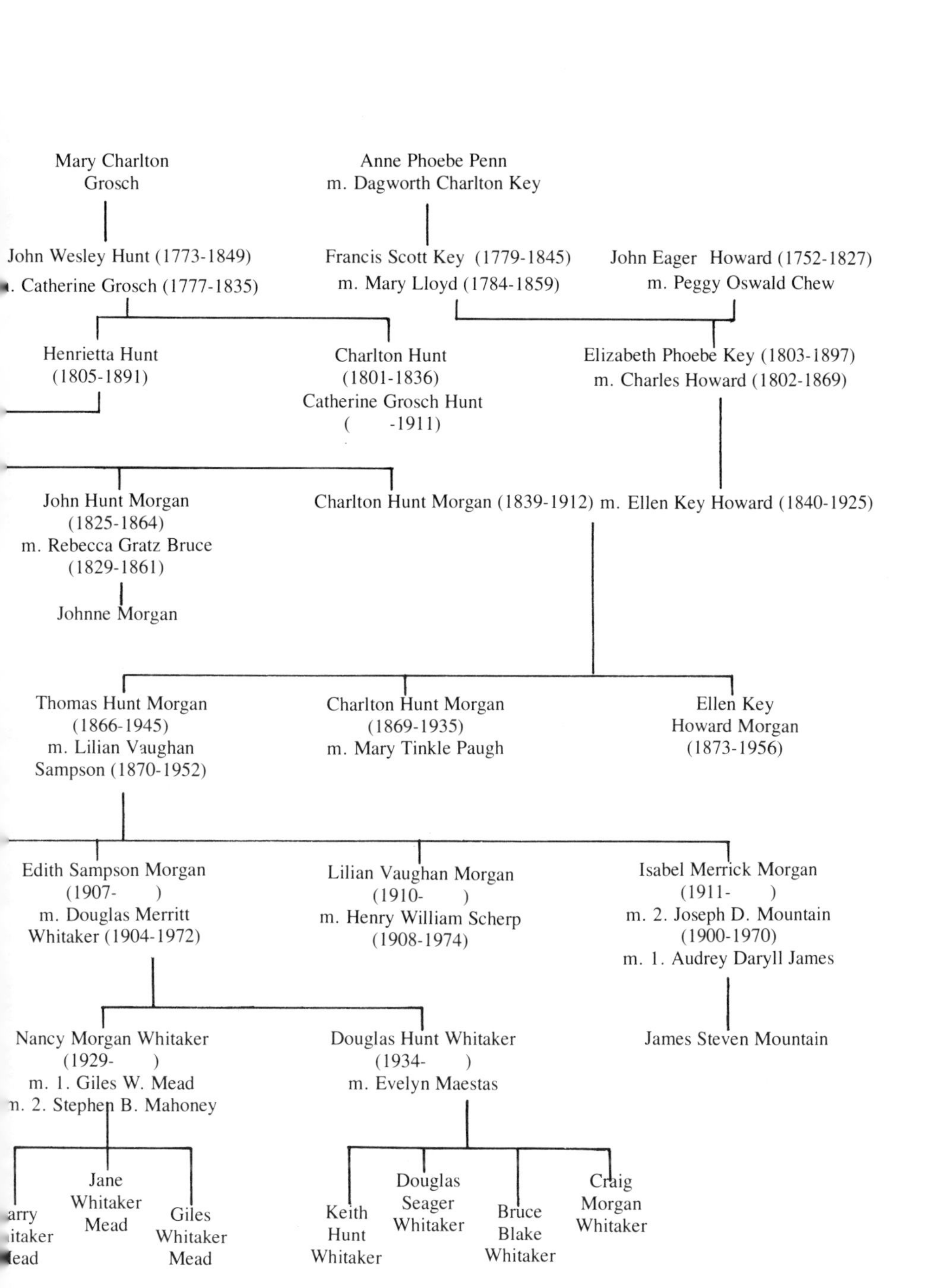

Mary Charlton
Grosch
Anne Phoebe Penn
m. Dagworth Charlton Key
John Wesley Hunt (1773-1849)
. Catherine Grosch (1777-1835)
Francis Scott Key (1779-1845)
m. Mary Lloyd (1784-1859)
John Eager Howard (1752-1827)
m. Peggy Oswald Chew
Henrietta Hunt
(1805-1891)
Charlton Hunt
(1801-1836)
Catherine Grosch Hunt
(-1911)
Elizabeth Phoebe Key (1803-1897)
m. Charles Howard (1802-1869)
John Hunt Morgan
(1825-1864)
m. Rebecca Gratz Bruce
(1829-1861)
Johnne Morgan
Charlton Hunt Morgan (1839-1912) m. Ellen Key Howard (1840-1925)
Thomas Hunt Morgan
(1866-1945)
m. Lilian Vaughan
Sampson (1870-1952)
Charlton Hunt Morgan
(1869-1935)
m. Mary Tinkle Paugh
Ellen Key
Howard Morgan
(1873-1956)
Edith Sampson Morgan
(1907-)
m. Douglas Merritt
Whitaker (1904-1972)
Lilian Vaughan Morgan
(1910-)
m. Henry William Scherp
(1908-1974)
Isabel Merrick Morgan
(1911-)
m. 2. Joseph D. Mountain
(1900-1970)
m. 1. Audrey Daryll James
Nancy Morgan Whitaker
(1929-)
m. 1. Giles W. Mead
n. 2. Stephen B. Mahoney
Douglas Hunt Whitaker
(1934-)
m. Evelyn Maestas
James Steven Mountain
arry
itaker
lead
Jane
Whitaker
Mead
Giles
Whitaker
Mead
Keith
Hunt
Whitaker
Douglas
Seager
Whitaker
Bruce
Blake
Whitaker
Craig
Morgan
Whitaker

APPENDIX 2

Intellectual Genealogy of T. H. Morgan

Morgan's intellectual ancestors stretch back on one side via W. K. Brooks to the naturalist tradition exemplified by Louis Agassiz. On the other side his ancestry stretches back, via H. N. Martin, to the strongly experimentalist tradition of Michael Foster and the European physiologists. Morgan's chief intellectual descendants all developed their mentor's strong experimentalist bias. The group on the lower left were from the Columbia University days and worked largely on breeding and cytological problems in *Drosophila*, influenced (in varying degrees) by both Morgan and E. B. Wilson. The group on the lower right were from Morgan's later career at the California Institute of Technology, and were influenced by Morgan toward genetic problems at the physiological —i.e., biochemical and molecular—levels. It was this group that broke with the use of *Drosophila* as a major experimental organism, and helped introduce microorganisms such as bacteria and fungi into genetics research.

The purpose of an intellectual genealogy is to sketch the major intellectual traditions that come to bear on the development of an individual's outlook. Morgan's genealogy shows clearly the naturalist and experimentalist traditions in his background. Both traditions continued to mix and merge in his approach to biological problems throughout his life. (Modified from A. H. Sturtevant, *A History of Genetics* [New York: Harper and Row, 1965], pp. 140, 143.)

A word or two of caution is in order about the interpretation of intellectual genealogies. They are useful primarily in a broad schematic way, sketching in the general and most obvious lines of influence. Clearly, they do not have the same concrete meaning a family genealogy has. Biologically speaking, there is no doubt that an offspring receives half his or her genetic makeup from one parent and half from the other. In contrast, it is not clear how much a given student received from the teachers who were his or her intellectual parents. In Morgan's case, for example, we have seen that he reacted against many of the ideas and methods of his most direct intellectual parent, W. K. Brooks, and appears to have been more positively influenced, in methodological directions at least, by his more indirect contact with H. N. Martin. The fact that a person studies with a particular teacher does not mean that he or she adopts all, or even any, of that teacher's ideas, methods, or philosophical outlook.

Another problem with intellectual genealogies (and the present one is no exception) is that usually they do not include influences of a nonofficial sort which often have a direct bearing on a given investigator's development. Exclusion of the influence of fellow students, later colleagues, or written materials is usually made for quite practical though arbitrary reasons (there is only so much that can be included on such charts for the sake of space and clarity). In Morgan's case the influence of contemporaries such as Hans Driesch and Jacques Loeb was extremely important, though because he did not study officially with either man, such influences are not shown on the chart. The intellectual genealogy of Morgan, like that of most scientists, is an oversimplified scheme which can be a useful summary once the details are fully known and understood. By itself, it is not a very reliable indicator of how influence spread.

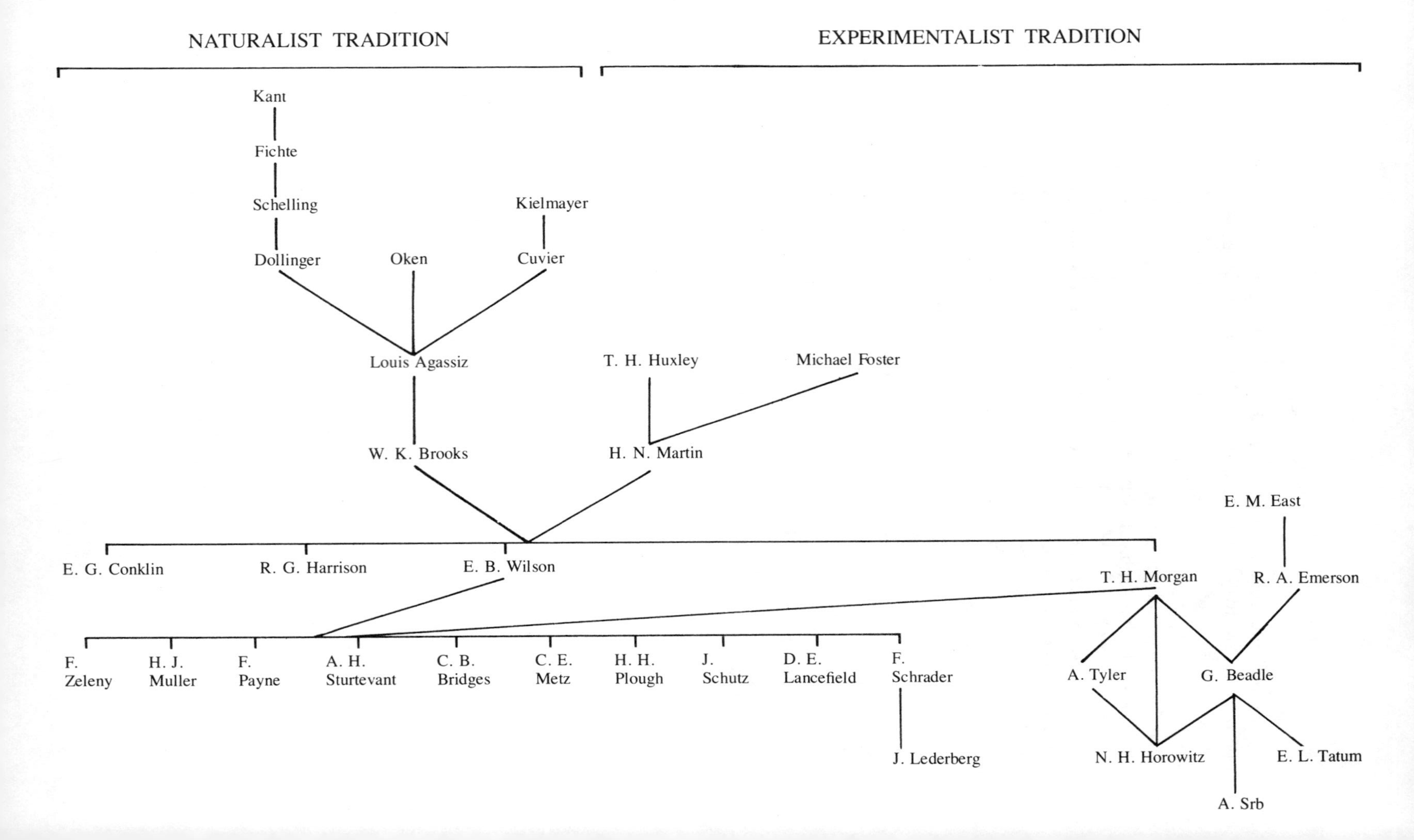

NATURALIST TRADITION
EXPERIMENTALIST TRADITION
Kant
Fichte
Schelling
Kielmayer
Dollinger
Oken
Cuvier
Louis Agassiz
T. H. Huxley
Michael Foster
W. K. Brooks
H. N. Martin
E. M. East
E. G. Conklin
R. G. Harrison
E. B. Wilson
T. H. Morgan
R. A. Emerson
F. Zeleny
H. J. Muller
F. Payne
A. H. Sturtevant
C. B. Bridges
C. E. Metz
H. H. Plough
J. Schutz
D. E. Lancefield
F. Schrader
A. Tyler
G. Beadle
J. Lederberg
N. H. Horowitz
E. L. Tatum
A. Srb

BIBLIOGRAPHICAL ESSAY

I. PRIMARY SOURCES

A. Published Writings of T. H. Morgan

T. H. Morgan was a prolific scholar. His total output consisted of twenty-two books and over 370 papers. Any attempt to list even a significant fraction of this work would be beyond the scope of this essay. A complete list of Morgan's published writings can be found in A. H. Sturtevant's biographical essay. "Thomas Hunt Morgan," in *Biographical Memoirs, National Academy of Sciences* 33 (1959): 283-325. It will be most helpful, here, to list only Morgan's major writings. For easiest reference the listing will be by general topic.

1. *Embryology*

A good view of the morphological studies to which Morgan was exposed as a student at Johns Hopkins can be found in two papers of 1891: his doctoral dissertation, "A Contribution to the Embryology and Phylogeny of the Pycnogonids," *Studies from the Biological Laboratory Johns Hopkins University*, vol. 5, no. 1 (Baltimore: Isaac Friedenwald, 1891), pp. 1-76; and "The Relationships of the Sea-Spiders," in *Biological Lectures Delivered at the Marine Biological Laboratory of Woods Hole in the Summer Session of 1890* (Boston: Ginn & Co., 1891), pp. 142-167. Morgan's early conversion to experimental embryology is mirrored in any number of papers beginning in 1893. Some of the most representative are: "Half-Embryos and Whole-Embryos from One of the First Two Blastomeres of the Frog's Egg," *Anatomische Anzeiger* 10 (1895): 623-628; "The Formation of One Embryo from Two Blastulae," *Roux' Archiv für Entwickelungsmechanik der Organismen* 2 (1895): 65-71; "The Fertilization of Non-Nucleated Fragments of Echinoderm Eggs," *Roux' Archiv* 2 (1895): 268-280; along the same lines are Morgan's two interesting papers written with Hans Driesch: "Zur Analysis der ersten Entwickelungsstadien des Ctenophoreneies. I. Von der Entwickelung einzelner Ctenophorenblastomeren. II. Von der Entwickelung ungefurchter Eier mit Protoplasmadefekten," *Roux' Archiv* 2 (1895): 204-215; 216-224. The strong influence of the Naples

Zoological Station on Morgan's thinking as an experimentalist can be found in his appreciative account, "Impressions of the Naples Zoological Station," *Science* 3 (1896): 16-18; and "The Enlargement of the Naples Station," *Science* 16 (1902): 993-994.

Following experimental paths back in the United States after a year at Naples required that Morgan shift from marine to freshwater or terrestrial organisms. His early papers on regeneration in the earthworm represent the application of principles of *Entwickelungsmechanik* to nonmarine forms: "Regeneration in *Allolobophora foetida*," *Science* 5 (1897): 570-586; and "Experimental Studies of the Regeneration of *Planaria maculata*," *Science* 7 (1898): 364-397. Morgan's work on regeneration from an experimental point of view can be found in two of his Woods Hole lectures, "Some Problems of Regeneration," in *Biological Lectures Delivered at the Marine Biological Laboratory of Woods Hole in the Summer Session of 1897 and 1898* (Boston: Ginn & Co., 1899), pp. 193-207, and "Regeneration: Old and New Interpretations," in *Biological Lectures . . . Summer Session, 1899* (Boston: Ginn & Co., 1900), pp. 185-208; in addition, his first major book, *Regeneration* (New York: Macmillan Co., 1901), admirably brings together many aspects of the problem of regeneration as it relates to the larger issues of development.

Of more general interest is Morgan's article "Developmental Mechanics," *Science* 7 (1898): 156-158, which discusses his views on the new methods of experimental, compared to the old methods of descriptive, embryology. An attempt to provide a mechanism for regeneration can be found in a series of articles on "organic polarity," a term referring to the differential distribution of substances along a gradient throughout an organism or any of its parts: "An Attempt to Analyze the Phenomena of Polarity in Tubularia," *Journal of Experimental Zoology* 1 (1904): 587-591; "An Analysis of the Phenomena of Organic 'Polarity,'" *Science* 20 (1904): 742-748; "Polarity and Regeneration in Plants," *Torrey Botanical Club Bulletin* 31 (1904): 227-230. Lengthy studies of the relationship between normal and abnormal development of the frog's egg appeared in 1904 and 1905 in *Roux' Archiv* 16 (1903), 18 (1904), and 19 (1905). Much work on regeneration and experimental embryology is discussed in the 1907 book *Experimental Zoology* (New York: Macmillan Co.), a masterful summary of much of the literature up to that time, including Morgan's evaluation of current work and future directions. Twenty years later Morgan summarized much of his own, as well as others', embryological work (especially that of the Spemann and Needham-Huxley-de Beer schools), in *Experimental Embryology* (New York: Columbia University Press, 1927). His last series of studies in the field of embryology were concerned with cross- and self-fertilization in Ascidians (especially *Ciona*), published in the *Biological Bulletin* 82 (1942): 161-171; 172-177; 455-460; 88 (1945): 50-62, and in the *Journal of Experimental Zoology* 90 (1942): 199-228; 95 (1944): 37-59; 97 (1944): 231-248; and 100 (1945): 407-416. (Morgan may be unique in the history of the *Journal of Experimental Zoology*, or of any scientific journal for that matter, in having published in both volume 1 and volume 100.)

2. *Evolution*

While Morgan's association with W. K. Brooks laid the groundwork for his serious interest in the larger problems of evolutionary theory, he claimed that it was through his studies on regeneration that he came face to face with Darwinian theory. To Morgan in the period before 1915, Darwinian theory was inadequate to explain the evolution of complex adaptations such as the ability to regenerate lost parts. This view is most thoroughly expressed in *Evolution and Adaptation* (New York: Macmillan Co., 1903), a strong indictment of the theory of natural selection. More succinct and limited criticisms are found in: "Darwinism in the light of Modern Criticism," *Harper's Magazine* 106 (1903): 476-479; "The Origin of Species through Selection Contrasted with Their Origin through the Appearance of Definite Variations," *Popular Science Monthly* 67 (1905): 54-65; "Fifty Years of Darwinism," *Nation* 89 (1909): 145-147; "Chance or Purpose in the Origin and Evolution of Adaptation," *Science* 31 (1910): 201-210. In contrast to the preceding articles, "For Darwin," *Popular Science Monthly* 74 (1909): 367-380 assesses the positive role Morgan felt Darwin's theory had played in the history of biology.

On the controversial issue of the inheritance of acquired characteristics, Morgan generally felt that the point had never been demonstrated in any creditable way. His views on this topic can be found in *Evolution and Adaptation*, chapter 7; "The Influence of Heredity and Environment in Determining the Coat Colors in Mice," *New York Academy of Science Annals* 21 (1911): 87-117; "Are Acquired Characters Inherited," *Yale Review* 13 (1924):

712-729; "The Apparent Inheritance of an Acquired Character and its Explanation," *American Naturalist* 64 (1930): 97-114.

After the work on *Drosophila* had gotten well underway, Morgan returned to the problem of natural selection, interpreted now in light of Mendelian genetics. *A Critique of the Theory of Evolution* (Princeton, N.J.: Princeton University Press, 1916; considerably revised as *Evolution and Genetics* in 1925), Morgan's Louis Clark Vauxeum Foundation Lectures at Princeton in 1915, represents a major turning point in his acceptance of Darwinian theory. Although still lacking a full understanding of either the species question, or the populational aspects of evolutionary thinking, Morgan dropped many of his previous objections (such as the effects of swamping of new variations) to Darwinian theory. More expanded and up-to-date views are presented in Morgan's Messenger Lectures at Cornell in 1931, published as *The Scientific Basis of Evolution* (New York: W. W. Norton, 1932; rev. ed., 1935). Although Morgan was apparently aware of the new work in population genetics in the late 1920s and early 1930s, his later writings on evolution never incorporate any of this approach.

3. *Heredity*

As with his initial interest in evolution, so Morgan's initial interest in heredity came largely through his embryological investigations. In particular, the problem of sex determination, equally as much an issue in heredity as in embryology, led Morgan to the question of how adult traits are determined during development. A summary of the contending schools of thought around the turn of the century is found in Morgan's article "Recent Theories in Regard to the Determination of Sex," *Popular Science Monthly* 64 (1903): 97-116. As an embryologist Morgan was initially suspicious of theories purporting to explain the development of sex by either internal factors (such as chromosomes) or external factors (such as diet and temperature) alone. His agnosticism is evident in several articles prior to 1909, for example, "Ziegler's Theory of Sex Determination and an Alternative Point of View," *Science* 22 (1905): 839-841; and "Sex Determining Factors in Animals," *Science* 25 (1907): 382-384. A substantial switch in his acceptance of the possible role of internal factors came with his own cytological studies of chromosome dispersal in parthenogenetic forms such as rotifers and aphids: "A Biological and Cytological Study of Sex Determination in Phylloxerans and

Aphids," *Journal of Experimental Zoology* 7 (1909): 239-352; Morgan's conversion is even more explicitly stated in "Chromosomes and Heredity," *American Naturalist* 44 (1910): 449-496.

Morgan's earliest views of Mendelism, always skeptical, begin with a single article in 1905: "The Assumed Purity of the Germ Cells in Mendelian Results," *Science* 22 (1905): 877-879; the same theme is reiterated in "Are the Germ-Cells of Mendelian Hybrids 'Pure'?" *Biologische Zentralblatt* 26 (1906): 289-296. Critical responses to Castle's experiments with mice and guinea pigs are voiced in two book reviews: "*Heredity of Coat Characters in Guinea Pigs and Rabbits*, by W. E. Castle (Review)," *Science* 21 (1905): 737-738; and "*Selection and Cross-Breeding in Relation to the Inheritance of Coat-pigments and Coat-patterns in Rats and Guinea Pigs*, by H. MacCurdy and W. E. Castle (Review)," *Science* 26 (1907): 751-752. A critical discussion of the work on "yellow mice" (what amounted to the discovery of a lethal gene) by the French Mendelian Lucien Cuénot is found in "Some Experiments in Heredity in Mice (Abstract)," *Science* 27 (1908): 493. Morgan's most scathing and direct attack on Mendelism is the 1909 article, "What Are 'Factors' in Mendelian Explanations?" *American Breeders' Association Report* 6 (1909): 365-368.

The first published paper on the white-eye *Drosophila* is "Sex Limited Inheritance in *Drosophila,*" *Science* 32 (1910): 120-122; in this article Morgan adopts a limited Mendelian explanation for the inheritance pattern of white eye. In another paper of the same year sex linkage is explained more fully: "The Method of Inheritance of Two Sex-Limited Characters in the Same Animal (Abstract)," *Society of Experimental Biology and Medicine Proceedings* 8 (1910): 17-19. A whole series of articles followed in the next several years: "The Origin of Nine-Wing Mutations in *Drosophila,*" *Science* 33 (1911): 496-499; "The Origin of Five Mutations in Eye Color in *Drosophila* and Their Modes of Inheritance," *Science* 33 (1911): 534-537; "The Explanation of a New Sex Ratio in *Drosophila,*" *Science* 36 (1912): 718-719. The early work on sex inheritance and sex linkage is well summarized in *Heredity and Sex* (New York: Columbia University Press, 1913), Morgan's Jessup Lectures at Columbia in 1913.

The first thorough analysis of lethal factors by Morgan is "Two Sex-Linked Lethal Factors in *Drosophila* and Their Influence on the Sex-Ratio," *Journal of Experimental Zoology* 17 (1914): 81-

122; a brief summary of the *Drosophila* work is given nontechnically in "The Mechanism of Heredity as Indicated by the Inheritance of Linked Characters," *Popular Science Monthly* 84 (1914): 5-16. The book that made such an impact on many of Morgan's contemporaries is *The Mechanism of Mendelian Heredity*, with A. H. Sturtevant, H. J. Muller, and C. B. Bridges (New York: Henry Holt & Co., 1915). In 1919 Morgan published an expanded and updated summary, *The Physical Basis of Heredity* (New York: Lippincott), as part of the Lippincott Monographs on Experimental Biology. A good description of the methods of mapping and determination of the linear and spatial arrangement of genes is found in: "The Evidence for the Linear Order of the Genes," with A. H. Sturtevant and C. B. Bridges, *Proceedings of the National Academy of Sciences* 6 (1920): 162-164. Morgan's most complete summary of the Mendelian-chromosome theory as developed in the first fifteen years of the work with *Drosophila* is in his Hepsa Ely Silliman Lectures at Yale, published as *The Theory of the Gene* (New Haven, Conn.: Yale University Press, 1926; rev. ed. 1928).

The central role of genetics in the study of other areas of biology is discussed in several articles and books: "Some Possible Bearings of Genetics on Pathology," the Middleton Goldsmith Lectures (Lancaster, Pa.: New Era Printing Co., 1922). Morgan made a relatively unsuccessful attempt to explore the relationship between genetics and embryology in the book, *Embryology and Genetics* (New York: Columbia University Press, 1934); on this same subject see the briefer summary, "Genetics and the Physiology of Development," *American Naturalist* 60 (1926): 489-515; and "The Modern Theory of Genetics and the Problem of Embryonic Development," *Physiological Review* 3 (1923): 603-627. In relation to other sciences, see "The Bearing of Mendelism on the Origin of Species," *Scientific Monthly* 16 (1923): 237-247; "Mendelian Heredity in Relation to Cytology," in *General Cytology*, ed. E. V. Cowdry (Chicago: University of Chicago Press, 1924); and Morgan's presidential address to the Sixth International Congress of Genetics in 1932: "The Rise of Genetics," *Science* 76 (1932): 261-267; also published in *Proceedings, Sixth International Congress of Genetics* 1 (1932): 87-103. The specific relationship between genetics and physiology is featured in Morgan's Nobel Lecture, "The Relation of Genetics to Physiology and Medicine," *Scientific Monthly* 41 (1935): 5-18.

4. *Method and Philosophy*

Morgan was not given to much overt philosophizing, so that many of his views on the philosophy of science must be gleaned from writings devoted primarily to other subjects. The importance of approaching problems from an experimental point of view is adequately summarized in Morgan's unsigned review of Jacques Loeb's *Comparative Physiology and Comparative Psychology* appearing in *The Independent* 53 (1901): 1,564. Morgan's strong opposition to vitalistic explanations emerges in his review of Otto Bütschli's *Mechanism and Vitalism* in *American Naturalist* 36 (1902): 154-156. Although appreciative of Driesch's efforts to explore the philosophical side of biology, Morgan's own strongly empirical philosophy is clearly stated in his review of Driesch's *The Science and Philosophy of the Organism* in *Journal of Philosophy, Psychology and Scientific Method* 6 (1909): 101-105.

Morgan's strong conviction that biology must be placed on the same footing as physics and chemistry is clearly stated in "The Relation of Physics to Biology," in *Physics and Its Relations* (Poughkeepsie, N.Y.: Vassar College, 1927), pp. 7-24. An article titled "The Genetics Controversy," by Lucretius Smith in *Soviet Russia Today* 8 (1940): 10-13 is erroneously attributed to Morgan in Sturtevant's bibliography.

B. Unpublished Writings of T. H. Morgan

Unfortunately Morgan did not save his correspondence in any systematic way; according to Sturtevant, he cleaned out his file cabinets every five years, disposing of all, or nearly all, items. Several archives contain materials collected by Morgan himself, but most have been gleaned from letters and manuscripts found in the collections of other people, mostly scientific colleagues. In describing the various archival sources used in preparation of this biography, I will give some indication of the extent and usefulness of each collection. Where more detailed descriptions of the collections have been published, the appropriate reference will be given.

There are two major collections of the papers of T. H. Morgan. One is at the American Philosophical Society (APS) in Philadelphia, and the other at the California Institute of Technology (Cal Tech). The APS collection consists of thirty-one items, mostly letters to Otto L. Mohr written between 1919 and 1945 (presented to the APS by Otto Mohr in 1963). More generally helpful has been the larger collection, the Thomas Hunt Morgan Papers, in the Archives of Cal Tech. Although most of the material at Cal Tech comes from Morgan's later years, a number of items from earlier periods are also included (for example, a copy of Morgan's valedictory address at Kentucky State College in 1886). The Cal Tech collection also includes a number of photographs and other items documenting family as well as scientific history.

The APS genetics collection, whose initial development was guided by L. C. Dunn, has become the most extensive and systematic archival source for the history of twentieth-century genetics—with a slant, it should be added, toward American genetics—anywhere in the world. Descriptions of many of these holdings can be found in *The Mendel Newsletter*, issued by the library of the APS. Letters to and from Morgan can be found in a number of collections now housed at the APS: the Bateson Papers (see William Coleman, "Bateson Papers," *Mendel Newsletter*, no. 2 [1968; 1-3], A. F. Blakeslee Papers, Charles B. Davenport Papers (see L. C. Dunn, "The Davenport Papers," *Mendel Newsletter*, no. 6 1970: 1-2), Milislav Demerec Papers, L. C. Dunn Papers, H. D. Goodale Papers, and the H. S. Jennings Papers. In addition, much information relating directly or indirectly to the work of the Morgan group can be found in the Papers of the Genetics Society of America, and the Papers of the Department of Genetics, University of California, Berkeley, both housed at the APS. Most interesting in the latter collection are notebooks and lecture notes from genetics courses at Berkeley as early as 1914 (the lecture notes preserved are those of E. B. Babcock, with whom Morgan spent part of his sabbatical year, 1920-1921).

Some valuable letters of the Morgan family, which throw light on T. H. Morgan's early days, are found in the Hunt-Morgan Family Papers, 1784-1949, at the Margaret I. King Library at the University of Kentucky. Of particular interest are some letters from T. H. Morgan and Lilian V. Morgan to Morgan's mother, Ellen Key Howard Morgan, written principally between 1904 and 1907. One letter (from Morgan to Ballard Thurston) is now held in the archives of the Filson Club, Louisville, Kentucky.

A large number of letters to and from Morgan are found in the E. G. Conklin Papers, Firestone Library, Princeton University (see Garland Allen, "The Conklin Papers," *Mendel Newsletter*, no. 1 [1968]: 3-4); a more complete description is found in Garland Allen and Dennis McCullough, "Notes on Source Materials:

the Edwin Grant Conklin Papers at Princeton University," *Journal of the History of Biology* 1 (1968): 325-331. Similarly, Morgan's correspondence with Ross Harrison was voluminous. Much of it has been preserved in the Harrison Papers, Sterling Library, Yale University, New Haven, Connecticut. An extensive collection of Morgan letters appears in the papers of Frank Zeleny at the University of Illinois, Champaign-Urbana (see M. J. Brichford, "University of Illinois Mendel and Zeleny Papers," *Mendel Newsletter*, no. 3 [1969]: 2). Not surprisingly, given the relationship between Morgan and H. J. Muller, relatively few letters are found in the Muller Collection, Lilly Library, Indiana University, Bloomington (see E. A. Carlson, "Indiana University: the Muller Archives," *Mendel Newsletter*, no. 4 [1969]: 1-2). A few letters between Morgan and E. B. Wilson are in the possession of Wilson's daughter, Nancy Lobb, of South Hadley, Massachusetts. Many letters between Morgan and Jacques Loeb are in the Loeb Papers, Library of Congress (see Nathan Reingold, "Jacques Loeb, the Scientist: His Papers and His Era," *Library of Congress Quarterly Journal of Current Acquisitions* 19 [1962]: 119-130). Letters from Morgan to his close friend Hans Driesch are now housed at the Leipzig University Library. A microfilm copy of these letters has been supplied to me by Professor Reinhard Mocek, and a copy of this microfilm has recently been deposited at the library of the APS.

A number of archival sources provide valuable information about Morgan's era, although containing little if any of his direct correspondence. Among these are the library of the Marine Biological Laboratory, Woods Hole, Massachusetts, which contains two collections: the institutional archives of the MBL (see Garland Allen, "The M.B.L. Library," *Mendel Newsletter*, no. 2 [1968]: 3-4) and the F. R. Lillie Papers (Garland Allen, "The F. R. Lillie Papers in Woods Hole, Massachusetts," *Mendel Newsletter*, no. 9 [1973]: 1-6). The Papers of Richard Goldschmidt at the University of California, Berkeley, contain much information on Goldschmidt's views of genetics, principally his opposition to the Morgan school in the period after his arrival in the United States in 1936 (see Garland Allen, "Goldschmidt Papers, University of California, Berkeley," *Mendel Newsletter*, no. 7 [1971]: 1-7). A number of Goldschmidt's letters are also in the personal files of Emeritus Professor Curt Stern, University of California, Berkeley.

Personal letters from Morgan to members of his family are in the private possession of several of his children and former associates. Isabel Morgan Mountain, Morgan's youngest daughter, has made available copies of her father's letters to her, written mostly between 1930 and 1945. Tove Mohr, of Fredrikstad, Norway, the wife of Otto L. Mohr, has supplied letters to her from T. H. Morgan, Lilian Morgan, the Morgan children (including Howard K. Morgan's account of the family's train trip across the continent in 1920), and other associates from the fly-room days, such as Calvin Bridges and H. J. Muller. Dr. Mohr also provided her diaries from the year at Columbia (1918-1919), and from visits with the Morgans both in Norway and in the United States (1932, 1933).

II. SECONDARY SOURCES

A. General Biology in Nineteenth and Early Twentieth Centuries

There are no modern histories of biology in general. Erik Nordenskiold's *The History of Biology* (New York: Tudor Publishing Co., 1928), covers much of the nineteenth- and early twentieth-century background in a rudimentary way. More directly useful as background for Morgan's era in biology is William Coleman's *Biology in the Nineteenth Century* (New York: John Wiley, 1971). This admirable book, while selectively treating certain trends in the nineteenth century, focuses thoroughly on evolutionary and hereditary theory and the growth of cytology. Garland E. Allen's *Life Science in the Twentieth Century* (New York: John Wiley, 1975; Cambridge, The University Press, 1978) describes the growth of genetics, evolutionary theory, and embryology, as well as general physiology and biochemistry, from the 1880s through the 1960s. Together, Coleman's and Allen's books cover certain highlights in the development of biology in general, and evolutionary and genetic theories in particular, over the past 150 years.

Because so much of Morgan's early research was related to the interaction between the morphological and experimental traditions, some further reading may be desired on these subjects. The general development of experimentalism has been treated in a very readable form in Elizabeth Gasking's *The Rise of Experimental Biology* (New York: Random House, 1970). Gasking traces experimentalism from the seventeenth through the end of the nineteenth centuries. A glimpse at what late nineteenth-century biologists thought morphology to be is found in Patrick Geddes's article, "Morphology," for the ninth edition of the *Encyclopedia Britannica* (1883), vol. 16, pp. 837-846. A more expanded treatment can be found in E. S. Russell's *Form and Function,* first published in 1916 and now available in a reprint edition with an introduction by William Coleman (New York: Johnson Reprint Corporation, 1967).

Valuable background to Morgan's own era can be found in the biography of his contemporary, William Morton Wheeler: *William Morton Wheeler, Biologist* (Cambridge, Mass.: Harvard University Press, 1970) by Mary Alice and Howard Ensign Evans. This is a useful biographical sketch of an investigator who was much involved in the study of evolution and heredity from the early 1900s through the early 1940s.

B. Evolutionary Theory, 1880-1910

Many historical studies have been devoted to the history of Darwinism, focusing mostly on factors contributing to the development of, and immediate reaction to, Darwin's theory of natural selection. Considerably less effort has been focused on the history of evolutionary theory after the 1870s or 1880s. Although dated, several early studies provide some insight into the spectrum of scientific reaction to Darwinism in the late nineteenth and early twentieth centuries—the very period in which Morgan was entering on his scientific career. Vernon L. Kellogg's *Darwinism Today* (New York: Henry Holt & Co., 1907), contains much useful material summarizing the enormous pro- and anti-Darwinian literature of the period up through 1905.

Although prejudiced toward the neo-Lamarckian viewpoint P. G. Fothergill's *Historical Aspects of Organic Evolution* (New York: Philosophical Library, 1953), and Yves Delage and Marie Goldsmith's *The Theories of Evolution*, translated by A. Tridon (New York: B. W. Huebsch, 1912) give a good picture of the debates surrounding the evolutionary significance of the inheritance of acquired characteristics.

On the prominence of neo-Lamarckism, against which Morgan reacted so strongly, see: Edward J. Pfeiffer, "The Genesis of American Neo-Lamarckism," *Isis* 56 (1965): 155-167; George

W. Stocking, "Lamarckism in American Social Science, 1890-1915," *Journal of the History of Ideas* 23 (1962): 239-256; and Nathaniel Brown, "Neo-Lamarckism versus Neo-Darwinism: A Controversy over the Inheritance of Acquired Characters" (B. A. thesis, Harvard University, 1961). Arthur Koestler's *The Case of the Midwife Toad* (London: Hutchinson, 1971), although biased heavily in favor of its central figure, presents a detailed historical account of the neo-Lamarckian controversy surrounding Austrian biologist Paul Kammerer in the 1920s. Kammerer was thought to have presented faked results purporting to demonstrate the supposed inheritance of an acquired trait. Frederick Churchill has prepared a valuable study of the notion of the inheritance of acquired characters in medical circles in the latter part of the nineteenth century: "Rudolf Virchow and the Pathologist's Criteria for the Inheritance of Acquired Characteristics," *Journal of the History of Medicine and Allied Sciences* 31 (1976): 117-148.

A specific example of opposition to Darwinian theory is found in the mutation theory of Hugo de Vries, popular between 1900 and 1910. De Vries's work is analyzed in several papers. Garland Allen's "Hugo de Vries and the Reception of the Mutation Theory," *Journal of the History of Biology* 2 (1969): 55-87, documents how de Vries's theory was accepted as an alternative to Darwin's concept of natural selection. Peter W. vander Pas, in "The Correspondence of Hugo de Vries and Charles Darwin," *Janus* 57 (1970): 173-213, describes de Vries's admiration for Darwin and the influence of Darwin on the young de Vries's interest in heredity and evolution. An up-to-date discussion of de Vries's work in historical perspective is found in Ralph Cleland's posthumous *Oenothera: Cytogenetics and Evolution* (New York: Academic Press, 1972), especially chapters 1-3.

More special aspects of the growth of evolutionary theory in the late nineteenth and early twentieth centuries are found in several shorter papers. Stephen Gould's "Dollo on Dollo's Law: Irreversibility and the Status of Evolutionary Laws," *Journal of the History of Biology* 3 (1970): 189-212, analyzes the pressing question of the 1920s: whether or not evolutionary change, like human historical change, is unique and thus nonrepeatable. In a very thorough and well-presented essay, "The Role of Isolation in Evolution: George J. Romanes and John T. Gulick," *Isis* 66 (1975): 483-503, John E. Lesch analyzes the growth of ideas about geographic, reproductive, and physiological isolation in natural speciation.

Studies of the growth of biometry and population genetics, as the converging interface between the fields of heredity and evolution, have become increasingly prominent in recent years. Although initially anti-Mendelian, the biometrical movement, through its younger workers, eventually overcame its bias and adopted the Mendelian scheme; the fusion of the two fields was accomplished in the late 1920s and early 1930s by R. A. Fisher, J.B.S. Haldane, Sewell Wright, and others in both England and the United States. The whole development of this field, from the earliest biometricians through classical population genetics in Great Britain has been discussed admirably by William Provine in *The Origins of Theoretical Population Genetics* (Chicago: University of Chicago Press, 1971). Provine includes some detailed discussion of the biometrical movement in chapters 1-3.

Specific studies of the biometrical movement itself include: Franz Weiling, "Quellen und Impulse in der Entwicklung der Biometrie," *Sudhoffs Archiv. Zeitschrift für Wissenschaftsgeschichte* 53 (1969): 306-325; R. G. Swinburne, "Galton's Law—Formulation and Development," *Annals of Science* 21 (1965): 15-31; P. Froggart and N. C. Nevin, "The 'Law of Ancestral Heredity' and the Mendelian-Ancestrian Controversy in England, 1889-1906," *Journal of Medical Genetics* 8 (1971): 1-36; the same authors' "Galton's 'Law of Ancestral Heredity': Its Influence on the Early Development of Human Genetics," *History of Science* 10 (1971): 1-27; Ruth Cowan, "Francis Galton's Contribution to Genetics," *Journal of the History of Biology* 5 (1972): 389-412; and, by the same author, "Galton and the Continuity of the Germ Plasm: A Biological Idea with Political Roots," *XIIe Congrès International d'Histoire des Sciences, Actes* (Paris: Albert Blanchard, Libraire Scientifique et Technique, 1968), pp. 181-186. All these papers give some insight into the hereditary problems that biometricians were attempting to solve prior to the rediscovery of, or outside the context of, Mendelian theory.

Of particular importance as background for the early development of Mendelian theory, is the lengthy debate between the biometricians and Mendelians in England between 1900 and 1910. Although coming from a very different direction, the biometricians advanced many of the same arguments against Mendelian theory as Morgan and others in the United States. The debate has been discussed in several sources. Provine devoted chapters 2 and 3 of *The Origins of Theoretical Population Genetics* to the controversy and its background. Several recent monographs approach the debate from different perspectives. A. G. Cock's "William Bateson, Mendelism and Biometry," *Journal of the History of Biology* 6 (1973): 1-36, explores Bateson's role (as an antibiometrician) in retarding the fusion of statistical and Mendelian thinking. Lyndsay A. Farrall, "Controversy and Conflict in Science: A Case Study—the English Biometric School and Mendel's Laws," *Social Studies of Science* 5 (1975): 269-301, provides a sociological analysis of the controversy (largely opposing the Mertonian view of conflict in science). A similar sociological analysis by D. A. MacKenzie and S. B. Barnes, "Historical and Sociological Analyses of Scientific Change: The Case of the Mendelian-Biometrician Controversy in England," is found in *Kölner Zeitschrift für Soziologie und Socialpsychologie*, in press. An older, unpublished work covering much of the same ground, but focusing more on the Mendelian side, is Morton Arnsdorf, "The Great Controversy: The Introduction of Mendelism into England" (B.A. thesis, Harvard University, 1962).

From the work of Provine and others, the impression is given that population genetics developed solely in England, with a small bit of help from Sewell Wright in the United States. That this was not the case is demonstrated in two very fine papers by Mark Adams on the Russian school of population genetics in the 1920s and 1930s: "The Founding of Population Genetics: Contributions of the Chetverikov School, 1924-1934," *Journal of the History of Biology* 1 (1968): 23-39; and "Toward a Synthesis: Population Concepts in Russian Evolutionary Thought, 1925-1935," *Journal of the History of Biology* 3 (1970): 107-129. A review of the historiography of population genetics is found in Ernst Mayr's essay review, "The Recent Historiography of Genetics," *Journal of the History of Biology* 6 (1973): 125-154.

C. The Study of Heredity

Works summarizing the general history of theories of heredity from antiquity up to around 1900 include: L. C. Dunn's *A Short History of Genetics* (New York: McGraw-Hill, 1965) and A. H. Sturtevant's *A History of Genetics* (New York: Harper & Row,

1965), Hans Stubbe's thorough *Kurze Geschichte der Genetik bis zur Wiederentdeckung der Vererbungsregeln Gregor Mendels*, 2d ed. (Jena: Gustav Fischer, 1965), translated as *History of Genetics from Prehistoric Times to the Rediscovery of Mendel's Laws* by T.R.W. Waters (Cambridge, Mass.: M.I.T. Press, 1973), contains a wealth of information, albeit in covering so much ground it necessarily is superficial in many aspects. Stubbe's book by and large supersedes the older *Beginnings of Plant Hybridization* by Conway Zirkle (Philadelphia: University of Pennsylvania Press, 1935) and *Plant Hybridization Before Mendel* by H. F. Roberts (Princeton, N.J.: Princeton University Press, 1929). Despite its title, Ernst Mayr's essay, "The Recent Historiography of Genetics," focuses exclusively on the history of population genetics, and ignores all recent work in the history of classical or biochemical and molecular genetics.

Historical investigations of many of the theories of heredity prominent in the late nineteenth century are sparse, despite the obvious importance of such theories in conditioning reaction to Mendel after 1900. L. C. Dunn has summarized some of the more well-known particulate theories of heredity in chapter 3 of his *A Short History of Genetics*; this chapter has been reprinted as "Ideas about Living Units, 1864-1909: A Chapter in the History of Genetics," *Perspectives in Biology and Medicine* 8 (1965): 335-346. Particularly valuable is Frederick Churchill's characteristically thorough "August Weismann and a Break from Tradition," *Journal of the History of Biology* 1 (1968): 91-112. Peter Vorzimmer has presented a detailed study of Darwin's theories of inheritance, on which many late nineteenth-century particulate theories were in some degree modeled: "Charles Darwin and Blending Inheritance," *Isis* 54 (1963): 371-390. A very useful, though unpublished, work is Gloria Robinson's "Theories of a Material Substance of Heredity: Darwin to Weismann" (Ph.D. dissertation, Yale University, 1969). Robinson's admirable thesis summarizes in detail various of the speculative schemes against which Morgan and his generation argued so strongly.

Much has been written about the genesis of Mendel's ideas and the factors surrounding their rediscovery by Correns, Tschermak, and de Vires in 1900. A useful source is Robert Olby's *Origins of Mendelism* (New York: Schocken Books, 1966); he assesses both Mendel's work and the impact of its rediscovery on the first decade of the twentieth century. In addition, much of the current research on Mendel's life, times, and work can be found in articles in the periodical, *Folia Mendeliana*, published by the Moravian Museum, Brno, Czechoslovakia. Also of interest is L. C. Dunn's "Mendel, His Work, and His Place in History," *Proceedings of the American Philosophical Society* 109 (1965): 189-198. The nature of the rediscovery of Mendel's work and its significance for the early twentieth century is traced in several articles. Among the best is Elizabeth Gasking's "Why Was Mendel's Work Ignored?" *Journal of the History of Ideas* 20 (1959): 60-84, which outlines many of the factors that may have contributed to the general indifference most plant breeders and evolutionists displayed toward Mendel between 1865 and 1900. Conway Zirkle, in "Some Oddities in the Delayed Discovery of Mendelism," *Journal of Heredity* 55 (1964): 65-72, suggests that statistical thinking was required before Mendel's ideas of segregation and random assortment could be adequately appreciated. The factors that may have led to the simultaneous rediscovery of Mendel by three different individuals are analyzed carefully by J. S. Wilkie in "Some Reason for the Rediscovery and Appreciation of Mendel's Work in the First Years of the Present Century," *British Journal for the History of Science* 1 (1962): 5-18.

The early years of the establishment of Mendelian ideas in the twentieth century has been the subject of several studies. R. C. Punnett's "Early Days of Genetics," *Heredity* 4 (1950): 1-10 provides an amusing and illuminating account of Bateson's struggle to establish Mendelian theory in England between 1900 and 1906. A similar, though less vibrant, account of the American scene around 1900 is found in W. E. Castle's "The Beginnings of Mendelism in America," in *Genetics in the Twentieth Century*, ed. L. C. Dunn (New York: McGraw-Hill, 1950), pp. 59-76. In the same volume Jay Lush's "Genetics and Animal Breeding," pp. 493-510, gives some indication of the degree to which animal breeders were prepared, from their own work, to appreciate the significance of Mendel's results. A. H. Sturtevant has provided an interesting analysis of the kinds of people who became attracted to Mendelism in the early years of the century: "The Early Mendelians," *Proceedings of the American Philosophical Society* 109 (1965): 199-204. In a very suggestive article, "Factors in the Development of Genetics in the United States: Some Suggestions," *Journal of the History of Medicine and Allied Sciences* 22 (1967): 27-46, Charles Rosenberg explores the reaction to Mendelism among three groups of workers: practical breeders, physicians, and university biologists. More recently Denis Buican has explored the development of classical genetics in France in "Sur le dévélopment de la génétique classique en France," *Revue Synth* 94 (1974): 231-241.

An important twentieth-century development was the convergence of Mendelian breeding results with cytological studies of chromosomes and their behavior. Some aspects of the cytological background prior to the work of the Morgan school in 1910 can be found in several articles. Among the older, but still valuable, works is John R. Baker's "The Cell Theory: A Restatement, History and Critique. Part V. The Multiplication of Nuclei," *Quarterly Journal of Microscopical Science* 96 (1955): 449-481. Frederick Churchill's "Hertwig, Weismann, and the Meaning of Reduction Division circa 1890," *Isis* (1970): 429-457, deals primarily with the problem of meiosis during germ-cell formation, as it was viewed during the later years of the nineteenth century. William Coleman's "Cell, Nucleus and Inheritance: An Historical Study," *Proceedings of the American Philosophical Society* 109 (1965): 124-158, traces the growth of the idea that the cell nucleus, and chromosomes in particular, control heredity.

The work of E. B. Wilson was extremely important in laying the groundwork for the later association of chromosomes with Mendelian genes. His work has been assessed in two sources: H. J. Muller's introduction to the reprint edition of Wilson's *The Cell in Development and Inheritance* (New York: Johnson Reprint Corporation, 1966) as well as his earlier essay, "Edmund B. Wilson—an Appreciation," *American Naturalist* 77 (1943): 142-172, are both thorough, detailed studies of all Wilson's major work, as well as laudatory testaments to Muller's appreciation of his mentor. More recently Alice Levine Baxter has published an excellent analysis of the factors that led Wilson to accept the chromosome theory earlier than many of his American contemporaries: "Edmund B. Wilson as a Preformationist: Some Reasons for His Acceptance of the Chromosome Theory," *Journal of the History of Biology* 9 (1976): 29-57. A general survey of Wilson's life and work is found in Garland Allen's biographical essay in the *Dictionary of Scientific Biography* (New York: Charles Scribner's Sons, 1976), vol. 14, pp. 423-436.

There are as yet few detailed studies of later (post-1915) developments in the Mendelian-chromosome theory. Among the best summaries are the one found in E. A. Carlson's *The Gene: A Critical History* (Philadelphia: W. B. Saunders, 1966) and the briefer ones found in Sturtevant's and Dunn's histories of genetics. A good summary of some critical developments in the Mendelian-chromosome theory between 1915 and 1950 is found in H. J. Muller's "The Development of the Gene Theory," in *Genetics in the Twentieth Century*, ed. L. C. Dunn. Also valuable are H.L.K. Whitehouse's *The Mechanism of Heredity* (London: Arnold, 1965) and John Moore's *Heredity and Development* (New York: Oxford University Press, 1963). Both books approach the problems of heredity from a historical point of view, and thus contain excellent summaries of some of the technical issues in the development of the Mendelian and chromosome theories.

Opposition to the Mendelian and chromosome theories has been studied in three papers. Frederick Churchill's outstanding account, "Wilhelm Johannsen and the Genotype Concept," *Journal of the History of Biology* 7 (1974): 4-30, discusses Johannsen's genotype-phenotype distinction and his opposition to the identification of genes with material bodies in the cell. Similar opposition to the materialistic implications of the Mendelian-chromosome theory was echoed by William Bateson and Richard Goldschmidt. William Coleman has thoroughly explored Bateson's idealism with regard to chromosomes in "Bateson and Chromosomes: Conservative Thought in Science," *Centaurus* 15 (1970): 228-314. Garland Allen has discussed Goldschmidt's numerous objections to the work of the Morgan school in "Richard Goldschmidt's Opposition to the Mendelian-Chromosome Theory," *Folia Mendeliana* 6 (1970): 229-303, and in a more expanded and detailed form in "Opposition to the Mendelian-Chromosome Theory: The Physiological and Developmental Genetics of Richard Goldschmidt," *Journal of the History of Biology* 7 (1974): 49-92.

D. The Study of Embryology

Although no complete histories of modern embryology (post-1880) have yet been written, a number of excellent shorter monographs exist, describing the growth of various aspects of this discipline. Of particular interest are studies dealing with the development of experimental embryology in the period from 1870 through the 1930s. A collection of essays by Jane Oppenheimer,

Essays in the History of Embryology and Biology (Cambridge, Mass.: M.I.T. Press, 1967), written primarily in the 1950s and 1960s, provides background for particular aspects of Morgan's work. The most relevant essays are: "Embryological Concepts in the Twentieth Century" (pp. 1-61); "Questions Posed by Classical Descriptive and Experimental Embryology" (pp. 62-91); "Ross Harrison's Contributions to Experimental Embryology" (pp. 92-116); and "Embryology and Evolution: Nineteenth Century Hopes and Twentieth Century Realities" (pp. 206-221). Oppenheimer has more recently prepared an insightful introduction to the second edition of *Foundations of Experimental Embryology*, ed. B. H. Willier and Jane M. Oppenheimer (New York: Macmillan Co., 1974). A collection of historical papers in experimental embryology (by Roux, Driesch, Spemann, etc.), this book, and its introduction, provide a detailed insight into the questions, methods, and procedures that revolutionized embryology from the 1880s on.

Frederick Churchill's studies have done much to illuminate the beginnings of experimental embryology, particularly in the work of Wilhelm Roux. His unpublished dissertation, "Wilhelm Roux and a Program for Embryology" (Ph.D. dissertation, Harvard University, 1968) gives a thorough account of the influences acting on Roux and the impact Roux made on his contemporaries. A less extensive treatment, with a slightly different emphasis, is found in Churchill's "Chabry, Roux, and the Experimental Method in Nineteenth-Century Embryology," in *Foundations of Scientific Method*, ed. R. N. Giere and R. S. Westfall (Bloomington, Ind.: Indiana University Press, 1973), pp. 161-205. A study of the transition from mechanistic to vitalistic models in embryology is traced through the work of Roux to that of Driesch in the same author's "From Machine-Theory to Entelechy: Two Studies in Developmental Teleology," *Journal of the History of Biology* 2 (1969): 165-185.

The philosophical and scientific aspects of the work of Roux and Driesch are given considerable attention by Reinhard Mocek, most notably in a comparative scientific biography, *Wilhelm Roux–Hans Driesch. Zur Geschichte der Entwicklungsphysiologie der Tiere* (Jena: Gustav Fischer, 1974). This is a detailed study of both Roux and Driesch in terms of their biological, philosophical, and methodological approaches to embryological questions.

E. The Life and Work of Thomas Hunt Morgan

1. *Biographical Studies*

The most complete biographical study to date is Ian Shine and Sylvia Wrobel, *Thomas Hunt Morgan, Pioneer of Genetics* (Lexington, Ky.: University of Kentucky Press, 1976). A popularized account prepared for the Kentucky Bicentennial Bookshelf series, this study admirably conveys the extent, ingenuity, and feeling of Morgan's life and work. The authors draw on all relevant modern research, including archival sources and oral history.

A number of obituary notices written by Morgan's professional colleagues shortly after his death cover the more conventional details of his career. The most complete of these is A. H. Sturtevant's account in *Biographical Memoirs, National Academy of Sciences* 33 (1959): 283-325. Sturtevant's memoir contains a complete bibliography of Morgan's published work. Other biographical sketches include: E. G. Conklin, "Thomas Hunt Morgan, 1866-1945," *Biological Bulletin* 93 (1947): 14-18; R. A. Fisher and Gavin de Beer, "Thomas Hunt Morgan, 1866-1945," *Obituary Notices of Fellows of the Royal Society* 5 (1947): 451-466 (also with a complete bibliography of Morgan's published works).

A lengthy account of Morgan the man and scientist is Garland E. Allen's article "Thomas Hunt Morgan" in *Dictionary of Scientific Biography* (New York: Charles Scribner's Sons, 1974), vol. 9, pp. 515-526; a shorter version by the same author appears in *Encyclopaedia Britannica*, 15th ed. (1974), vol. 10, pp. 440-442. George Beadle's article on Morgan for the *Dictionary of American Biography* (1969) is short and direct. Bernard Jaffee's older essay "Thomas Hunt Morgan (1866-1945): American Science Comes of Age," in *Men of Science in America* (New York: Simon & Schuster, 1946) discusses both Morgan's personal life and scientific work. Jaffee is overenamored of scientists in general, and his account perpetuates numerous myths about science and about Morgan in particular (for example, the notion that Morgan "discovered" the use of *Drosophila* for breeding experiments). A strong feature of Jaffee's essay, however, is the discussion he provides of Lilian Morgan's biological contributions, principally her role in the *Drosophila* work.

Among the more personal accounts of Morgan's life work are

both published and unpublished sources. Concerning Morgan's early life, two essays have been useful: Wendell H. Stephenson, "Thomas Hunt Morgan: Kentucky's Gift to Biological Science," *The Filson Club Historical Quarterly* 20 (1946): 97-106; and Herbert Parkes Riley, "Thomas Hunt Morgan," *Transactions of the Kentucky Academy of Sciences* 53 (1974): 1-8. Howard K. Morgan's unpublished essay, "Notes on Thomas Hunt Morgan's Life," prepared for the dedication of the Thomas Hunt Morgan Elementary School in Seattle, Washington in 1953, contains interesting insights and anecdotes of his father's life. Lilian V. Morgan prepared a somewhat shorter account, "Some Chronological Notes on the Life of Thomas Hunt Morgan" (date unknown), now contained in the Morgan Papers, Archives of the California Institute of Technology. "Personal Recollections of T. H. Morgan," Tove Mohr's essay written for the centennial celebration of Morgan's birth in Lexington, Kentucky, September 1966, is warm, insightful, and humorous. An extensive typescript, "Notes on T. H. Morgan," prepared by Isabel Morgan Mountain (Morgan's youngest daughter) for the author contains many personal reminiscences and details of family life that have been most useful. Two unpublished personal accounts by Morgan's colleagues and close friends, E. G. Conklin and A. H. Sturtevant, give further insight into Morgan's personality as a research worker: Conklin's "Some Personal Recollections of Thomas Hunt Morgan, 1866-1945," and Sturtevant's "Personal Recollections of Thomas Hunt Morgan," are both housed in the Morgan Papers, Archives of the California Institute of Technology.

2. *Morgan's Scientific Work*

Some aspects of Morgan's scientific work are discussed in each of the biographical sketches or obituary notices listed above. However, in recent years a number of monographs dealing specifically with one or another aspect of Morgan's scientific work have appeared. A number of these are by Garland Allen: "Thomas Hunt Morgan and the Problem of Sex Determination," *Proceedings of the American Philosophical Society* 110 (1966): 48-57 deals with Morgan's opposition to both the Mendelian and the chromosome theories because of their inability to explain the inheritance of sex. "Thomas Hunt Morgan and the Problem of Natural Selection," *Journal of the History of Biology* 1 (1968): 113-139, describes Morgan's opposition to the Darwinian theory before 1910 and his modified views on natural selection between 1910 and 1935. "The Introduction of *Drosophila* into the Study of Heredity and Evolution, 1900-1910," *Isis* 66 (1975): 322-333, traces the many paths that led Morgan to the use of the fruit fly as a laboratory organism.

The development of the *Drosophila* work, especially between 1910 and 1915, has been described in two articles: A. H. Sturtevant's account, "The Fly Room," published as chapter 6 of his *A History of Genetics*, conveys in slightly romanticized ways the excitement and cooperative attitudes prevailing among the early group of workers before 1915. Garland Allen has described the development of the *Drosophila* work, with special emphasis on the confluence of breeding results and cytology, in the introduction to the reprint edition of Morgan, Sturtevant, Muller, and Bridges, *The Mechanism of Mendelian Heredity* (New York: Johnson Reprint Corporation, 1972), pp. v-xxv. E. A. Carlson has presented a view of the workings of the *Drosophila* group somewhat opposed to Sturtevant's: "The *Drosophila* Group," *Genetics* 79 (1975): 15-27; a slightly modified version of this same article appears as "The *Drosophila* Group: The Transition from the Mendelian Unit to the Individual Gene," *Journal of the History of Biology* 7 (1974): 31-48; and in an earlier paper describing H. J. Muller's contributions to the development of the gene theory, "An Unacknowledged Founding of Molecular Biology: H. J. Muller's Contributions to Gene Theory, 1910-1936," *Journal of the History of Biology* 4 (1971): 149-170. An exchange of views on Carlson's treatment of Muller, and specifically on the relationship between Muller and Morgan, is found in a brief letter, "The Role of Hermann J. Muller in the Drosophila Group," *Genetics* 81 (1975): 222, a-b; and Carlson's reply, p. 222, c-d.

Morgan's response to opponents of the Mendelian-chromosome theory, such as Richard Goldschmidt, is described in Garland Allen's, "T. H. Morgan, Richard Goldschmidt, and the Opposition to Mendelian Theory, 1900-1940," *Biological Bulletin* 139 (1970): 412-413. An expanded version is found in "Opposition to the Mendelian-Chromosome Theory: The Physiological and Developmental Genetics of Richard Goldschmidt," *Journal of the History of Biology* 7 (1974): 49-92. Morgan's views on movements such as eugenics are briefly discussed in Allen's "Genetics, Eugenics and Class Struggle," *Genetics* 79 (1975): 29-45.

Some recent works dealing with aspects of the philosophy of genetics include David Hull, *Philosophy of Biological Science*

(Englewood Cliffs, N.J.: Prentice-Hall, 1974), especially chapter 1. Particularly enlightening and helpful to my own thinking has been Kenneth Schaffner's "Approaches to Reduction," *Philosophy of Science* 34 (1967): 137-147; highly stimulating is Nils Roll-Hansen's book, *Reduction and Biological Form in Historical Perspective,* of which chapter 5 is devoted specifically to reductionism in classical genetics (in press). Edward Manier's "Genetics and the Philosophy of Biology," *Proceedings of the American Catholic Philosophical Association* (Washington, D.C., 1965), pp. 124-133, deals with broader philosophical problems in classical genetics than simply reductionism. Studies dealing more specifically with Morgan include Manier's "The Experimental Method in Biology: T. H. Morgan and the Theory of the Gene," *Synthese* 20 (1969): 185-205, and Garland Allen's "T. H. Morgan and the Emergence of a New American Biology," *Quarterly Review of Biology* 44 (1969): 168-188.

INDEX

LIBRARY OF CONGRESS CATALOGING IN PUBLICATION DATA

Allen, Garland E
Thomas Hunt Morgan.

Bibliography: p.
Includes index.
1. Morgan, Thomas Hunt, 1866-1945. 2. Geneticists
—United States—Biography.
QH429.2.M67A44 575.1'092'4 [B] 77-85526
ISBN 0-691-08200-6